Gene and Cell Delivery for Intervertebral Disc Degeneration

GENE AND CELL THERAPY SERIES

Series Editors
Anthony Atala & Graça Almeida-Porada

PUBLISHED TITLES

Gene and Cell Delivery for Intervertebral Disc Degeneration
Raquel M. Gonçalves and Mário Adolfo Barbosa

*Regenerative Medicine Technology: On-a-Chip Applications for Disease Modeling,
Drug Discovery and Personalized Medicine*
Sean V. Murphy and Anthony Atala

Therapeutic Applications of Adenoviruses
Philip Ng and Nicola Brunetti-Pierri

Cellular Therapy for Neurological Injury
Charles S. Cox, Jr

Placenta: The Tree of Life
Ornella Parolini

Gene and Cell Delivery for Intervertebral Disc Degeneration

Edited by
Raquel M. Gonçalves
Mário Adolfo Barbosa

CRC Press is an imprint of the
Taylor & Francis Group, an **informa** business

CRC Press
Taylor & Francis Group
6000 Broken Sound Parkway NW, Suite 300
Boca Raton, FL 33487-2742

© 2018 by Taylor & Francis Group, LLC
CRC Press is an imprint of Taylor & Francis Group, an Informa business

No claim to original U.S. Government works

Printed on acid-free paper

International Standard Book Number-13: 978-1-4987-9940-9 (Hardback)

This book contains information obtained from authentic and highly regarded sources. Reasonable efforts have been made to publish reliable data and information, but the author and publisher cannot assume responsibility for the validity of all materials or the consequences of their use. The authors and publishers have attempted to trace the copyright holders of all material reproduced in this publication and apologize to copyright holders if permission to publish in this form has not been obtained. If any copyright material has not been acknowledged please write and let us know so we may rectify in any future reprint.

Except as permitted under U.S. Copyright Law, no part of this book may be reprinted, reproduced, transmitted, or utilized in any form by any electronic, mechanical, or other means, now known or hereafter invented, including photocopying, microfilming, and recording, or in any information storage or retrieval system, without written permission from the publishers.

For permission to photocopy or use material electronically from this work, please access www.copyright.com (http://www.copyright.com/) or contact the Copyright Clearance Center, Inc. (CCC), 222 Rosewood Drive, Danvers, MA 01923, 978-750-8400. CCC is a not-for-profit organization that provides licenses and registration for a variety of users. For organizations that have been granted a photocopy license by the CCC, a separate system of payment has been arranged.

Trademark Notice: Product or corporate names may be trademarks or registered trademarks, and are used only for identification and explanation without intent to infringe.

Library of Congress Cataloging-in-Publication Data

Names: Gonçalves, Raquel M., editor. | Barbosa, Mário A., editor.
Title: Gene and cell delivery for intervertebral disc degeneration / editors: Raquel M. Gonçalves and Mário Adolfo Barbosa.
Other titles: Gene and cell therapy series.
Description: Boca Raton : Taylor & Francis/CRC Press, 2018. | Series: Gene and cell therapy series | Includes bibliographical references.
Identifiers: LCCN 2017056453 | ISBN 9781498799409 (hardback : alk. paper)
Subjects: | MESH: Intervertebral Disc Degeneration--therapy | Low Back Pain--therapy | Genetic Therapy--methods | Stem Cell Transplantation--methods
Classification: LCC RD771.B217 | NLM WE 740 | DDC 617.5/64--dc23
LC record available at https://lccn.loc.gov/2017056453

Visit the Taylor & Francis Web site at
http://www.taylorandfrancis.com

and the CRC Press Web site at
http://www.crcpress.com

Contents

Series Preface vii
Preface ix
Editors xi
Contributors xiii

Chapter 1 Intervertebral Disc Degeneration in Clinics: Therapeutic Challenges 1
Pedro Santos Silva, Paulo Pereira, and Rui Vaz

Chapter 2 Animal Models and Imaging of Intervertebral Disc Degeneration ...19
Marion Fusellier, Johann Clouet, Olivier Gauthier, Catherine Le Visage, and Jerome Guicheux

Chapter 3 Intervertebral Disc Whole Organ Cultures: How to Choose the Appropriate Model 67
Sebastian Wangler, Zhen Li, Sibylle Grad, and Marianna Peroglio

Chapter 4 Adult Stem Cells for Intervertebral Disc Repair 103
Esther Potier and Delphine Logeart-Avramoglou

Chapter 5 Materials for Cell Delivery in Degenerated Intervertebral Disc 137
Joana Silva-Correia, Joaquim Miguel Oliveira, and Rui Luís Reis

Chapter 6 Cell Recruitment for Intervertebral Disc 155
Catarina Leite Pereira, Sibylle Grad, Mário Adolfo Barbosa, and Raquel M. Gonçalves

Chapter 7 Immunomodulation in Degenerated Intervertebral Disc 183
Graciosa Q. Teixeira, Mário Adolfo Barbosa, and Raquel M. Gonçalves

Chapter 8 Gene Delivery for Intervertebral Disc 231
Gianluca Vadalà, Luca Ambrosio, and Vincenzo Denaro

Index 255

Series Preface

Gene and cell therapies have evolved in the past several decades from a conceptual promise to a new paradigm of therapeutics, able to provide effective treatments for a broad range of diseases and disorders that previously had no possibility of cure.

The fast pace of advances in the cutting-edge science of gene and cell therapy, and supporting disciplines ranging from basic research discoveries to clinical applications, requires an in-depth coverage of information in a timely fashion. Each book in this series is designed to provide the reader with the latest scientific developments in the specialized fields of gene and cell therapy, delivered directly from experts who are pushing forward the boundaries of science.

In this volume of the Gene and Cell Therapy book series, *Gene and Cell Delivery for Intervertebral Disc Degeneration*, the editors have assembled a remarkable team of outstanding investigators and clinicians, each one of whom is an expert in a specific area of IVD, to give us an integrated approach to the most current and controversial aspects of IVD and the forefront research that is set to reform the way IVD management/treatment is approached.

This highly innovative and timely book brings together aspects pertaining to developmental and stem cell biology of the Nucleus Pulposus; cell recruitment, chemoattractants, immunology, and inflammation of disc degeneration; molecular-, cellular-, and biomaterials-based therapies targeting the degenerated disc; and the current therapeutic challenges that clinicians face when treating patients with IVD.

We would like to thank the volume editors, Raquel Gonçalves and Mário Barbosa, and all the authors, all of whom are remarkable experts, for their valuable contributions. We would also like to thank our senior acquisitions editor, Dr. C.R. Crumly, and the CRC Press staff for all their efforts and dedication to the Gene and Cell Therapy book series.

Anthony Atala
Wake Forest Institute for Regenerative Medicine

Graça Almeida-Porada
Wake Forest Institute for Regenerative Medicine

Preface

Low back pain (LBP) is the global leading disorder in terms of number of years lived with disability. It is also a social problem with a heavy economic burden, and tendency to increase as long as the population ages. Intervertebral disc (IVD) degeneration is one of the major causes of LBP, for which common therapeutic interventions are not efficient. The current clinical approaches, either conservative or nonconservative, are determined by the degree, severity, and persistence of pain, but the outcome of these solutions is often transient, some of them affecting patients' mobility, and others causing adjacent IVD degeneration, which in the end leads to chronic LBP symptoms in many patients.

A search for alternative therapies for LBP and particularly IVD degeneration has been encouraged, with special focus on cell-based therapies. However, contrary to many other tissues, the IVD has an avascular nature, maintained under hypoxia, low-glucose and is highly pressurized, which turns degenerated IVD into a hostile environment for cell survival.

Furthermore, cellular characterization in the IVD, and particularly in the nucleus pulposus, remains controversial, meaning that its molecular and cellular signature is not consensual among the scientific community, mainly due to a lack of specific markers and species variability. This impacts directly on the knowledge about the regenerative potential of this tissue by itself.

Overall, this book aims to contribute to increasing the knowledge on cellular and molecular therapies for degenerated IVD and associated LBP. The most relevant issues include the *ex vivo* and *in vivo* models of IVD degeneration, the types of cells, and cell sources for treating degenerated IVD, the current and alternative routes of therapies for degenerated IVD, the vehicles for cell delivery into degenerated IVD, and the intradiscal molecular therapies for degenerated IVD.

Finally, it is the goal of this book to approach current controversial aspects of IVD research and bring together the most recent advances in the field of molecular and cell therapies for degenerated IVD.

Editors

Raquel M. Gonçalves has a degree in chemical engineering and a PhD in biotechnology (University of Lisbon, Portugal). She gained expertise in human hematopoietic stem cell and mesenchymal stem cell expansion, with a period abroad in the University of Nevada, Reno, United States. Since 2009, she has been an assistant investigator and has been dedicated to intervertebral disc research. Presently, at I3S (Instituto de Investigação e Inovação em Saúde/INEB, Institute of Biomedical Engineering, University of Porto), she develops her work at the Microenvironments for New Therapies Group, whose main goal is to dissect the microenvironment elements that contribute to reestablish homeostasis upon disease and/or injury and to bioengineer therapeutic strategies to modulate host response. In this group, she has been involved in the establishment of IVD *ex vivo* and *in vivo* models, high-throughput tools to characterize IVD cells and extracellular matrix, and immunomodulatory strategies and stem cell recruitment to degenerated IVD. In parallel, she is involved in science dissemination activities at Porto high schools and in teaching at the University of Porto.

Mário Adolfo Barbosa is a full professor at Instituto de Ciências Biomédicas Abel Salazar (ICBAS), University of Porto, Portugal. For nearly 30 years, biomaterials science and technology has been the topic of Mário's research. He is internationally recognized for his contributions to biomaterials science, particularly in cell–biomaterial interactions. Mário was one of the founding members of the Instituto de Engenharia Biomédica (INEB, http://www.ineb.up.pt), created in June 1989. From 2000 to 2012, he was the scientific coordinator of the institute and its president from 2000–2006 and 2010–2012. Presently, he is the scientific coordinator of the I3S (Instituto de Investigação e Inovação em Saúde), of the University of Porto and leader of the Microenvironments for New Therapies Group at i3S. His research interests focus on the modulation of the microenvironment, in particular the inflammatory response to improve tissue repair and/or regeneration.

Contributors

Luca Ambrosio
Department of Orthopedic
and Trauma Surgery
Campus Bio-Medico University
of Rome
Rome, Italy

Mário Adolfo Barbosa
Instituto de Engenharia Biomédica
(INEB)
Instituto de Inovação e Investigação
em Saúde (I3S)
Instituto de Ciências Biomédicas
Abel Salazar (ICBAS)
University of Porto
Porto, Portugal

Johann Clouet
INSERM (Institut National de la
Santé et de la Recherche Médicale)
Laboratory for Regenerative Medicine
and Skeleton (RMeS)
School of Dental Surgery
and
Department of Pharmacy
CHU Nantes
Nantes, France

Vincenzo Denaro
Department of Orthopedic
and Trauma Surgery
Campus Bio-Medico University
of Rome, Italy
Rome, Italy

Marion Fusellier
INSERM (Institut National de la
Santé et de la Recherche Médicale)
Laboratory for Regenerative Medicine
and Skeleton (RMeS)
School of Dental Surgery
and
Department of Diagnostic Imaging
CRIP, Oniris College of Veterinary
Medicine
Food Science and Engineering
Nantes, France

Olivier Gauthier
INSERM (Institut National de la
Santé et de la Recherche Médicale)
Laboratory for Regenerative Medicine
and Skeleton (RMeS)
School of Dental Surgery
and
Department of Surgery
CRIP, Oniris College of Veterinary
Medicine
Nantes, France

Raquel M. Gonçalves
Instituto de Engenharia Biomédica
(INEB)
Instituto de Inovação e Investigação
em Saúde (I3S)
Instituto de Ciências Biomédicas
Abel Salazar (ICBAS)
University of Porto
Porto, Portugal

Sibylle Grad
AO Research Institute Davos
Davos, Switzerland

Jerome Guicheux
INSERM (Institut National de la
 Santé et de la Recherche Médicale)
Laboratory for Regenerative Medicine
 and Skeleton (RMeS)
School of Dental Surgery
Nantes, France

Catherine Le Visage
INSERM (Institut National de la Santé
 et de la Recherche Médicale)
Laboratory for Regenerative Medicine
 and Skeleton (RMeS)
School of Dental Surgery
Nantes, France

Zhen Li
AO Research Institute Davos
Davos, Switzerland

Delphine Logeart-Avramoglou
Laboratory for Osteo-Articular
 Bioengineering and Bioimaging
CNRS (Centre National de la
 Recherche Scientifique)
Paris Diderot University - Sorbonne
 Paris Cité
École nationale vétérinaire d'Alfort
Paris, France

Joaquim Miguel Oliveira
3B's Research Group–University
 of Minho
ICVS/3B's–PT Government
 Associated Laboratory
The Discoveries Centre for
 Regenerative and Precision Medicine
Barco GMR, Portugal

Catarina Leite Pereira
Instituto de Engenharia Biomédica
 (INEB)
Instituto de Inovação e Investigação
 em Saúde (I3S)
Instituto de Ciências Biomédicas
 Abel Salazar (ICBAS)
University of Porto
Porto, Portugal

Paulo Pereira
Department of Neurosurgery
Centro Hospitalar São João
and
Faculty of Medicine
University of Porto
and
Neurosciences Center
Hospital CUF Porto
Porto, Portugal

Marianna Peroglio
AO Research Institute Davos
Davos, Switzerland

Esther Potier
Laboratory for Osteo-Articular
 Bioengineering and Bioimaging
CNRS (Centre National de la
 Recherche Scientifique)
Paris Diderot University - Sorbonne
 Paris Cité
École nationale vétérinaire d'Alfort
Paris, France

Rui Luís Reis
3B's Research Group–University
 of Minho
ICVS/3B's–PT Government
 Associated Laboratory
The Discoveries Centre for
 Regenerative and Precision Medicine
Barco GMR, Portugal

Contributors

Pedro Santos Silva
Department of Neurosurgery
Centro Hospitalar São João
and
Faculty of Medicine
University of Porto
and
Neurosciences Center
Hospital CUF Porto
Porto, Portugal

Joana Silva-Correia
3B's Research Group-University
  of Minho
ICVS/3B's-PT Government Associated
  Laboratory
Barco GMR, Portugal

Graciosa Q. Teixeira
Instituto de Engenharia Biomédica
  (INEB)
Instituto de Investigação e Inovação
  em Saúde (I3S)
Instituto de Ciências Biomédicas
  Abel Salazar (ICBAS)
Universidade do Porto
Porto, Portugal

and

Institute for Orthopedic Research
  and Biomechanics
University of Ulm
Ulm, Germany

Gianluca Vadalà
Department of Orthopedic and Trauma
  Surgery
Campus Bio-Medico University
  of Rome
Rome, Italy

Rui Vaz
Department of Neurosurgery
Centro Hospitalar São João
and
Faculty of Medicine
University of Porto
and
Neurosciences Center
Hospital CUF Porto
Porto, Portugal

Sebastian Wangler
AO Research Institute Davos
Davos, Switzerland

and

Graduate School for Cellular
  and Biomedical Sciences
University of Bern
Bern, Switzerland

1 Intervertebral Disc Degeneration in Clinics
Therapeutic Challenges

Pedro Santos Silva, Paulo Pereira, and Rui Vaz

CONTENTS

1.1 Introduction 1
1.2 Importance of Disc Degeneration and LBP 2
1.3 Clinical Manifestations of Disc Degeneration 3
1.4 Intervertebral Disc Degeneration in Imaging Studies 4
1.5 Relation between Disc Degeneration and Pain 6
1.6 Prognosis 7
1.7 Therapeutic Challenges for Discogenic LBP 7
    1.7.1 Conservative Treatment 7
    1.7.2 Percutaneous Techniques 8
    1.7.3 Operative Treatment 9
    1.7.4 Regenerative Techniques 10
1.8 Therapeutic Challenges for Sciatica 11
1.9 Final Remarks 13
References 13

1.1 INTRODUCTION

For spine surgeons, lumbar degenerative disc disease (DDD) is an everyday challenge. In the last decades, this condition has raised significant questions and controversies that are far from being solved. From this contextual uncertainty, we can point two main sides of this problem that are relevant for interventional treatments: disc herniation as a cause of *radicular compression* and disc degeneration as a source of *discogenic pain.*

    Lumbar discectomy is by far the most common surgical treatment for intractable sciatica caused by lumbar disc herniation. Although it has been an option for many years, this procedure remains an act of aggression to the disc itself. Like a dentist extracting a tooth, the spine surgeon treats a dislocated part of the intervertebral disc by removing it. There is no procedure in our days than can resolve a lumbar radicular compression in a physiologic and reconstructive way.

Degenerative changes in lumbar discs can be associated with low back pain (LBP) in a selected group of patients, but the definition and diagnosis of discogenic pain remain a disputed concept. Beyond the diagnosis, the treatment of patients who are presumed to have discogenic pain involves a spectrum of more or less destructive procedures, none of which, including percutaneous techniques and lumbar fusion, is established as the standard of care.

For patients, the major setback is the lack of adequate treatments that can restore the biologic and mechanical intervertebral disc structure and function. This difficulty leads to several different treatment proposals with disappointing results and high rates of disability.

1.2 IMPORTANCE OF DISC DEGENERATION AND LBP

It would be no exaggeration to say that lumbar DDD is ubiquitous in the aging population. In a recent report (Armbrecht et al. 2017) of a prospective cohort of 10,132 individuals aged more than 50 years, all the participants had some degree of radiologic DDD, and moderate or severe DDD was present in 47% of the cases. Following the same trend, pain located in the lumbar area is an almost universal experience during human life. LBP has an enormous health and economic burden and has been the leading cause of years lived with disability in the world, in the last two decades (Global Burden of Disease Study Collaborators 2015).

According to Waddell (2005), LBP can be related to a *specific pathology*, *nerve root pain*, and *nonspecific causes* (Table 1.1). Specific causes are potentially severe spinal pathologies (fractures, tumors, and infection) and correspond to only 1%–2% of cases.

In this classification, about 5% of patients have nerve root pain (associated with disc herniation or vertebral canal stenosis) (Waddell 2005). The estimated incidence of sciatica in Western countries is 5:1000 (Cherkin et al. 1994), and while most cases

TABLE 1.1
Possible Causes of LBP

Specific LBP	Tumor
	Infection
	Fracture
Nonspecific LBP	Myofascial syndrome
	DDD
	Facet syndrome
	Instability
	Sacroiliitis
Nerve root pain	Lumbar disc herniation
	Vertebral canal stenosis
	Foraminal stenosis

of radicular pain resolve spontaneously, about 30% of patients can have persistent symptoms (Weber, Holme, and Amlie 1993).

Hence, by far, the vast majority of cases (85%–95%) correspond to nonspecific LBP, a widely heterogeneous group, where accepted diagnosis criteria are absent and the treatment is mainly empirical and unproved. In this category, a wide variety of pain generators have been included, such as the following:

- Intervertebral disc degeneration (discogenic pain)
- Muscle and ligaments (myofascial syndrome, low back strain)
- Facet joint osteoarthritis (facet syndrome)
- Abnormal movement (segmental instability)
- Sacroiliac joint (sacroiliitis)

In a study based on computed tomography (CT) and discography, about 40% of cases of nonspecific LBP were attributed to intervertebral discs (Schwarzer et al. 1995). Despite the advances with magnetic resonance imaging (MRI) and its association with provocative discography, there is no consensus on how to diagnose discogenic pain and the contribution of this entity to LBP.

1.3 CLINICAL MANIFESTATIONS OF DISC DEGENERATION

In clinical practice, the differential diagnosis between specific, nonspecific LBP, and radicular pain is a primary concern. Specific causes of LBP are fractures, tumors, infection, and inflammatory diseases.

A *fracture* is commonly suspected when acute back pain that worsens with loading begins after trauma; however, elderly women with osteoporosis can suffer vertebral fractures without a history of significant trauma. In adults older than 50 years, a slow-onset and progressive pain that wakes the patient at night suggests a *tumor*, particularly if there is cancer history. LBP in the presence of fever can be associated to infectious causes. Ankylosing spondylitis, psoriatic spondylitis, or Reiter syndrome are inflammatory diseases that affect the spine, mostly in young adults. These rheumatologic conditions are characterized by an inflammatory back pain that is more significant in the morning and is accompanied by stiffness with limitation of spine movements.

Nonspecific LBP can be related to the mechanical structures of the spine, such as the vertebrae and their articulations (intervertebral discs and facet joints), apposed joints (hip or sacroiliac joint), muscles, and ligaments. Injury or inflammation of these structures can cause a mechanical type of pain, which exacerbates with movement and exertion and tends to increase to the end of the day, and there is some relief with resting. The pain can radiate to the groin, buttocks, and thigh, which is call referred pain, a pain perceived at a different location from the site of the painful stimulus; this form of pain typically has a proximal radiation above the knee and is less well localized than a pain originating from a nerve root. There are some clinical aspects that can suggest an intervertebral disc origin for a LBP (Tonosu et al. 2017): a *discogenic pain* the can be triggered after sustained loading (sitting too long) and lumbar flexion, especially with knee extension (washing one's face), and can radiate

TABLE 1.2
Common Lumbar Nerve Root Compression Syndromes

Nerve Root	Pain/Numbness	Motor Weakness
L4	Inner side of the leg and foot	Extension of the knee
L5	Anterior and outer side of the leg and dorsal side of the foot	Dorsiflexion of the foot
S1	Outer and plantar side of the foot	Plantar flexion of the foot

to the anterior thigh. This condition should be differentiated from facet syndrome, a condition that results from facet joint osteoarthritis, in which the back pain may exacerbate with lumbar extension or rotation and radiate to the posterior thigh.

Radicular pain, also known as nerve root pain or sciatica, is caused by nerve root compression, usually in the lateral recess of the vertebral canal or in the intervertebral foramen. This compression can be originated by osteophytes, facet joint, or yellow ligament hypertrophies, but the most common cause is a lumbar disc herniation. Radicular pain runs from the lumbar region to the lower limb, is usually unilateral and sharp, and sometimes is associated with numbness. The topography of the pain is usually well defined and depends on the affected nerve root and its respective dermatome. Since the compression is more common in L4, L5, or S1 nerve roots, the pain typically radiates below the knee, affecting the foot. Neurological examination can reveal motor weakness and sensitive alterations, depending on the myotome or dermatome of the compressed root (Table 1.2). The straight leg raising test, also called *Lasègue's* test, is positive when the radicular pain is recreated by lifting the patient's leg while the knee is extended.

1.4 INTERVERTEBRAL DISC DEGENERATION IN IMAGING STUDIES

Lumbar DDD findings in imaging studies are well recognized, and there are several classifications of disc degeneration for radiographs, CT scans, and MRI. There are three progressive markers of DDD in radiography images and CT scans: disc height loss, osteophyte formation, and sclerosis of vertebral bodies (Lane et al. 1993; Wilke et al. 2006). The progression of the disease also leads to intervertebral disc vacuum sign, intervertebral disc space collapse, degeneration of the facet joints, subluxation (spondylolisthesis), and deformity in sagittal or coronal planes. The neural elements can be compressed by stenosis in the vertebral canal, lateral recess, or intervertebral foramen (Thalgott et al. 2004).

Pfirrmann et al. (2001) developed the most widely used classification of MRI findings of lumbar DDD. It is a grading system that describes the degenerative process based on T2-weighted MRI sequences. The higher the grade is, the greater the severity of degeneration. Throughout the degeneration process, the nucleus pulposus loses its water content and its hyperintense signal on T2-weighted images, the distinction between nucleus and annulus is lost, and there is a reduction in the disc height (Table 1.3).

Degeneration of vertebral body endplates and subchondral bone on MRI was classified into three types (Modic et al. 1988). Modic type I changes are hypointense on T1-weighted and hyperintense on T2-weighted images, and indicate bone marrow edema. These are thought to represent acute changes of the vertebral body and can be related to discogenic back pain (Weishaupt et al. 2001). Type II changes are hyperintense on both T1 and T2 sequences and represent chronic degenerative changes with bone marrow replacement by fat. Modic type III changes are hypointense on both T1 and T2 sequences, indicating sclerotic vertebral endplates. The intervertebral disc can herniate through a disruption in the vertebral body endplates, causing an intravertebral disc herniation (Schmorl nodes).

Lumbar disc herniation can assume various forms, depending on the volume of the dislocated tissue and the integrity of the annulus. In 2014, an American consensus (Fardon et al. 2014) defined disc herniation as a localized or focal displacement of disc material beyond the limits of the intervertebral disc space. This disc material may be nucleus, cartilage, fragmented ring apophysis, or annular tissue. A disc *bulge or bulging* is the presence of disc tissue extending beyond the edges of the ring apophyses, throughout the circumference of the disc. This is not considered a form of herniation. A disc herniation is called *protrusion*: if the greatest distance between the edges of the disc material presenting outside the disc space is less than the distance between the edges of the base of that herniated disc material. When any distance between the edges of herniation is greater than its base, an *extrusion* is present. When no continuity exists between the herniation and the disc space, the extrusion is subclassified as *sequestration*. The term *migration* is used to describe the displacement of disc material away from the site of extrusion. Another classification of disc herniations, as *contained* or *uncontained*, depends on the displaced material being covered or not by annulus fibers, and/or the posterior longitudinal ligament.

TABLE 1.3
Pfirrmann Classification of Lumbar Disc Degeneration on MRI

Grade I	Disc is homogeneous, with a bright hyperintense white signal intensity and normal disc height.
Grade II	Disc structure is inhomogeneous, with a hyperintense white signal, with or without horizontal gray bands; distinction between nucleus and annulus is clear; and the disc height is normal.
Grade III	Disc is inhomogeneous, with an intermediate gray signal intensity; distinction between nucleus and annulus is unclear; and the disc height is normal or slightly decreased.
Grade IV	Disc is inhomogeneous, with a hypointense dark gray signal intensity; distinction between nucleus and annulus is lost; and the disc height is normal to moderately decreased.
Grade V	Disc is inhomogeneous, with a hypointense black signal intensity; distinction between nucleus and annulus is lost; and the disc space is collapsed.

In 2006, a lumbar degenerative disease severity score was proposed (Mirza et al. 2006), based on nine imaging features that evaluate disc and endplate degeneration on MRI, disc height loss, osteophytes, disc herniation type, stenosis, spondylolisthesis, instability, and deformity. This system is very complete and is scored from 0 to 39 points, and there was an excellent interrater agreement for this severity scale.

Provocative *discography* is an invasive diagnostic procedure that involves a pressurized injection of fluid into an intervertebral disc to elicit pain. It has been developed to identify the cases where the disc was the primary source of back pain. Dye extravasation from the injected annulus site indicates annular fissure, which may be interpreted as an abnormal finding; however, typical pain reproduction is essential to classify the discogram as positive. Currently, this procedure is less commonly used and the test is considered controversial due to its low specificity and the concerns that it can increase the risk of clinical disc problems by inducing iatrogenic degenerative changes (Cuellar et al. 2016).

1.5 RELATION BETWEEN DISC DEGENERATION AND PAIN

There are some controversies about the relation between pain and pathologic findings of the lumbar discs in imaging studies. Several studies revealed abnormal MRI disc findings in asymptomatic subjects: disc protrusion (25%–50%), disc degeneration (25%–70%), signal changes in the vertebral body endplates (10%), and annular fissures (14%–33%) (Boden et al. 1990; Carragee, Paragioudakis, and Khurana 2000; Jensen et al. 1994). On the other hand, these alterations of intervertebral discs, annulus, and vertebral endplates on MRI findings have been associated with pain intensity during provocative discography. Despite disc degeneration having a significant relation with age in asymptomatic individuals, in younger ages (younger than 50 years old), a strong association was found between disc degeneration and LBP, and similar findings were reported for disc bulges (Brinjikji et al. 2015). Furthermore, Pfirrmann grades ≥3 are strongly associated with a history of previous LBP (Tonosu et al. 2017). Posterior *annular tears* on discography and a high-intensity zone on T2-weighted MRI are likely to produce pain since its prevalence is higher in symptomatic patients, but the validity of these signs is limited because their prevalence in asymptomatic individuals is also elevated, so they are not clinically reliable as pain predictors (Carragee, Paragioudakis, and Khurana 2000; Ito et al. 1998). Regarding *endplate abnormalities*, moderate and severe type 1 and 2 Modic changes were related to pain during discography (Weishaupt et al. 2001).

A meta-analysis of 14 high-quality case control studies including more than 3000 individuals (Brinjikji et al. 2015) demonstrated that MRI findings of disc bulge (odds ratio [OR]: 7.54), degeneration (OR: 2.24), extrusion (OR: 4.38), protrusion (OR: 2.65), Modic 1 changes (OR: 4.01), and spondylolysis (OR: 5.06) are more prevalent in adults up to the age of 50 with back pain, when compared with asymptomatic individuals; annular fissures, high-intensity zones, spondylolisthesis, and central canal stenosis demonstrated no association with LBP.

Concerning radiographic abnormalities, a systematic review of the literature found a positive association between radiographic disc space narrowing and LBP (Raastad et al. 2015).

1.6 PROGNOSIS

The prognosis of LBP is variable, with a great proportion of patients undergoing remission but also with high rates of recurrence. Back pain episodes are typically transient, with improvements seen within a few weeks to a few months. Episode remission at 1 year ranges from 54% to 90% and recurrence at 1 year is estimated in a range from 24% to 80% (Hoy et al. 2010).

The prognosis of *sciatica* is good, and most patients will experience improvement in pain and disability in the short run without treatment. However, about 30% of patients refer persistent significant symptoms at 1 year (Weber, Holme, and Amlie 1993). Another study suggested that recovery from sciatica is less frequent than expected: 55% of patients still had symptoms of sciatica 2 years later, and 53%, after 4 years (25% who had recovered after 2 years had relapsed again by 4 years) (Tubach, Beaute, and Leclerc 2004).

1.7 THERAPEUTIC CHALLENGES FOR DISCOGENIC LBP

Interventional treatments for LBP are still controversial and should be reserved only for patients who failed to improve with time and appropriate conservative management. However, clear diagnostic criteria for discogenic pain are not established. Pain topography, characteristics, and worsening factors may suggest the anterior column as a pain generator. Spine imaging may show disc degeneration that can be related to pain, but these features are also found in asymptomatic individuals. In the last decades, several procedures to treat lumbar DDD emerged, with meaningful mechanisms of action and potentially favorable outcomes. In most of them, however, after some initial promising reports, randomized trials failed to prove unquestionable efficacy of the treatments. It seems that unclear and inadequate selection of patients is a determinant factor for failure of treatment and waning in generalized use of the techniques.

1.7.1 CONSERVATIVE TREATMENT

Guidelines for the management of LBP (Chetty 2017) strongly recommend conservative treatments as the first approach. Initial management includes medication and paracetamol, nonsteroidal anti-inflammatory drugs, and muscle relaxants that may be used for short-term treatment. For severe pain, stronger analgesics, such as opioids, are next in the recommendations.

For chronic pain, a different approach is proposed, and medication may include opioids and anti-depressants. *Cognitive behavioral therapy* (CBT) is recommended for chronic pain in most guidelines.

The efficacy of *passive physiotherapy modalities*, such as traction, ultrasound, massage, acupuncture, transcutaneous electrical nerve stimulation, and heat and cool therapies is unclear, with conflicting results in the literature. *Exercises* are usually indicated, which include aerobic, muscle conditioning, and back exercise classes.

1.7.2 Percutaneous Techniques

Several techniques have been developed to treat the initial stages of lumbar DDD, having in common the insertion and manipulation of catheters or electrodes within the disc space. They are appealing for the patient and for the physician because they are minimally aggressive and have much lower complication rates than operative treatment, particularly considering spine fusion. Whether they are injections or ablative techniques, delivering some type of energy, all these procedures produce some degree of destruction of the disc.

Thermal annular procedures involve delivering energy to the posterior annulus fibrosus. The rationale is that coagulation of nerve fibers occur in the annulus and that denaturation of collagen fibers results in shrinking of the annulus and promotes the healing of annular tears (Lu et al. 2014). As mentioned, there is little evidence that annular fissures are related to discogenic pain.

In intradiscal electrothermal therapy (IDET), a flexible electrode is steered in a circumferential fashion inside the annulus fibrosus of the disc and then it is heated to 90°C. Two randomized controlled trials (RCTs) (Freeman et al. 2005; Pauza et al. 2004) compared IDET with sham procedures and reported poor results with the procedure. One study showed no significant benefit from IDET over placebo, and in the other study, substantial numbers of patients benefited from the sham treatment, so the apparent efficacy of IDET was considered to be related to nonspecific factors and not to the procedure itself.

Another form of thermal annular procedure is percutaneous intradiscal radiofrequency thermocoagulation (PIRFT). The efficacy of this technique was assessed in two RCTs and there were no significant differences between sham and treated groups (Barendse et al. 2001; Kvarstein et al. 2009).

A third form of thermal annular procedure, *biacuplasty*, involves the use of two cooled radiofrequency electrodes placed on the posterolateral sides of the annulus fibrosus. One RCT compared biacuplasty with sham procedure and reported clinical benefits in the intervention group at 6 months after the treatment (Kapural et al. 2013). Another multicenter RCT compared biacuplasty with conventional medical management and reported superior performance in the procedure group in all study outcomes (Desai et al. 2016).

All in all, regarding thermal annular procedures, IDET and PIRFT are likely ineffective for patients with discogenic pain, and intradiscal biacuplasty showed some promising results, but further studies are needed to confirm its effectiveness (Lu et al. 2014).

Electrothermal ablation of ramus communicans, a possible neuropathway for discogenic pain, was evaluated in only one study in 2004, with promising results, but no additional studies were published (Oh and Shim 2004).

Intervertebral Disc Degeneration in Clinics

Methylene blue can be used to chemically ablate nerve endings. The intradiscal injection of methylene blue was evaluated in an RCT (Peng et al. 2010): patients who underwent the treatment reported significantly better outcome scores than the sham group. In a small prospective clinical series of 15 patients (Kallewaard et al. 2016), 40% of the patients claimed at least 30% pain relief. An additional retrospective observational study (Zhang et al. 2016) stated that intradiscal methylene blue might be an effective therapy in short-term follow-up.

Two RCT studies investigated the clinical success of *intradiscal steroid injections*. One study with 1-year follow-up failed to detect significant differences between intradiscal steroid and saline injections (Khot et al. 2004). The other study found that patients who received intradiscal injections of steroid or steroid plus an anti-inflammatory herbal had significantly improved outcomes at their 3 and 6 month follow-up, compared to the saline injection (Cao et al. 2011). A recently published RCT (Nguyen et al. 2017) compared the results of an intradiscal steroid injection performed during discography versus discography alone. One month after the intervention, the percentage of responders (reduction in LBP intensity) was higher in the steroid than control group (55.4% versus 33.3%), but the groups did not differ in pain intensity at 12 months.

1.7.3 Operative Treatment

Lumbar segment arthrodesis, or *fusion*, is the reference treatment for patients with lumbar DDD who failed all other treatments and in whom the intensity of pain and reduction of quality of life are so severe that surgical intervention is considered. Lumbar fusion can be performed using several techniques. Posterolateral fusion involves promoting bone fusion between the adjacent facet joints and transverse processes, usually using screws and rods as an internal fixation device. Interbody fusion requires a radical discectomy, removal of the cartilaginous endplates of the adjacent vertebrae, and usually the insertion of a cage and bone or a bone substitute inside the disc space to promote interbody bone growth. In most cases, a construct with pedicle screws and rods is also used. The variations of interbody fusion techniques depend on the approach to the disc space: posterior, posterolateral, lateral, or anterior, and besides the arthrodesis, decompression of neural elements may also be performed during surgery, depending on the clinical picture and the option of the surgeon. Minimally invasive surgical techniques were developed to reduce the soft tissue damage related to the approach, and these techniques have been reported to reduce the blood loss and the need for analgesic medications during the postoperative period and to decrease the length of hospitalization and complication rates (Khan et al. 2015).

The rationale for fusion in patients with discogenic pain is to remove the pain generator, and reduce the nociceptive input from loading of the disc and the facet joints from painful motion.

There is limited evidence for the use of fusion techniques in patients with discogenic back pain. Lumbar fusion has been compared to nonoperative management. A systematic review and meta-analysis of five RCTs included data from 707 patients (Bydon et al. 2014): there was an overall improvement of 7.39 points in the

Oswestry Disability Index in favor of lumbar fusion, but this difference was not statistically significant. In 2014, a guideline update for the performance of fusion procedures for degenerative disease of the lumbar spine was published (Eck et al. 2014) regarding fusion for intractable LBP without stenosis or spondylolisthesis. This study found a Level II evidence supporting the use of either intensive rehabilitation programs, incorporating cognitive therapy, or lumbar fusion.

Although lumbar fusion may benefit selected patients with discogenic pain, the fusion of a lumbar segment could lead to accelerated degeneration of adjacent disc segments. A recent study (Cho et al. 2014) reported that 66.8% of the patients have radiographic evidence of adjacent segment degeneration and 6.4% require a second operation at least 2 years after surgery.

Lumbar arthroplasty or *total disc replacement* was developed as a motion-preserving technique to lower the rate of adjacent segment disease, while keeping the rationale of lumbar fusion of removing the pain generator. There are various devices in the market, all of them including articular surfaces that tolerate loading and conserve the range of motion. Typically, they are inserted through an anterior approach. Most authors agree that adequate selection is the most important factor affecting arthroplasty outcomes and that ideal candidates are young patients with relatively preserved disc height, without any significant deformity, instability, or facet degeneration; in these patients, total disc replacement can be a suitable alternative to lumbar fusion (Salzmann et al. 2017). A review of five meta-analysis, about lumbar arthroplasty versus fusion, concludes that lumbar total disc replacement may be an effective technique for the treatment of selected patients with lumbar DDD, with at least equivalent results to lumbar fusion in the short-term; however, long-term studies are needed to address clinical outcomes, complications, and adjacent segment disease rates (Ding et al. 2017).

Surgery for discogenic pain may be only marginally superior to best conservative treatment, with the addition of significant complications and cost. The main limitations of operative treatment for DDD are the lack of pathoanatomical diagnosis in most patients and the absence of good-quality literature. Despite that, there has been an increase in the number of patients treated with spinal fusion for nonspecific LBP over the last two decades, which raised concerns about financial conflicts of interest among spine surgeons (Dhillon 2016).

1.7.4 REGENERATIVE TECHNIQUES

Novel technologies with regenerative objectives have been proposed as an alternative to ablative procedures or operative treatment. The challenge is to stop or reverse disc degeneration. The use of biomolecular strategies, cell transplantation, and tissue-engineering technology is under investigation to attempt that purpose.

Biomolecular strategies are suitable for early degrees of disc degeneration, when cell growth and anabolic responses may be stimulated. Recombinant proteins and genes have been used to regenerate the expression of target molecules, facilitating the production of extracellular matrix. Members of the families of bone morphogenetic proteins and transforming growth factor were shown to increase proteoglycan content and disc height in several *in vivo* studies (Moriguchi et al. 2016). Gene vector systems are being developed to regulate the transcription of growth factors, extracellular matrix degrading enzymes, and chondrocyte transcription factors (Woods et al. 2011).

Platelet-rich plasma has a high content of a variety of multifunctional growth factors. A recent RCT (Tuakli-Wosornu et al. 2016) studied intradiscal injection of platelet-rich plasma in 47 patients, and significant improvements in functional outcome were observed in the treatment group up to 1-year follow-up.

Introducing *stem cells* in the intervertebral disc has emerged as an attractive strategy for patients who have discs with intermediate structural damage, where the disc cell content is reduced. Several *in vivo* studies showed that these mesenchymal stem cells maintained viability and proliferate, and they can be induced to a chondrogenic pathway and then produce proteoglycans and collagen, increasing extracellular matrix and disc height (Moriguchi et al. 2016).

Intervertebral disc injection of autologous mesenchymal bone marrow cells was studied in 10 patients (Orozco et al. 2011), and the feasibility of the procedure was confirmed, with good clinical results despite unrecovered disc height.

Furthermore, after promising results in a canine model, a human trial was designed (Eurodisc) involving an autologous disc chondrocyte transplant into postdiscectomy patients. The study reported that in the intervention group, LBP was decreased and disc height was preserved at 2 years follow-up (Hohaus et al. 2008).

The impaired nutrient supply in degenerated discs is an obstacle to the feasibility of cell therapy. Furthermore, any injection inside the disc may induce additional degeneration.

Tissue-engineering technology is being developed for the treatment of advanced stages of disc degeneration. Scaffolds can be combined with cells, growth factors, and mechanical conditioning. The goal is an intervertebral disc construction *in vitro*, which can be implanted *in vivo*. Many studies have evaluated tissue-engineered components and whole-disc constructs, but no clinical study in lumbar spine was done (Moriguchi et al. 2016).

1.8 THERAPEUTIC CHALLENGES FOR SCIATICA

Most individuals with nerve root pain caused by intervertebral lumbar disc herniation have a high likelihood of recovery, spontaneously or with *conservative management*. As for discogenic back pain treatment, the initial approach can include pharmacological interventions and nonpharmacological strategies, such as physical therapy. In case of persistent pain or neurologic deficits, invasive treatments are the next step.

The evidence for caudal, interlaminar, and transforaminal *epidural glucocorticoid injections* in managing pain associated with lumbar disc herniation is good. These modalities may be an alternative to surgery, particularly in patients with contained disc herniations or moderate spinal stenosis (Manchikanti et al. 2013).

Lumbar discectomy is the standard surgical treatment for disc herniation. There is general agreement to indicate surgery in patients with good correlation between clinical picture and imaging studies, with progressive neurologic deficits or selected patients with persistent sciatica. There is still some controversy about the effectiveness of surgery in relation to conservative management. Studies that compared the two forms of treatment concluded that early surgery can offer a better short-term relief of sciatica (6–12 weeks) and there were no significant differences between surgery and conservative care at 1 and 2 years, but the evidence for this is of very low quality (Jacobs et al. 2011). The fast recovery and return to work can make the option for surgery cost-effective compared to prolonged conservative care (van den Hout et al. 2008). On the other hand, aggressive discectomy can exacerbate the degeneration process (Fakouri, Shetty, and White 2015), leading to more disability and increased healthcare costs in the long-term.

The goal of surgery is nerve root pain relief after removal of the herniated material, reducing the nerve root irritation or compression. The classical surgical technique is a partial disc removal, to reduce the risk of recurrence. However, the current trend in cases of sequestrations or when small tears in the annulus fibrosus are found intraoperatively is to remove only the herniated material, without invading the disc. This technique, called sequestrectomy, can provide similar pain reduction and reherniation rates compared to microdiscectomy (Fakouri, Shetty, and White 2015).

Surgery is usually performed with the aid of a magnification system, such as an operating microscope. As in lumbar fusion, minimally invasive surgical techniques can be used, and they have proved to be safe and effective by achieving similar degrees of disc removal and clinical long-term outcomes. As muscle trauma is reduced in the approach, decreased blood loss and shorter hospital stays have been reported (Clark et al. 2017; He et al. 2016). Relevant clinical questions remain about surgical treatment of lumbar disc herniation, namely, about the appropriate criteria for patient selection, timing for surgery, and long-term implications of discectomy for the spine degenerative process.

Some percutaneous intradiscal treatments are suitable for contained herniations. These techniques intend to produce a central lesion in the nucleus pulposus, with the purpose of reducing the volume of the disc and the intradiscal pressure. The goal is to retract the contained herniation toward the center of the disc, decreasing nerve root compression.

Chemonucleolysis consists of an intradiscal injection of proteolytic enzymes to dissolve the nucleus pulposus. Chemonucleolysis with chymopapain was widely used and researched some decades ago, but it is not in use today due to the neurotoxicity and anaphylactic reactions induced by the enzyme (Koknel Talu and Derby 2008). Intradiscal injection of collagenase or condroitinase ABC was also researched in former studies, as well as, more recently, matrix metalloproteinase and ethanol gel (Knezevic et al. 2017). *Ozone* is a strong oxidizing agent that reacts with the proteoglycans in the nucleus pulposus, which also results in a form of nucleolysis. There is low-quality evidence, mostly grounded in observational studies with problems in inclusion criteria, that intradiscal ozone therapy can produce positive results with low morbidity rates in patients with radicular pain and disc herniation (Magalhaes et al. 2012).

Other forms of percutaneous disc decompression produce the lesion of nucleus with application of energy. *Percutaneous laser disc decompression* (PLDD) uses an optic fiber inserted in the nucleus pulposus and a laser is activated, producing vaporization of nucleus water content. One RCT (Brouwer et al. 2017) compared microdiscectomy to this technique, followed by surgery if needed, in patients with disc herniations smaller than one-third of the spinal canal. The clinical outcomes were similar at 1- and 2-year follow-ups, there was a faster recovery in microdiscectomy group, and 52% of the patients submitted to PLDD needed surgery afterward. The authors concluded that surgery could be avoided in 48% of those patients who were originally candidates for surgery.

Nucleotome and *Dekompressor* are devices that induce a lesion in the nucleus by mechanical action, followed by aspiration. There is limited evidence, mostly based on observational studies or RCT with deficiencies, of their efficacy in selected patients with radicular pain, with a low incidence of complications (Ong, Chua, and Vissers 2016).

Nucleoplasty, or percutaneous disc coagulation therapy, uses plasma light to create channels in the nucleus pulposus, with the purpose of decreasing the intradiscal pressure. This technology can dissolve tissue at relatively low temperatures, sparing surrounding tissues. Currently, nucleoplasty is perhaps the most popular percutaneous disc decompression technique. One RCT concluded for the superiority of nucleoplasty compared to repeated steroid epidural injections (Gerszten et al. 2010), and numerous observational studies show favorable results after nucleoplasty, with comparable outcomes to microdiscectomy in cases of contained herniations (Ong, Chua, and Vissers 2016).

1.9 FINAL REMARKS

Lumbar DDD plays a major role in human pain and disability worldwide. When conservative treatment fails, there are no valid interventional techniques than can solve discogenic or radicular pain in a physiologic and reconstructive method. The evidence is fair for diagnosis and invasive treatment of discogenic back pain. Lumbar disc regenerative techniques are the main goal of current research in lumbar DDD.

REFERENCES

Armbrecht, G., D. Felsenberg, M. Ganswindt et al. 2017. "Degenerative inter-vertebral disc disease osteochondrosis intervertebralis in Europe: Prevalence, geographic variation and radiological correlates in men and women aged 50 and over." *Rheumatology (Oxford)*. doi: 10.1093/rheumatology/kex040.

Barendse, G.A., S.G. van Den Berg, A.H. Kessels, W.E. Weber, and M. van Kleef. 2001. "Randomized controlled trial of percutaneous intradiscal radiofrequency thermocoagulation for chronic discogenic back pain: Lack of effect from a 90-second 70 C lesion." *Spine (Phila Pa 1976)* 26 (3):287–92.

Boden, S.D., D.O. Davis, T.S. Dina, N.J. Patronas, and S.W. Wiesel. 1990. "Abnormal magnetic-resonance scans of the lumbar spine in asymptomatic subjects. A prospective investigation." *J Bone Joint Surg Am* 72 (3):403–8.

Brinjikji, W., F.E. Diehn, J.G. Jarvik et al. 2015. "MRI findings of disc degeneration are more prevalent in adults with low back pain than in asymptomatic controls: A systematic review and meta-analysis." *AJNR Am J Neuroradiol* 36 (12):2394–9. doi: 10.3174/ajnr.A4498.

Brouwer, P.A., R. Brand, M.E. van den Akker-van Marle et al. 2017. "Percutaneous laser disc decompression versus conventional microdiscectomy for patients with sciatica:

Two-year results of a randomised controlled trial." *Interv Neuroradiol* 23 (3):313–24. doi: 10.1177/1591019917699981.

Bydon, M., R. De la Garza-Ramos, M. Macki et al. 2014. "Lumbar fusion versus nonoperative management for treatment of discogenic low back pain: A systematic review and meta-analysis of randomized controlled trials." *J Spinal Disord Tech* 27 (5):297–304. doi: 10.1097/BSD.0000000000000072.

Cao, P., L. Jiang, C. Zhuang, Y. Yang et al. 2011. "Intradiscal injection therapy for degenerative chronic discogenic low back pain with end plate Modic changes." *Spine J* 11 (2):100–6. doi: 10.1016/j.spinee.2010.07.001.

Carragee, E.J., S.J. Paragioudakis, and S. Khurana. 2000. "2000 Volvo Award winner in clinical studies: Lumbar high-intensity zone and discography in subjects without low back problems." *Spine (Phila Pa 1976)* 25 (23):2987–92.

Cherkin, D.C., R.A. Deyo, J.D. Loeser, T. Bush, and G. Waddell. 1994. "An international comparison of back surgery rates." *Spine (Phila Pa 1976)* 19 (11):1201–6.

Chetty, L. 2017. "A critical review of low back pain guidelines." *Workplace Health Safety* 2165079917702384. doi: 10.1177/2165079917702384.

Cho, T.K., J.H. Lim, S.H. Kim et al. 2014. "Preoperative predictable factors for the occurrence of adjacent segment degeneration requiring second operation after spinal fusion at isolated L4–L5 level." *J Neurol Surg* 75 (4):270–5. doi: 10.1055/s-0033-1349331.

Clark, A.J., M.M. Safaee, N.R. Khan, M.T. Brown, and K.T. Foley. 2017. "Tubular microdiscectomy: Techniques, complication avoidance, and review of the literature." *Neurosurg Focus* 43 (2):E7. doi: 10.3171/2017.5.FOCUS17202.

Cuellar, J.M., M.P. Stauff, R.J. Herzog et al. 2016. "Does provocative discography cause clinically important injury to the lumbar intervertebral disc? A 10-year matched cohort study." *Spine J* 16 (3):273–80. doi: 10.1016/j.spinee.2015.06.051.

Desai, M.J., L. Kapural, J.D. Petersohn et al. 2016. "A prospective, randomized, multicenter, open-label clinical trial comparing intradiscal biacuplasty to conventional medical management for discogenic lumbar back pain." *Spine (Phila Pa 1976)* 41 (13):1065–74. doi: 10.1097/BRS.0000000000001412.

Dhillon, K.S. 2016. "Spinal fusion for chronic low back pain: A 'magic bullet' or wishful thinking?" *Malays Orthop J* 10 (1):61–8.

Ding, F., Z. Jia, Z. Zhao et al. 2017. "Total disc replacement versus fusion for lumbar degenerative disc disease: A systematic review of overlapping meta-analyses." *Eur Spine J* 26 (3):806–15. doi: 10.1007/s00586-016-4714-y.

Eck, J.C., A. Sharan, Z. Ghogawala et al. 2014. "Guideline update for the performance of fusion procedures for degenerative disease of the lumbar spine. Part 7: Lumbar fusion for intractable low back pain without stenosis or spondylolisthesis." *J Neurosurg Spine* 21 (1):42–7. doi: 10.3171/2014.4.SPINE14270.

Fakouri, B., N.R. Shetty, and T.C. White. 2015. "Is sequestrectomy a viable alternative to microdiscectomy? A systematic review of the literature." *Clin Orthop Relat Res* 473 (6):1957–62. doi: 10.1007/s11999-014-3904-3.

Fardon, D.F., A.L. Williams, E.J. Dohring et al. 2014. "Lumbar disc nomenclature: Version 2.0: recommendations of the combined task forces of the North American Spine Society, the American Society of Spine Radiology and the American Society of Neuroradiology." *Spine J* 14 (11):2525–45. doi: 10.1016/j.spinee.2014.04.022.

Freeman, B.J., R.D. Fraser, C.M. Cain, D.J. Hall, and D.C. Chapple. 2005. "A randomized, double-blind, controlled trial: Intradiscal electrothermal therapy versus placebo for the treatment of chronic discogenic low back pain." *Spine (Phila Pa 1976)* 30 (21):2369–77; discussion 2378.

Gerszten, P.C., M. Smuck, J.P. Rathmell et al. 2010. "Plasma disc decompression compared with fluoroscopy-guided transforaminal epidural steroid injections for symptomatic contained lumbar disc herniation: A prospective, randomized, controlled trial." *J Neurosurg Spine* 12 (4):357–71. doi: 10.3171/2009.10.SPINE09208.

Global Burden of Disease Study Collaborators. 2015. "Global, regional, and national incidence, prevalence, and years lived with disability for 301 acute and chronic diseases and injuries in 188 countries, 1990–2013: A systematic analysis for the Global Burden of Disease Study 2013." *Lancet* 386 (9995):743–800. doi: 10.1016/S0140-6736(15)60692-4.

He, J., S. Xiao, Z. Wu, and Z. Yuan. 2016. "Microendoscopic discectomy versus open discectomy for lumbar disc herniation: A meta-analysis." *Eur Spine J* 25 (5):1373–81. doi: 10.1007/s00586-016-4523-3.

Hohaus, C., T.M. Ganey, Y. Minkus, and H.J. Meisel. 2008. "Cell transplantation in lumbar spine disc degeneration disease." *Eur Spine J* 17 (Suppl 4):492–503. doi: 10.1007/s00586-008-0750-6.

Hoy, D., P. Brooks, F. Blyth, and R. Buchbinder. 2010. "The epidemiology of low back pain." *Best Pract Res Clin Rheumatol* 24 (6):769–81. doi: 10.1016/j.berh.2010.10.002.

Ito, M., K.M. Incorvaia, S.F. Yu et al. 1998. "Predictive signs of discogenic lumbar pain on magnetic resonance imaging with discography correlation." *Spine (Phila Pa 1976)* 23 (11):1252–8; discussion 1259–60.

Jacobs, W.C., M. van Tulder, M. Arts et al. 2011. "Surgery versus conservative management of sciatica due to a lumbar herniated disc: A systematic review." *Eur Spine J* 20 (4):513–22. doi: 10.1007/s00586-010-1603-7.

Jensen, M.C., M.N. Brant-Zawadzki, N. Obuchowski et al. 1994. "Magnetic resonance imaging of the lumbar spine in people without back pain." *N Engl J Med* 331 (2):69–73. doi: 10.1056/NEJM199407143310201.

Kallewaard, J.W., J.W. Geurts, A. Kessels et al. 2016. "Efficacy, safety, and predictors of intradiscal methylene blue injection for discogenic low back pain: Results of a multi-center prospective clinical series." *Pain Pract* 16 (4):405–12. doi: 10.1111/papr.12283.

Kapural, L., B. Vrooman, S. Sarwar et al. 2013. "A randomized, placebo-controlled trial of transdiscal radiofrequency, biacuplasty for treatment of discogenic lower back pain." *Pain Med* 14 (3):362–73. doi: 10.1111/pme.12023.

Khan, N.R., A.J. Clark, S.L. Lee et al. 2015. "Surgical outcomes for minimally invasive vs open transforaminal lumbar interbody fusion: An updated systematic review and meta-analysis." *Neurosurgery* 77 (6):847–74; discussion 874. doi: 10.1227/NEU.0000000000000913.

Khot, A., M. Bowditch, J. Powell, and D. Sharp. 2004. "The use of intradiscal steroid therapy for lumbar spinal discogenic pain: A randomized controlled trial." *Spine (Phila Pa 1976)* 29 (8):833–6; discussion 837.

Knezevic, N.N., S. Mandalia, J. Raasch, I. Knezevic, and K.D. Candido. 2017. "Treatment of chronic low back pain—New approaches on the horizon." *J Pain Res* 10:1111–23. doi: 10.2147/JPR.S132769.

Koknel Talu, G., and R. Derby. 2008. "Chemonucleolysis in low back pain." *Agri* 20 (2):8–13.

Kvarstein, G., L. Mawe, A. Indahl et al. 2009. "A randomized double-blind controlled trial of intra-annular radiofrequency thermal disc therapy—A 12-month follow-up." *Pain* 145 (3):279–86. doi: 10.1016/j.pain.2009.05.001.

Lane, N.E., M.C. Nevitt, H.K. Genant, and M.C. Hochberg. 1993. "Reliability of new indices of radiographic osteoarthritis of the hand and hip and lumbar disc degeneration." *J Rheumatol* 20 (11):1911–8.

Lu, Y., J.Z. Guzman, D. Purmessur et al. 2014. "Nonoperative management of discogenic back pain: A systematic review." *Spine (Phila Pa 1976)* 39 (16):1314–24. doi: 10.1097/BRS.0000000000000401.

Magalhaes, F.N., L. Dotta, A. Sasse, M.J. Teixera, and E.T. Fonoff. 2012. "Ozone therapy as a treatment for low back pain secondary to herniated disc: A systematic review and meta-analysis of randomized controlled trials." *Pain Physician* 15 (2):E115–29.

Manchikanti, L., S. Abdi, S. Atluri et al. 2013. "An update of comprehensive evidence-based guidelines for interventional techniques in chronic spinal pain. Part II: Guidance and recommendations." *Pain Physician* 16 (2 Suppl):S49–283.

Mirza, S.K., R.A. Deyo, P.J. Heagerty et al. 2006. "Towards standardized measurement of adverse events in spine surgery: Conceptual model and pilot evaluation." *BMC Musculoskelet Disord* 7:53. doi: 10.1186/1471-2474-7-53.

Modic, M.T., P.M. Steinberg, J.S. Ross, T.J. Masaryk, and J.R. Carter. 1988. "Degenerative disk disease: Assessment of changes in vertebral body marrow with MR imaging." *Radiology* 166 (1 Pt 1):193–9. doi: 10.1148/radiology.166.1.3336678.

Moriguchi, Y., M. Alimi, T. Khair et al. 2016. "Biological treatment approaches for degenerative disk disease: A literature review of *in vivo* animal and clinical data." *Global Spine J* 6 (5):497–518. doi: 10.1055/s-0036-1571955.

Nguyen, C., I. Boutron, G. Baron et al. 2017. "Intradiscal glucocorticoid injection for patients with chronic low back pain associated with active discopathy: A randomized trial." *Ann Intern Med* 166 (8):547–556. doi: 10.7326/M16-1700.

Oh, W.S., and J.C. Shim. 2004. "A randomized controlled trial of radiofrequency denervation of the ramus communicans nerve for chronic discogenic low back pain." *Clin J Pain* 20 (1):55–60.

Ong, D., N.H. Chua, and K. Vissers. 2016. "Percutaneous disc decompression for lumbar radicular pain: A review article." *Pain Pract* 16 (1):111–26. doi: 10.1111/papr.12250.

Orozco, L., R. Soler, C. Morera et al. 2011. "Intervertebral disc repair by autologous mesenchymal bone marrow cells: A pilot study." *Transplantation* 92 (7):822–8. doi: 10.1097/TP.0b013e3182298a15.

Pauza, K.J., S. Howell, P. Dreyfuss et al. 2004. "A randomized, placebo-controlled trial of intradiscal electrothermal therapy for the treatment of discogenic low back pain." *Spine J* 4 (1):27–35.

Peng, B., X. Pang, Y. Wu, C. Zhao, and X. Song. 2010. "A randomized placebo-controlled trial of intradiscal methylene blue injection for the treatment of chronic discogenic low back pain." *Pain* 149 (1):124–9. doi: 10.1016/j.pain.2010.01.021.

Pfirrmann, C.W., A. Metzdorf, M. Zanetti, J. Hodler, and N. Boos. 2001. "Magnetic resonance classification of lumbar intervertebral disc degeneration." *Spine (Phila Pa 1976)* 26 (17):1873–8.

Raastad, J., M. Reiman, R. Coeytaux, L. Ledbetter, and A.P. Goode. 2015. "The association between lumbar spine radiographic features and low back pain: A systematic review and meta-analysis." *Semin Arthritis Rheum* 44 (5):571–85. doi: 10.1016/j.semarthrit.2014.10.006.

Salzmann, S.N., N. Plais, J. Shue, and F.P. Girardi. 2017. "Lumbar disc replacement surgery successes and obstacles to widespread adoption." *Curr Rev Musculoskelet Med* 10 (2):153–9. doi: 10.1007/s12178-017-9397-4.

Schwarzer, A.C., C.N. Aprill, R. Derby et al. 1995. "The prevalence and clinical features of internal disc disruption in patients with chronic low back pain." *Spine (Phila Pa 1976)* 20 (17):1878–83.

Thalgott, J.S., T.J. Albert, A.R. Vaccaro et al. 2004. "A new classification system for degenerative disc disease of the lumbar spine based on magnetic resonance imaging, provocative discography, plain radiographs and anatomic considerations." *Spine J* 4 (6 Suppl):167S–172S. doi: 10.1016/j.spinee.2004.07.001.

Tonosu, J., H. Oka, K. Matsudaira et al. 2017. "The relationship between findings on magnetic resonance imaging and previous history of low back pain." *J Pain Res* 10:47–52. doi: 10.2147/JPR.S122380.

Tuakli-Wosornu, Y.A., A. Terry, K. Boachie-Adjei et al. 2016. "Lumbar intradiskal platelet-rich plasma (PRP) injections: A prospective, double-blind, randomized controlled study." *PM R* 8 (1):1–10; quiz 10. doi: 10.1016/j.pmrj.2015.08.010.

Tubach, F., J. Beaute, and A. Leclerc. 2004. "Natural history and prognostic indicators of sciatica." *J Clin Epidemiol* 57 (2):174–9. doi: 10.1016/S0895-4356(03)00257-9.
van den Hout, W.B., W.C. Peul, B.W. Koes et al. 2008. "Prolonged conservative care versus early surgery in patients with sciatica from lumbar disc herniation: Cost utility analysis alongside a randomised controlled trial." *BMJ* 336 (7657):1351–4. doi: 10.1136/bmj.39583.709074.BE.
Waddell, G. 2005. "Subgroups within 'nonspecific' low back pain." *J Rheumatol* 32 (3):395–6.
Weber, H., I. Holme, and E. Amlie. 1993. "The natural course of acute sciatica with nerve root symptoms in a double-blind placebo-controlled trial evaluating the effect of piroxicam." *Spine (Phila Pa 1976)* 18 (11):1433–8.
Weishaupt, D., M. Zanetti, J. Hodler et al. 2001. "Painful lumbar disk derangement: Relevance of endplate abnormalities at MR imaging." *Radiology* 218 (2):420–7. doi: 10.1148/radiology.218.2.r01fe15420.
Wilke, H.J., F. Rohlmann, C. Neidlinger-Wilke et al. 2006. "Validity and interobserver agreement of a new radiographic grading system for intervertebral disc degeneration: Part I. Lumbar spine." *Eur Spine J* 15 (6):720–30. doi: 10.1007/s00586-005-1029-9.
Woods, B.I., N. Vo, G. Sowa, and J.D. Kang. 2011. "Gene therapy for intervertebral disk degeneration." *Orthop Clin North Am* 42 (4):563–74, ix. doi: 10.1016/j.ocl.2011.07.002.
Zhang, X., J. Hao, Z. Hu, and H. Yang. 2016. "Clinical evaluation and magnetic resonance imaging assessment of intradiscal methylene blue injection for the treatment of discogenic low back pain." *Pain Physician* 19 (8):E1189–95.

2 Animal Models and Imaging of Intervertebral Disc Degeneration

Marion Fusellier, Johann Clouet, Olivier Gauthier, Catherine Le Visage, and Jerome Guicheux

CONTENTS

2.1 Introduction 20
2.2 *In Vivo* Models of DDD 21
2.3 Characteristics of the IVD Among Species 29
    2.3.1 Anatomy of the IVD: Comparison of Various Animal Models 29
    2.3.2 Cellular Population of the IVD 32
    2.3.3 Biomechanics of the IVD 33
2.4 DDD Induction Technique 34
    2.4.1 Animal Models of Spontaneously Occurring DDD 35
        2.4.1.1 Degeneration Related to Aging 35
        2.4.1.2 Inherited Lesions 35
        2.4.1.3 Genetic Models 37
    2.4.2 Experimentally Induced Disc Disease 38
        2.4.2.1 Mechanical Models 38
            2.4.2.1.1 Bipedalism 38
            2.4.2.1.2 Forceful Bending 38
            2.4.2.1.3 Compression 38
            2.4.2.1.4 Instability (Fusion and Facetectomy) 39
        2.4.2.2 Structural Models 40
            2.4.2.2.1 Physical Models 40
            2.4.2.2.2 Enzymatic Models: Chemonucleolysis 43
            2.4.2.2.3 Endplate Injury 44
2.5 Surgical Approach to the IVD 45
2.6 Potential Applications of the Different Models 45
2.7 Imaging of IVD Degeneration and Perspectives 46
    2.7.1 Physiopathology Context 46
    2.7.2 X-Ray Imaging 47
    2.7.3 Nuclear Medicine 49
    2.7.4 Magnetic Resonance Imaging 49

2.8   New Developments in MRI 51
2.9   Conclusion 53
References 53

2.1 INTRODUCTION

Low back pain (LBP) is commonly associated with a progressive age-related degeneration of intervertebral discs (IVDs). IVD degeneration is based on a cascade of degenerative events that ultimately lead to the degradation of the disc's extracellular matrix (ECM) and a loss of its mechanical function. The incidence of LBP is rising exponentially, notably because of the current demographic changes with an increasingly aged population, making LBP a major public health concern in Western industrialized societies (Hoy et al. 2014). Given the significant contribution of IVD alteration to LBP, research has recently focused on improving the understanding of the molecular mechanisms that govern the onset of degenerative disc disease (DDD) as well as on the development of clinically relevant therapeutic strategies. To elucidate such complex biochemical and cell-mediated processes and investigate innovative therapeutic strategies, models that closely mimic human disease are required. *In vitro* cell and organ cultures have notably been developed to test hypotheses without the influence of confounding factors. Unfortunately, given the complexity of the parameters (biochemical, biomechanical, etc.) acting simultaneously in the IVD, these models remain of limited value and their use is restricted to mechanistic analysis.

Considering that it is often necessary to compare *in vitro* findings to the reality of the whole body, complex models that more faithfully replicate the *in vivo* situation have been contemplated. Establishing preclinical animal models is an essential step before translation to humans. In addition, these *in vivo* models are critical because of ethical restrictions and safety issues that make it difficult to obtain human biological samples. They are therefore being developed for both fundamental research, notably to address issues regarding the etiopathogenesis and pathophysiological conditions that lead to DDD, and translational research purposes to assess the clinical relevance of novel regenerative strategies. Disc degeneration involves all parts of the disc, the *nucleus pulposus* (NP) as well as the *annulus fibrosus* (AF) and endplates. Nevertheless, it is acknowledged that DDD originates from the NP, where the degenerative cascade is initiated very early in adults. With respect to the initiating role of NP in the onset of DDD, this chapter will mainly focus on animal models of NP degeneration.

Over the past few years, many different preclinical models have been developed and numerous species have been studied. Many protocols of DDD induction have also been tested in order to obtain the optimal model that would closely mimic the different stages of IVD degeneration, including chemical, cellular, and biomechanical modifications. Since no animal model can mimic the human disc degeneration per se, selection of the most appropriate animal model implies a compromise related to the experimental purposes pursued.

2.2 IN VIVO MODELS OF DDD

The choice of a DDD model entails a balance between many requirements: size of the animal, morphology of the IVD and spine, mechanical and biochemical properties, time for the onset of degeneration, cost of the model, and housing concerns.

The ideal model should meet the following criteria:

- First, it must conform to ethics guidelines; i.e., it must follow the 3Rs Principle: replace the animal with nonanimal methods whenever possible; reduce the number of animals used, particularly using the most refined, reproducible, and sensitive methods to minimize the variability associated with each experimental manipulation; and refine animal experimentation by reducing pain, distress, or lasting harm and improving animal welfare.
- From the anatomical point of view, the size and anatomy of the model are very important because the IVD height should be sufficient to mimic the challenging nutritional environment found in human discs (O'Connell, Vresilovic, and Elliott 2007). The size should also be sufficiently large to evaluate the surgical devices that are developed for human surgery, allowing transposition of the surgical approach.
- The biomechanical properties of the IVD must be close to the human IVD. Quadrupedal models of DDD seem intuitively farther from the human situation than bipedal models. Nevertheless, since lumbar muscle and ligament tensions are the main contributors to the forces applied to IVDs, the influence of this parameter is reduced.
- The disease process studied must be adequately modeled: biochemical and cellular compositions should be close to those in humans (Moon et al. 2013) before and after degeneration. Moreover, time for the onset of degeneration should be as short as possible to reduce costs, and the model conditions, either spontaneous or induced, must be reproducible.
- High cost and housing concerns, notably for large-animal models, explain the wide use of small animals (i.e. rodents and rabbits).
- In the end, the animal models that are used vary greatly in size from the smallest ones to the largest ones (Table 2.1). Rabbits became the most widely used animal as an IVD degeneration model because of their relatively modest cost in terms of purchase and housing and because they are the smallest animals in which the annular puncture model is possible without using sophisticated microsurgical techniques. Nevertheless, there is a tendency toward larger models (dog, sheep, goat, pig, or monkey), in particular for models specifically designed for assessing cell-based therapies (Oehme et al. 2015).

TABLE 2.1
Different Animal Models of IVD Degeneration

Species	Presence of Notochordal Cells in Adulthood (Yes or No)	Onset of Degeneration	Type of Model	Comments	References
Mediterranean sand rat	Yes	2–12 months	Spontaneous		Gruber, Ingram, and Hanley 2006
		18 months	Spontaneous Diabetes		Adler, Schoenbaum, and Silberberg 1983
		6 weeks			Ziv et al. 1992
		3–30 months			Moskowitz et al. 1990
		6 months			Silberberg 1988a
		18 months			Ziran et al. 1994
Mouse	Yes	2.5 months	Spontaneous	Kyphoscoliosis	Mason and Palfrey 1984; Venn and Mason 1986
		3 months	Genetic model	Coll IX	Boyd et al. 2008; Kimura et al. 1996
		6 months		Myostatin knockout	Hamrick, Pennington, and Byron 2003
		4 weeks		Cmd aggrecan knockout mice	Watanabe and Yamada 2002
		3 months		Progressive ankylosis	Sweet and Green 1981
		2 weeks		Mutation in Mecom	Juneja et al. 2014
		1 month		Caveolin-1-null mice	Bach et al. 2016
		5–6 months		Murine progeroid syndrome	Vo et al. 2012
		2–6 months		Neurofibromatosis type 1 mice	

(*Continued*)

TABLE 2.1 (CONTINUED)
Different Animal Models of IVD Degeneration

Species	Presence of Notochordal Cells in Adulthood (Yes or No)	Onset of Degeneration	Type of Model		Comments	References
		1 month	Mechanical	Tail bending	Col2a1 gene for collagen II	Rhodes et al. 2015; Wang et al. 2011
		1 week		Dynamic compression		Sahlman et al. 2001
		1 week				Court et al. 2001
						Walsh and Lotz 2004
		3–12 months		Bipedalism		Higuchi, Abe, and Kaneda 1983
		6–12 months		Instability		Miyamoto, Yonenobu, and Ono 1991
		2–8 weeks	Structural	Needle puncture		Martin et al. 2013; Ohnishi et al. 2016
Rat	Yes	4–5 months	Spontaneous	HLA-B27		Taurog et al. 1993
		1–2 months	Mechanical	Static compression		Iatridis et al. 1999; Yurube et al. 2010
		0–14 days		Dynamic compression		Ching et al. 2003, 2004
		1–3 months		Bipedalism		Cassidy et al. 1988
		7–28 days	Structural	Needle puncture		Hsieh et al. 2009; Issy et al. 2013

(Continued)

TABLE 2.1 (CONTINUED)
Different Animal Models of IVD Degeneration

Species	Presence of Notochordal Cells in Adulthood (Yes or No)	Onset of Degeneration	Type of Model	Comments	References
		7–28 days	AF incision		Rousseau et al. 2007
		2 weeks	Chemonucleolysis	Chondroitinase	Norcross et al. 2003
		4–12 weeks			Boxberger et al. 2008
					Clouet et al. 2011; Leung et al. 2008
Rabbit	Yes	6–30 months	Spontaneous Aging		Kroeber et al. 2002
		1 month	Mechanical Static compression		
		6–9 months	Fusion		Phillips, Reuben, and Wetzel 2002
		6 months	Facetectomy		Sullivan, Farfan, and Kahn 1971
		3 months	Torsion + facetectomy		Hadjipavlou et al. 1998
		2–6 weeks	Structural AF incision		Lipson and Muir 1981; Masuda et al. 2004
		2–8 weeks	Needle puncture		Masuda et al. 2004
		3 weeks			Sobajima et al. 2005
		4–9 weeks			Kim, Chun, and Lee 2015
		2–6 weeks			Chai et al. 2016
		2 weeks	Nucleotomy		Sakai et al. 2005
		20–40 days			Lucas et al. 2012

(Continued)

TABLE 2.1 (CONTINUED)
Different Animal Models of IVD Degeneration

Species	Presence of Notochordal Cells in Adulthood (Yes or No)	Onset of Degeneration	Type of Model	Comments	References
		60–90 days	Laser		Fusellier et al. 2016; Lucas et al. 2012
		6 days	Chemonucleolysis	Chymopapain	Kiester et al. 1994
		10 days	Chemonucleolysis	Chondroitinase	Takahashi et al. 1996
		1 week			Henderson, Stanescu, and Cauchoix 1991
		2–6 days		Hyaluronidase	Fusellier et al. 2016
		3–6 months	Ischemic lesion of endplate	Pingyangmycin	Wei et al. 2015
Chondrodystrophic dogs	No	3–7 years	Spontaneous		Bergknut et al. 2011
		6 months	Mechanical Fusion		Taylor et al. 1976
		4 weeks	Structural Nucleotomy		Serigano et al. 2010
		1 month to 1–7 months	Chemonucleolysis	Chymopapain	Melrose et al. 1996
		1–8 days			Atilola et al. 1988
		7–21 days		Chondroitinase	Fry et al. 1991
Nonchondrodystrophic dogs	Yes	1–4 years	Spontaneous Kyphoscoliosis		Faller et al. 2014
		4–12 weeks	Structural AF incision		Chen et al. 2015

(Continued)

TABLE 2.1 (CONTINUED)
Different Animal Models of IVD Degeneration

Species	Presence of Notochordal Cells in Adulthood (Yes or No)	Onset of Degeneration	Type of Model	Comments	References
		6 weeks	Partial nucleotomy		Ganey et al. 2009
		2 weeks	Chemonucleolysis	Chymopapain	Bradford, Cooper, and Oegema 1983
		3 weeks			Wakano et al. 1983
		2 weeks			Spencer, Miller, and Schultz 1985
		1 week			Lü et al. 1997
		2 weeks			Kudo, Sumi, and Hashimoto 1993
		2 weeks		Collagenase	Spencer, Miller, and Schultz 1985
		2 weeks			Bromley et al. 1980
		1 week		Chondroitinase	Lü et al. 1997
Goat	No	2 months	Structural	Annulotomy/drill bit	Zhang et al. 2011
		12–26 weeks	Chemonucleolysis	Chondroitinase	Hoogendoorn et al. 2008
Sheep	No	4–7 years	Spontaneous	Spontaneous aging	Nisolle et al. 2016
		4–12 months	Structural	Partial annulotomy	Osti, Vernon-Roberts, and Fraser 1990

(Continued)

TABLE 2.1 (CONTINUED)
Different Animal Models of IVD Degeneration

Species	Presence of Notochordal Cells in Adulthood (Yes or No)	Onset of Degeneration	Type of Model	Comments	References
		6 months	AF incision		Freeman et al. 2016
		3–12 months			Melrose et al. 2002
		3 months			Schollum et al. 2010
		12–48 weeks	Nucleotomy		Guder et al. 2009; Reitmaier et al. 2014; Russo et al. 2016; Vadalà et al. 2013, 2015
		1 week	Chemonucleolysis	Chondroitinase	Sasaki et al. 2001
Pig	Yes	3 months	AF incision Structural		Kääpä et al. 1994
		3–24 weeks	Nucleotomy		Acosta et al. 2011; Omlor et al. 2009
		3 months	Perforation of endplate		Holm et al. 2004
		7 months			Cinotti et al. 2005
Macaque	Yes	22.3 years	Spontaneous Spontaneous aging		Nuckley et al. 2008
Baboon	Yes	14–15.7 years	Spontaneous Spontaneous aging		Lauerman et al. 1992; Platenberg et al. 2001
		4 weeks	Chemonucleolysis Structural	Chymopapain	Zook and Kobrine 1986
		4 weeks		Collagenase	Zook and Kobrine 1986

(Continued)

TABLE 2.1 (CONTINUED)
Different Animal Models of IVD Degeneration

Species	Presence of Notochordal Cells in Adulthood (Yes or No)	Onset of Degeneration	Type of Model	Comments	References
Rhesus monkey	Yes	7 days to 14 months	Structural	Annulotomy and collagenase	Stern and Coulson 1976
		6 weeks	Chemonucleolysis	Chymopapain	Sugimura et al. 1996
		6 weeks		Chondroitinase	Sugimura et al. 1996
		7 days to 14 months		Collagenase	Stern and Coulson 1976
		1–3 months	Ischemic lesion of endplate	Pingyangmycin	Zhou et al. 2013b
		3–15 months		Bleomycin	Wei et al. 2014
Cow	No	3 weeks	Structural	Chymopapain	Roberts et al. 2008
		Immediate	Chemonucleolysis	Collagenase	Antoniou et al. 2006; Kalaf, Sell, and Bledsoe 2014
		16 h			Mwale et al. 2008
		3 weeks		Trypsin	Roberts et al. 2006

2.3 CHARACTERISTICS OF THE IVD AMONG SPECIES

2.3.1 ANATOMY OF THE IVD: COMPARISON OF VARIOUS ANIMAL MODELS

Despite the widespread use of animals as models of DDD, limited data are available on the dimensions of the IVD in different species, particularly on IVD and NP volume. Nevertheless, IVD size and height are important parameters to be considered for at least three reasons. First, the large size of human IVDs and the largely avascular nature of IVDs affect the diffusion of nutritive solutes in IVDs. Discs of most animal models are smaller than those of humans, with great variations among species. The second reason is practical: difficulties in the use of small-animal models come from the small size of their IVD, which makes the cell volume injected disproportionate to the volumes intended to be used in humans, notably when considering the injection of stem cells and/or biomaterials. Finally, large-animal models enable the use of the same surgical ancillaries and techniques as those used in humans.

In mammals, IVDs classically present the same organization: they consist of three anatomical and distinct structures. The central and gelatinous part of the IVD is called *nucleus pulposus*. It is a highly hydrated and proteoglycan-rich tissue. NP is surrounded by a ring made of fibrous tissue called the annulus fibrosus, which is attached to the vertebral endplates by fiber bundles. The anatomy of the lumbar spine in mammals is similar from one animal to another, but the size differs considerably from rodents to large herbivores.

Regarding the spine anatomy, in thoracic and lumbar regions, large animals have a greater mean pedicle height than human pedicles do. Except in sheep, the mean pedicle width is similar between animals and humans when considering baboons, sheep, pigs, calves, and deer (Sheng et al. 2010). Moreover, human lumbar vertebral body width and depth are greater than in animals, but the mean vertebral body height is lower than that of all animals (Sheng et al. 2010).

A closer examination of the IVD shows that lumbar discs of mammals are all bean shaped and are thinnest in the dorsal region and thickest in the ventral region. Conversely, tail discs are more rounded. IVD sizes in the main animal models are given in Table 2.2. It should be stressed that the values reported in the literature are quite variable even for the same species. These discrepancies are likely related to the differences in age or breed of the animals used. In addition, the techniques used for evaluating disc size range from histology to microradiography, computed tomography (CT), and magnetic resonance imaging (MRI), which intrinsically could also explain the variations observed.

In mice, it has been shown that disc thickness increases from birth until 4 months and then decreases with aging (Cao et al. 2017). As compared to humans, rabbits have a lumbar IVD that is more elliptical in shape with a smaller dorsoventral length/lateral length ratio (O'Connell, Vresilovic, and Elliott 2007). In sheep lumbar IVDs, AF water and collagen contents are quite comparable to those of humans (Reid et al. 2002). The sheep lumbar IVD is obviously thinner than the human IVD (Mageed et al. 2013a), and anterior disc height in the sheep lumbar spine is as much as 5 mm lower than the 11- to 16-mm height in humans (Wilke et al. 1997). Nevertheless, lumbar IVDs have similar gross anatomical features in sheep and humans.

TABLE 2.2
Comparison of IVD Geometry among Animal Models and Humans

	Disc Height (SD) (mm)						NP Volume	Disc	
Species	Ventral Height	Dorsal Height	Dorsoventral Width (mm)	Lateral Width (mm)	Disc Surface (mm^2)	NP Area (mm^2)	(mm^3)	Localization	References
Mouse	0.37–0.38	0.22–0.23	0.98–1	1.35–1.47	1.04–1.17			Lumbar discs	Elliott and Sarver 2004
	0.31 (0.03)		1.08	1.82	1.61	0.43		Lumbar discs	Beckstein et al. 2008
	0.75	0.4					0.58	Lumbar discs	Cao et al. 2017
	0.31 (0.03)		1.24	1.84	1.81	0.33		Lumbar discs	O'Connell, Vresilovic, and Elliott 2007
Rat	0.70–0.72	0.52–0.57	2.95–3.17	3.77–4.25	8.28–10.7			Lumbar discs	Elliott and Sarver 2004
	0.77 (0.04)		3.16	4.49	11.85	3.34		Lumbar discs	Beckstein et al. 2008
	0.99		3.23	3.8				Lumbar discs	Bowles et al. 2012
	1.02–1.31							Lumbar discs	Jaumard et al. 2015
	0.93 (0.24)		4.36	5.79	20.4	5		Lumbar discs	O'Connell, Vresilovic, and Elliott 2007
Rabbit	2.4 (0.23)		7.46	13.81	90	22.6		Lumbar discs	Beckstein et al. 2008
	1.42 (0.39)		6.59	12.7	73.4	18		Lumbar discs	O'Connell, Vresilovic, and Elliott 2007
Sheep	2.6–3.3							Lumbar discs	Mageed et al. 2013a
	4.9 (0.4)		18.3 (1.8)	24.9 (1.7)				Lumbar discs	Monaco, Dewitte-Orr, and Gregory 2016
	3.40 (0.46)		19.78	30.03	511	193		Lumbar discs	Beckstein et al. 2008

(Continued)

TABLE 2.2 (CONTINUED)
Comparison of IVD Geometry among Animal Models and Humans

	Disc Height (SD) (mm)								
Species	Ventral Height	Dorsal Height	Dorsoventral Width (mm)	Lateral Width (mm)	Disc Surface (mm^2)	NP Area (mm^2)	NP Volume (mm^3)	Disc Localization	References
Goat	3.93 (0.07)		21.6	34.5	676	267		Lumbar discs	O'Connell, Vresilovic, and Elliott 2007
	7.1–8.2				380.6–404.9			Lumbar discs	Paul et al. 2017
Pig	7.8 (1.3)		25 (2.3)	38 (3.8)				Lumbar discs	Monaco, Dewitte-Orr, and Gregory 2016
	5.46 (0.71)		25.56	37.13	872	232		Lumbar discs	Beckstein et al. 2008
							799 (212)		Kasch et al. 2012
Calf	6.09 (1.32)		32.75	39.33	1100			Lumbar discs	Beckstein et al. 2008
Cow	12.4 (0.4)		24.1 (1.2)	23.7 (1.7)				Tail discs	Monaco, Dewitte-Orr, and Gregory 2016
	9.18 (0.65)		32	33.3	857	231		Tail discs	Beckstein et al. 2008
	6.90 (0.35)		27.8	28.9	622	176		Tail discs	O'Connell, Vresilovic, and Elliott 2007
Human	7.9–10							Lumbar discs	Mageed et al. 2013a
	10.9 (2.7)				1560			Lumbar discs	Elliott and Sarver 2004
	10.91 (0.83)		37.67	55.38	1925	598		Lumbar discs	Beckstein et al. 2008
	10.27 (2.23)	6.07 (1.97)						Lumbar discs	Miao et al. 2013
	11.3 (0.3)		37.2	55.9	1727	479		Lumbar discs	O'Connell, Vresilovic, and Elliott 2007

Variations can be observed and are related to differences in age, sex, or strain of the sheep studied. When comparing the same vertebrae, sheep lumbar vertebral body volumes are as much as 48% smaller than human volumes (Mageed et al. 2013a), with vertebral bodies longer than wider, unlike in humans (Mageed et al. 2013b).

Bovine IVDs exhibit the same general organization as found in humans and are biconvex in the lumbar segments (Cotterill et al. 1986). In calves, the lumbar spine is a similar size to the human lumbar spine, but the IVDs contribute only 10% of the total length of the spine, in contrast to 25% in humans. Moreover, when studying biomechanical parameters, the presence of weak cartilaginous growth plate can affect the expected results. The tail IVD of cows is one of the thickest and it has a circular shape (Monaco, Dewitte-Orr, and Gregory 2016) and the NP area compared to the disc area is one of the closest to human IVD (O'Connell, Vresilovic, and Elliott 2007).

Data related to human IVD geometry are not fully comparable and are strongly dependent on age, sex, and anatomical level. Among all species, humans have the thickest IVD with the highest disc area (O'Connell, Vresilovic, and Elliott 2007). Thus, the cross-sectional area of the ovine disc is about 630 mm^2, only half the size of the human disc with around 1380 mm^2 (Schmidt and Reitmaier 2012). The distance between the NP and the adjacent vertebral body is therefore much closer than in humans, and in smaller species such as rodents, results may be biased because they do not suffer from the nutritional deficiencies observed in human IVD. Surprisingly, after analyzing the geometries of IVDs (disc height, width, and NP size), O'Connell, Vresilovic, and Elliott (2007) found that mouse and rat lumbar IVDs and mouse tail IVDs are the closest representation of the human lumbar IVD geometry. Nevertheless, large-animal models seem to better mimic the challenging nutritional environment of human IVDs.

Dogs, sheep, goats, and pigs are the most widely used large-animal models because of their relatively large size, which intuitively appears to be more appropriate when surgical procedures need to be tested. Despite a number of differences with human anatomy, sheep vertebra size and volume are comparable to those in humans (Wilke et al. 1997), making the use of the same implants and instruments possible. This has probably contributed to making sheep the animal model of choice to test vertebral implants (Benneker et al. 2012).

2.3.2 Cellular Population of the IVD

There are differences between animal species and humans in terms of the cell population that constitutes the NP. This difference may be of scientific relevance, because notochordal cells play a pivotal role in the cellular dialog with nucleopulpocytes (Colombier et al. 2014), with a great influence on the metabolism of proteoglycans (Aguiar, Johnson, and Oegema 1999), and hyaluronan production (Stevens et al. 2000) and notochordal cells can synthetize matrix. In contrast to the human situation, the persistence of notochordal cells in the NP of certain animal models could likely influence the experimental data generated, notably those obtained after treatment. In particular, this could modify the response to growth factors by providing signal-responsive cells that are absent in the human IVD, and they could contribute to a self-repair process of the IVD that is not efficient in humans. In human IVD, notochordal cells are present at birth, but their number rapidly decreases before the age of 10 years and they are completely absent in the NP of adults (Alini et al. 2008). Interestingly, notochordal cells are present in the NP of most of the species used as models of DDD such as mice, rats, rabbits, and nonchondrodystrophic dogs and pigs (Omlor et al. 2009). Mouse and rat notochordal cells remain in the NP throughout life and rabbits until at least 12 months of age (Hunter, Matyas, and Duncan 2004). These nonchondrodystrophic species are less pertinent as models due to their lower incidence of spontaneous disc degeneration. Horses have no notochordal cells at birth (Yovich, Powers, and Stashak 1985). As has been described in humans, a rapid decrease in the number of notochordal cells is observed after birth in the NP of sheep and goats (Daly et al. 2016; Hoogendoorn et al. 2008), as well as chondrodystrophic dogs (Bergknut et al. 2012) and cows. It is notable that in sheep and goats, NP proteoglycan contents and whole-disc biomechanics are superimposable on these features in humans (Beckstein et al. 2008). These species thus exhibit greater relevance than others, notably for experiments intended to decipher the IVD degenerative process or to assess the efficacy of cell-based regenerative therapies.

2.3.3 Biomechanics of the IVD

The question of quadrupedal locomotion and the influence on IVD biomechanics frequently arises when looking for an animal model of DDD. With the exception of monkeys and bipedal mice and rats, all animal models of DDD are quadrupedal. Nevertheless, the vertical position seems to have a significant influence on the mechanical load to which human lumbar IVDs are exposed. This could limit the use of quadrupedal models. However, muscle contraction and ligament tension are significant contributors to the load to which IVDs are exposed (Wilke et al. 2003), and the influence of the weight of the upper body on the lumbar spine is lessened by the influence of the loading action of muscles and ligaments, as shown in Figure 2.1.

100: man lying prone
120: man lying laterally
350–950: man jogging
360: rabbit
460: man sitting unsupported
500: man relaxed standing
530–650: man walking
500–800: ovine lying
830: man sitting with maximum flexion
750–1000: ovine standing
1100: man standing bent forward
2300: man lifting a 20 kg weight with round flexed back

Pression kPa

FIGURE 2.1 Comparison of intradiscal pressure in human (Wilke et al. 1999), rabbit (Guehring et al. 2006), and ovine (Reitmaier et al. 2013) lumbar disc.

It would even appear that the load exerted by these structures on the lumbar IVDs of large animals by these structures may be greater than that observed in humans because of the increased difficulty of stabilizing a horizontal spinal column compared to a vertically balanced spine. Thus, the mechanical performance of IVDs is very similar across species (in calf, pig, baboon, sheep, rabbit, rat, and mouse lumbar IVDs), after normalization by the geometric parameters of disc height and area (Beckstein et al. 2008). Similarly, although tail IVDs are suspected of not being subjected to the same loading pressure as human IVDs, the swelling pressure of bovine tail IVDs is reported to be similar to that of human IVDs (Oshima et al. 1993).

2.4 DDD INDUCTION TECHNIQUE

Research focusing on IVD degeneration or experimental therapeutics has been conducted on models that more or less mimic the natural pathophysiological process of human IVD degeneration. A great variety of approaches have been used to induce IVD degeneration in both large- and small-animal models (Table 2.1).

Two main categories of models are commonly described, as shown in Figure 2.2: spontaneous and experimentally induced models. Spontaneous models include animals that, like humans, spontaneously develop a DDD in direct relationship with aging, genetic alteration, or specific breeding conditions. Experimentally induced models are subdivided into models based on mechanical or structural lesions. Mechanical lesions are generally induced by modifications of the forces applied to the IVD. Structural lesions are mainly related to direct alterations of the IVD by means of injury, inflammation, or chemical lesions.

The time course and extent of IVD degeneration strongly differ depending on the induction method. Ideally, the onset of degeneration should be rapid, to limit the cost of housing, and should be reproducible.

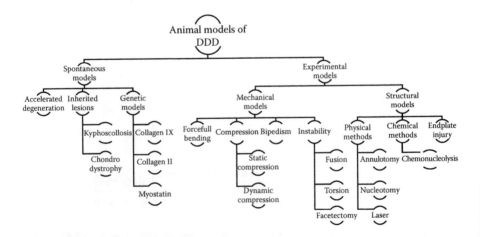

FIGURE 2.2 Different animal models of DDD.

2.4.1 Animal Models of Spontaneously Occurring DDD

2.4.1.1 Degeneration Related to Aging

Like humans, several animal species exhibit spontaneous degeneration. Mediterranean sand rats that develop extensive DDD associated with diabetes when fed with a standard laboratory diet (Silberberg 1988b) have been thoroughly studied. The discal lesions develop between 2 and 18 months of age and are extended from the lumbar IVD to the tail (Moskowitz et al. 1990). They are concomitant with the diffuse idiopathic skeletal hyperostosis frequently associated with diabetes mellitus. The modifications of the IVDs are biomechanic, biochemical, morphologic, and cellular alterations and consist of loss of notochordal cells, end-plate sclerosis, and osteophytosis along with a decrease in hydration and osmotic pressure (Gruber et al. 2002; Ziv et al. 1992). Spontaneous degeneration unrelated to diabetes has also been shown in aging rabbits (Clouet et al. 2011; Leung et al. 2008; Lucas et al. 2012), sheep (Nisolle et al. 2016), as well as baboons and macaques (Lauerman et al. 1992; Nuckley et al. 2008; Platenberg et al. 2001).

2.4.1.2 Inherited Lesions

Some species may also present inherited lesions that lead to IVD degeneration. The most frequent cases are kyphoscoliosis and chondrodystrophy.

In dogs and mice, hereditary kyphoscoliosis can induce early degeneration. Kyphoscoliosis is an abnormal conformation of the vertebral column associated with spinal deformity that leads to dorsal (kyphosis) and lateral (scoliosis) curvature. It has been associated with a decrease in cell content, a loss of distinction between NP and AF, and a decrease in proteoglycan content and cervicothoracic IVD size (between the fifth cervical and the second thoracic vertebrae) in 80-day-old homozygous recessive mice (ky/ky) (Mason and Palfrey 1984; Venn and Mason 1986). In these mice, a homozygous variant in the kyphoscoliosis peptidase gene (ky) induces a neuromuscular disorder that leads to kyphoscoliosis because of the weakness of postural muscles. The ky transcript has been detected only in skeletal muscle and heart, and ky protein belongs to a family of transglutaminase-like proteins (Blanco 2001; Blanco et al. 2004). The same effect is frequently observed in canine IVDs adjacent to a vertebral malformation (Faller et al. 2014), as illustrated in Figure 2.3. This malformation often affects the thoracic or thoracolumbar spine.

Chondrodystrophy is an endochondral ossification defect with disturbance in chondrogenesis that leads to impairment of ossification and growth of long bones. It results in a short-limbed disproportionate dwarfism and, in some cases, in vertebral abnormalities. This anomaly has been fully described in dogs and is even selected in some popular breeds such as Dachshund, Basset Hound, Bulldogs, Beagle, Lhassa Apso, Pekingese, Cavalier King Charles Spaniel, Basset Hounds, and Welsh Corgis (Brisson 2010; Cherrone et al. 2004; Griffin et al. 2009; Priester 1976). As seen in Figure 2.4, these dogs tend to exhibit early onset of IVD degeneration at all levels, with a high incidence of disc herniation at the thoracolumbar junction. DDD in chondrodystrophic breeds is of multigene etiology and is termed chondroid metamorphosis. It is characterized by premature loss of proteoglycans, with an increased metalloproteinase 2 activity and a decrease in glycosaminoglycan (GAG) content.

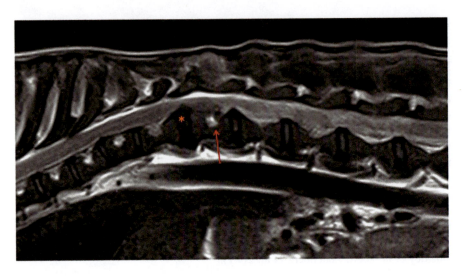

FIGURE 2.3 T2-weighted sagittal MRI of the thoracolumbar spine of a dog with kyphoscoliosis. In case of kyphoscoliosis, the adjacent disc to the malformation exhibits early degeneration, as observed with a decreased MRI T2 signal (*) that is adjacent to a malformed vertebra inducing kyphoscoliosis (red arrow).

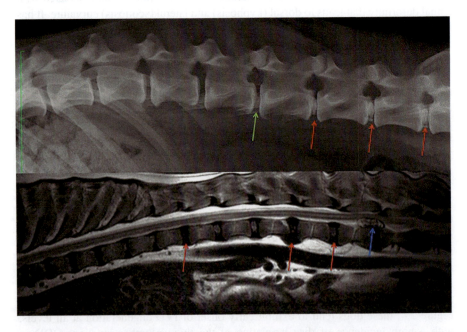

FIGURE 2.4 Lumbar lateral radiography and thoracolumbar T2-weighted sagittal MRI of the spine of a 5-year-old chondrodystrophic dog showing multiple IVD degeneration with a decrease in IVD height (green arrow), MRI T2 signal, and calcification (red arrows) and even disc herniation (blue arrow).

Notochordal cell replacement by chondrocyte-like cells in chondrodystrophic breeds is similar to what is observed in human DDD (Bergknut et al. 2011, 2012; Hunter, Matyas, and Duncan 2003; Johnson, Da Costa, and Allen 2010). Conversely, in nonchondystrophic breeds, notochordal cells are still present in the IVD in adulthood. On MRI, marked similarities have been observed between the different stages of DDD progression in dogs and humans (Bergknut et al. 2011).

2.4.1.3 Genetic Models

Numerous genetically modified models have allowed investigation into the contribution of genetics in disc degeneration:

- Deletion mutation for collagen IX leads to mucous degeneration of IVD in mice (Boyd et al. 2008; Kimura et al. 1996), and collagen IX gene polymorphisms increase the risk of disc generation in the human population (Paassilta et al. 2001).
- Mice with a knockout allele of the *Col2a1* gene encoding for type II collagen (Sahlman et al. 2001) exhibit premature endplate calcification and IVD degeneration.
- Myostatin (GDF8) acts as a negative regulator of skeletal muscle growth and myostatin knockout ($Mstn^{-/-}$) mice have increased weight and a higher trabecular bone mineral density than normal mice. Their IVD showed degenerative changes with a decreased proteoglycan content in the hyaline end plates and inner AF (Hamrick, Pennington, and Byron 2003).
- Many other genetic models have been described with DDD, including, for example, Cmd aggrecan knockout mice (Watanabe and Yamada 2002), progressive ankylosis mice (Sweet and Green 1981), a Mecom mutation in mice (Juneja et al. 2014), caveolin-1-null mice (Bach et al. 2016), neurofibromatosis type 1 mice (Rhodes et al. 2015; Wang et al. 2011), murine progeroid syndrome (Vo et al. 2012), and HLA-B27 rats (Hammer et al. 1990; Taurog et al. 1993).

Spontaneous models of DDD are particularly interesting in providing models of disc degeneration with processes close to those observed in humans. One of the limitations frequently reported for the use of such models is the small number of species in which spontaneous degeneration occurs. We believe that this is a relative limitation to our opinion because of the wide variety of the species concerned. Indeed, spontaneous degeneration has been thoroughly described in Mediterranean sand rats (Gruber, Ingram, and Hanley 2006; Gruber et al. 2002; Moskowitz et al. 1990; Pokharna et al. 1994; Tapp et al. 2008; Ziv et al. 1992), chondrodystrophoid dog species (Bergknut et al. 2011, 2012; Brisson 2010; Cherrone et al. 2004; Griffin et al. 2009; Priester 1976), and rabbits (Clouet et al. 2011; Leung et al. 2008) and has also been observed in sheep (Nisolle et al. 2016) and nonhuman monkey (Lauerman et al. 1992; Nuckley et al. 2008; Platenberg et al. 2001). Although these spontaneous models are particularly advantageous because of their closeness to the human situation, the main limitation to their use is the long and sometimes unpredictable course of degeneration, with inconsistent onset.

2.4.2 Experimentally Induced Disc Disease

Unlike spontaneous models, experimental models require a direct intervention on the IVD or spine to induce degeneration. They are classically divided into two groups: mechanical models that induce IVD degeneration by modification of the mechanical stress applied to the IVD, and structural models that are the result of a direct lesion of IVDs or their adjacent structures such as endplates.

2.4.2.1 Mechanical Models

In agreement with epidemiological data that suggest a link between IVD degeneration and any abnormal loading condition, including immobilization (Battié, Videman, and Parent 2004; Stokes and Iatridis 2004), mechanical models of DDD result from an altered mechanical loading applied to the IVD. It can be the result of bending (Court et al. 2001), static or dynamic disc compression (Ching et al. 2003, 2004; Iatridis et al. 1999; Kroeber et al. 2002; Lotz et al. 1998; Walsh and Lotz 2004), postural changes such as bipedalism (Cassidy et al. 1988; Higuchi, Abe, and Kaneda 1983), hyperactivity such as running (Puustjärvi et al. 1993, 1994; Saamanen et al. 1993), fusion (Bushell et al. 1978; Phillips, Reuben, and Wetzel 2002; Taylor et al. 1976), and spinal distraction or instability (Hadjipavlou et al. 1998; Kroeber et al. 2005).

2.4.2.1.1 Bipedalism

Considering that evolution toward the erected posture could increase the mechanical stress applied to the lumbar spine, bipedal mice and rats have been created by forelimb clipping or amputation (Cassidy et al. 1988; Higuchi, Abe, and Kaneda 1983). Mouse models exhibit an acceleration of age-related disc degeneration (Higuchi, Abe, and Kaneda 1983). Postural models such as bipedalism were thought to increase the time spent in the standing posture. Unfortunately, a recent study has shown that "bipedal" rats spend less time in the erected posture than their quadruped counterparts do (Bailey et al. 2001). Disc degeneration may therefore not be the consequence of the erected posture in the bipedal rat. Moreover, bipedal models today present an ethical issue because amputation induces a significant behavioral stress and is no longer acceptable.

2.4.2.1.2 Forceful Bending

One of the first mechanical models of DDD was the forceful bending of the rat tail by Lindblom in 1957 (Lindblom 1957). Rat tails were fixed in bent shapes. This model was recently taken over in mice instrumented with an external device applying bending to the IVD between coccygeal vertebrae for 1 week (Court et al. 2001). IVD degeneration was evidenced with decreased cellularity in the AF and decreased aggrecan and type II collagen gene expression.

2.4.2.1.3 Compression

The effects of compression have also been reported. Static compression has been investigated in rat tails through the use of an Ilizarov-type apparatus, used commonly as external distractor for limb reconstruction (Iatridis et al. 1999). This resulted in thinning of the IVD with a decrease in axial compliance and angular laxity associated with an increase in proteoglycan content. Degeneration induced by static compression was further confirmed by MRI in another study (Yurube et al. 2010) with a progressive increase in matrix metalloproteinase 3 (MMP-3) in NP cells correlated with radiological degenerative scores. In the rabbit (Kroeber et al. 2002), compressive static force also induces an irreversible decrease in IVD height associated with a decrease in cell density in the AF and endplates after 28 days.

A rat tail model of dynamic or cyclic compression has been developed, suggesting that IVD degeneration can depend on both the total load exposure (magnitude and amplitude) and the frequency of loading (Ching et al. 2003, 2004). Compressive stress was applied to a caudal IVD via pins inserted in caudal vertebral bodies for 14 days. The dynamic compression was applied for 1 hour in a square wave pattern between 440 and 940 kPa at three different frequencies (0.5, 1.5, and 2.5 Hz) (Ching et al. 2004). The effect of cyclic loading was less than the effect of static loading, because the decrease in IVD height was smaller in the cyclical loading group. Nevertheless, while a significant reduction in total proteoglycan content was observed in the IVD submitted to static compression, the effect of cyclic loading did not seem to be limited to the IVD alone but rather extended to adjacent levels. The differential effects of dynamic loading depending on frequency and stress, have been confirmed in a murine tail model of dynamic loading (Walsh and Lotz 2004).

2.4.2.1.4 Instability (Fusion and Facetectomy)

Instability models also depend on the effects of altered loading on IVDs. Various types of model exist: vertebral fusion, torsional injury, and resection of the spinal process or facet joint. These models induce progressive IVD degeneration that initiates within 3–12 months depending on the mechanism of instability and the species. These models were used for many years but are no longer widely used.

The IVD degeneration induced by fusion has been studied in both rabbits (Phillips, Reuben, and Wetzel 2002) and dogs (Bushell et al. 1978). Vertebral fusion, or arthrodesis, aims at surgically fusing vertebrae together by means of bone grafting and devices in order to eliminate movement. In rabbits, progressive degeneration was observed in adjacent treated IVDs as early as 3 months after spine fusion. Radiographs showing narrowing of the IVD space and endplate sclerosis highlighted the degenerative process. These modifications accompanied degradation of the structure of the IVD characterized by proliferation of chondrocytes and loss of notochordal cells in the NP. Fusion in immature beagles (Taylor et al. 1976) also led to a decrease in collagen and a rise in noncollagenous protein content in fused and adjacent IVDs. Nevertheless, it should be noted that in greyhounds, this model failed to induce IVD degeneration (Bushell et al. 1978).

Facetectomy is the removal of a caudal articular process (facet) of a vertebra in order to induce joint destabilization. It can be unilateral or bilateral. Torsional injury after facetectomy in the rabbit has been shown to induce progressive (around 3 months) DDD with a decrease in NP volume accompanied by increased phospholipase A2 levels, indicating inflammation that was not observed with facetectomy alone (Hadjipavlou et al. 1998). Yet facetectomy alone in the lumbar spine of immature white rabbits (Sullivan, Farfan, and Kahn 1971) triggers a decrease in IVD height with thinning of the posterior AF and disorganization of the NP after 6 months.

Mechanical instability by resection of the spinal process in mice also seemed effective in inducing disc degeneration in the cervical spine with proliferation of cartilaginous tissue and fissures in the AF, shrinkage of the NP, disc herniation, and osteophytic formation (Miyamoto, Yonenobu, and Ono 1991). It is interesting to note that instability is less efficient in inducing degeneration in the lumbar spine than in the cervical spine. This is probably due to the different muscular environments that can counteract the induced destabilization.

All these instability models allow the induction of progressive IVD degeneration with many of the features observed in the clinical condition, but several criteria may explain their limited usage: surgical procedures are invasive, complex, and often require expensive equipment, and the time course of progression of the degeneration is uncertain, sometimes relatively long (around 9–12 months), leading to significant costs in animal housing. The main problem of mechanical models is their lack of reproducibility, which probably explains why they have been virtually abandoned.

2.4.2.2 Structural Models

Currently, structural models are the most widely used and concern numerous animal species ranging from mice to sheep. These models consist in inducing degeneration by directly compromising the structural integrity of the IVD by physical methods (including annular injury models [scalpel incision or needle puncture], nucleotomy, and laser heating of the NP), chemical models (chemonucleolysis), or lesions of the vertebral endplates.

2.4.2.2.1 Physical Models

2.4.2.2.1.1 Annulotomy Annulotomy consists in an incision into the AF. Damage to the AF is the most extensively used group of models. These models have been tested in a variety of animal species, including mice (Martin et al. 2013), rats (Hsieh et al. 2009; Issy et al. 2013; Rousseau et al. 2007), rabbits (Kim, Chun, and Lee 2015; Masuda et al. 2004; Sobajima et al. 2005), dogs (Chen et al. 2015), pigs (Kääpä et al. 1994; Pfeiffer et al. 1994), and sheep (Freeman et al. 2016). Many techniques have been described and lesions of the AF can be induced by a clear annular laceration model with a scalpel, a drill bit (Kääpä et al. 1994; Lipson and Muir 1980, 1981; Melrose et al. 2002; Olsewski et al. 1996; Osti, Vernon-Roberts, and Fraser 1990; Rousseau et al. 2007; Schollum et al. 2010; Zhang et al. 2011), or a more superficial injury with needle puncture (Hsieh et al. 2009; Issy et al. 2013; Kim et al. 2005; Martin et al. 2013; Masuda et al. 2004; Michalek, Funabashi, and Iatridis 2010; Sobajima et al. 2005). In small species, a retroperitoneal approach is required for lumbar discs, making the procedure relatively invasive, but for rabbits (Chai et al. 2016; Kim et al. 2005) or larger animals, annular puncture can easily be guided percutaneously by fluoroscopy or CT (Zhou et al. 2013a).

The depth of the lesion varies from partial-thickness annular injury to full-thickness annular stab with NP involvement. With a full-thickness injury, IVD degeneration develops quickly, but the risk of NP herniation is high, making this model less appropriate for studying the progressive onset of degeneration. The larger the lesion is, the earlier the degeneration takes place. Therefore, in superficial laceration or puncture, the modification can be relatively slow and can take months to be apparent, as shown in Table 2.1 (Lipson and Muir 1981; Osti, Vernon-Roberts, and Fraser 1990; Sobajima et al. 2005; Zhang et al. 2011). Nevertheless, this active degeneration may be more relevant for studying the pathophysiological processes.

In puncture models, the degree of degeneration depends on the ratio of the injury size to the disc size (Elliott et al. 2008; Masuda et al. 2004). In mice, the ratio of the needle diameter to the height of the punctured IVD needs to exceed 0.4 to induce significant degeneration. The location of the needle is also of importance and the needle needs to reach the NP with a laterodorsal position (Ohnishi et al. 2016). In rabbits, the use of 16-G and 18-G needles for puncture seems to produce a predictable and slowly progressive disc narrowing apparent 2 weeks after injury (Masuda et al. 2004). After puncture with a 21-G needle, degeneration does not differ from nonpunctured discs at 8 weeks (Masuda et al. 2004). The degeneration induced by needle puncture in rabbit lumbar discs is reproducible and can be quantitatively assessed *in vivo* by radiographs and MRI (Sobajima et al. 2005). Histological analysis confirms these results and reveals a loss of notochordal cells in the NP, replaced by chondrocytic cells. In ovine models of annular injury, MRI also shows a decrease in IVD height and an increase in MRI Pfirrmann scoring corroborated by histological injury scoring (Freeman et al. 2016).

2.4.2.2.1.2 Nucleotomy Nucleotomy consists in the removal of NP tissue by aspiration. The gold standard is an aspiration with a 10-ml syringe (Kim, Chun, and Lee 2015). Nucleotomy is a common surgical procedure used to treat disc herniation, but it has also been used as an *in vivo* animal model of disc degeneration, not only to study the physiological process of degeneration but also mostly to validate new cell-based therapeutic strategies (Acosta et al. 2011; Ganey et al. 2009; Sakai et al. 2005; Serigano et al. 2010; Vadalà et al. 2015), aiming at replacing NP, particularly cell therapy and tissue engineering. Nucleotomy makes it possible to create a cavity in the NP that allows injection of large volumes of engineered constructs based on reparative cells that may or may not be associated with biomaterials. Two approaches can be used to perform nucleotomy. The classic approach, and the most frequently used one, consists in an anterolateral approach toward the NP through the AF (transannular approach) under fluoroscopic guidance. This approach has the previously mentioned limitations of annular lesions. A transpedicular approach that keeps the AF intact has also been recently developed (Russo et al. 2016; Vadalà et al. 2013). It consists in targeting the NP after introduction of a wire in the caudal vertebra through the pedicle and the inferior endplate to the NP. *In vitro* studies showed partial nucleotomy even with an intact AF induces destabilization (Vadalà et al. 2013), but in contrast, *in vivo* experiments in an ovine model suggested decreased flexibility in absence of dorsal structure lesions (Reitmaier et al. 2014). After partial nucleotomy in ovine lumbar IVD, degeneration has been highlighted by MRI and histological analysis (Guder et al. 2009). This progressive degeneration was also observed by MRI in Goettingen minipigs that underwent partial nucleotomy (Omlor et al. 2009). The intensity of the MRI T2 signal was accompanied by a decrease in notochordal cell density and aggrecan content in NP after immunohistological analysis (hematoxylin-eosin and Alcian-blue stainings and cytokeratin-8 immunostaining). Studies remain difficult to compare, however, given the diversity of routes used and the volume of NP removed.

It has been shown that nucleotomy can be complete (Vadalà et al. 2015) or partially (Ganey et al. 2009; Guder et al. 2009; Omlor et al. 2009; Sakai et al. 2005; Serigano et al. 2010), and sadly, the amount of aspired NP is sometimes not even reported in publications. Moreover, nucleotomy is closer to hernia than DDD and therefore could be more useful in evaluating treatment of discal avulsion.

2.4.2.2.1.3 Laser Laser has long been used in spinal surgery for disc ablation (Dickey et al. 1996) or to perform experimental nucleotomy (Sato et al. 2003). The use of a diode laser has been recently described in an experimental model of IVD degeneration in adult rabbits (Lucas et al. 2012). Laser-induced degeneration was evidenced by a decrease in IVD height on X-ray and a decrease in the MRI T2 signal (as shown in Figure 2.5), as well as an increase in histological Boos scoring that confirmed IVD ECM degeneration. The IVD degeneration induced by laser treatment was histologically comparable to the spontaneous degenerative process. This method seems to induce a more progressive degeneration than aspiration-induced IVD degeneration (Lucas et al. 2012) and chemonucleolysis (Fusellier et al. 2016).

2.4.2.2.1.4 Advantages and Limitations These physical models are numerous and therefore do not exhibit the same advantages and limitations. They are relevant because it is believed that AF lesions play an essential role in the degeneration of human IVDs. These models could therefore reproduce the primitive lesion initiating the DDD. Nevertheless, the surgical approach needs to be considered. The retroperitoneal approach to the IVD is invasive because of the need of a laparotomy or lumbotomy. The percutaneous route with fluoroscopic or CT guidance is seductive because it is less invasive (Zhou et al. 2013a). Among injury models, a full-thickness stab causes a risk of immediate NP avulsion. This model is therefore inadequate to study progressive degeneration processes and to test methods of early regeneration. The aspiration technique is not reliable and is unreproducible in rabbits because of difficulties reaching the center of the nucleus and removing a controlled amount of disc material (Kim et al. 2005). Mechanical nucleotomy by transpedicular approach seems to be more reproducible and has the advantage of keeping AF intact. This model is suitable for testing the intradiscal injection of engineered constructs but is not likely a model of degeneration (Vadalà et al. 2015). Following superficial stab or needle puncture, degeneration is slower and progressive and involves an active process.

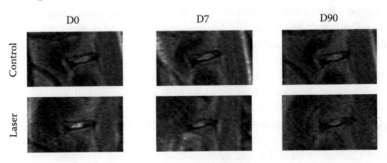

FIGURE 2.5 Progressive decrease of MRI T2 signal of rabbit IVDs after laser treatment compared with control treatment.

The degree of degeneration after annular lesions remains dependent on the size of the injury, however (Elliott et al. 2008; Masuda et al. 2004). This makes it a good model for studying pathophysiological processes and regeneration methods. Unfortunately, the time course of the degenerative process is difficult to control. The laser method seems to be more reproducible, but additional studies are needed to confirm its potential.

2.4.2.2.2 Enzymatic Models: Chemonucleolysis

Chemonucleolysis consists in inducing an enzymatic degradation of the NP. This method was initially considered as a treatment of choice for sciatica secondary to lateral IVD herniation (Chicheportiche et al. 1997). The goal of chemonucleolysis is to mimic the selective degradation of proteoglycans that occurs in disc degeneration. The digestion of GAG chains can be achieved by chymopapain (Atilola et al. 1988; Bradford, Cooper, and Oegema 1983; Kiester et al. 1994; Melrose et al. 1996; Roberts et al. 2008; Spencer, Miller, and Schultz 1985; Wakano et al. 1983; Zook and Kobrine 1986), chondroitinase ABC (Boxberger et al. 2008; Hoogendoorn et al. 2007; Norcross et al. 2003; Sasaki et al. 2001; Sugimura et al. 1996; Takahashi et al. 1996), hyaluronidase (Antoniou et al. 2006; Fusellier et al. 2016), and trypsin, while collagen fibers are digested by collagenase (Antoniou et al. 2006; Bromley et al. 1980; Kalaf, Sell, and Bledsoe 2014; Spencer, Miller, and Schultz 1985; Stern and Coulson 1976; Zook and Kobrine 1986) and trypsin (Mwale et al. 2008; Roberts et al. 2008). The main advantage of the enzymatic model is the capacity to control the degree of degeneration as a function of the dose of enzyme delivered in the NP. Although numerous models are described, the most widely used enzymes are chondroitinase ABC and chymopapain, particularly in the evaluation of regenerative strategies of the IVD. The concentrations of enzyme used to induce DDD are variable among studies, making critical comparisons very difficult.

Chymopapain is a proteolytic enzyme derived from the papaya latex that selectively cleaves the noncollagenous protein connections of proteoglycans in a dose-dependent manner. It has been shown to provoke degeneration in an *in vitro* organ culture model (Chan et al. 2013). Chymopapain treatment has been tested on rabbits (Gullbrand et al. 2016) and dogs (Kudo, Sumi, and Hashimoto 1993; Lü et al. 1997). It induces severe degeneration with rapid proteoglycan decrease and concomitant disc space narrowing and spine instability. It has been shown that NP contained degenerated notochordal cells after chymopapain treatment. Later, the center of the NP was replaced by fibrocartilage tissue (Kudo, Sumi, and Hashimoto 1993). The degeneration induced by chymopapain is greater than chondroitinase-induced degeneration in the dog (Lü et al. 1997) but is strongly dependent on the dose.

Chondroitinase ABC has also been used to produce IVD degeneration in animal models such as rats (Boxberger et al. 2008; Norcross et al. 2003), rabbits (Henderson, Stanescu, and Cauchoix 1991), sheep (Sasaki et al. 2001), goats (Hoogendoorn et al. 2007), monkeys (Sugimura et al. 1996), and dogs (Fry et al. 1991). Chondroitinase acts by specific degradation of the chondroitin and dermatan sulfate side chains of the proteoglycans. Models of intradiscal chondroitinase ABC injection *in vivo* have shown mild progressive degenerative changes characterized by a loss of proteoglycan content in NP (Fry et al. 1991; Norcross et al. 2003), a progressive decrease in IVD height (Fry et al. 1991; Gullbrand et al. 2017; Henderson, Stanescu, and Cauchoix 1991; Sasaki et al. 2001), and an increased Pfirrmann grading on MRI (Gullbrand et al. 2017). Biomechanical dysfunction with an increase in spine segment range of motion (Boxberger et al. 2008; Gullbrand et al. 2017) and decrease in intradiscal pressure (Sasaki et al. 2001) have also been reported. This enzymatic method is of particular relevance because chondroitinase ABC selectively degrades proteoglycans in a dose-dependent manner, and it is a reproducible, fast, and inexpensive method. Studies using this method have demonstrated IVD structural changes similar to those observed in human degeneration with a progressive reduction in IVD height, loss of proteoglycans in NP, fibrosis of the NP matrix, and disorganization of the AF. Similarly, changes observed by MRI (values of T2 and T1ρ and Pfirrmann grading) are correlated with the severity of degeneration. Nevertheless, proteoglycan restoration over time has been observed with low enzyme doses. Conversely, high doses can directly induce AF disruption in animal models. Moreover, enzymatic degradation by chondroitinase cannot mimic the natural DDD since there is no extracellular chondroitinase in mammals. Lastly, due to the avascular nature of IVD, it remains difficult to totally rule out the possibility that enzymes may persist for a long period of time. This persistence of enzymes or their degradation products with residual activities (Oegema et al. 1992) could jeopardize the regenerative therapies tested and probably affect the behavior of the cells of the surrounding tissues.

2.4.2.2.3 Endplate Injury

As the IVD is largely avascular, its nutrient supply comes mostly through endplate diffusion. Conversely, vertebral endplates and subchondral bone function as a barrier that prevents substances such as enzymes from damaging IVD ECM. These mechanisms explain why endplate sclerosis could be an initiating factor in IVD degeneration (Bernick and Cailliet 1982; Benneker et al. 2005) and have therefore been exploited to induce IVD degeneration. Endplate injuries can be either mechanical (physical) or ischemic (i.e., induced by reduction of blood flow).

Mechanical injuries of the endplate have been developed in porcine lumbar IVDs by perforations of the endplate with a drill bit (Holm et al. 2004) or with a Kirschner wire (Cinotti et al. 2005). Both methods induce a late IVD degeneration (3–7 months) with a decrease in cellularity and proteoglycan content of the NP (Holm et al. 2004) and a decrease in the MRI T2 signal (Cinotti et al. 2005). According to Cinotti et al. (2005), the extent of degenerative changes was related to the severity of endplate injuries. Ischemic lesions of the vertebral endplate can be induced by pingyangmycin or bleomycin injection in subchondral bone, which generates a blockade of the blood pathway through the endplate. This model has been tested in rhesus monkeys (Wei et al. 2014; Zhou et al. 2013b) and rabbits (Wei et al. 2015), in which it induces slowly (around 3 months) progressive IVD degeneration with a decrease in IVD height and T2 signal. These methods are attractive because they are based on mechanisms that are associated with DDD in humans. Lesions of the endplate could partially reproduce those observed in human DDD and notably the reduction in the subchondral bone vascularization that induces slow degeneration with a significant decrease in cell density and GAG in NP, disorganization of fibrocartilage lamellae in the AF, and IVD height loss. Finally, this method, while being invasive, is effective and provides long-lasting effects even though their reproducibility has not yet been fully assessed.

2.5 SURGICAL APPROACH TO THE IVD

One important problem to solve when attempting to induce IVD degeneration is how to hit either the AF or the NP. This issue was covered by investigating the different ways to approach the IVD.

Traditionally, the IVD is largely exposed by a wide surgical route such as an anterior approach to the lumbar spine after laparotomy (Boxberger et al. 2008; Sobajima et al. 2005) or a retroperitoneal approach (Kim et al. 2005; Masuda et al. 2004; Moss et al. 2013). These surgical procedures are invasive and time-consuming and require surgical experience. Moreover, they can be debilitating for animals and induce complications. Minimally invasive approaches have therefore been developed.

One alternative route is the percutaneous approach with fluoroscopy guidance (Kim, Chun, and Lee 2015). It consists in a transannular access commonly described to induce IVD degeneration by puncture (Xi et al. 2013) or aspiration, which could be also used as a way of injection (Prologo et al. 2012) or laser introduction. It has been used to perform discography and is rarely prescribed today because of the development of CT and MRI. This approach is relevant because it is a simple way to access the IVD and the procedure is fast. Moreover, this approach is noninvasive and painless. Nevertheless, this technique is difficult to use, notably to target the caudal lumbar levels because of the sterically hindered iliac bone, which in most species covers the posterolateral part of the spine. Furthermore, lesions of the AF may induce subsequent DDD, as mentioned in Section 2.4.2.2.1 on the physical IVD model of degeneration.

Another recently described way to access the IVD is the transpedicular approach (Vadalà et al. 2013). It consists in targeting the NP after having placed a wire in the caudal vertebra through the pedicle and the inferior endplate to the NP. This method is typically used in spinal surgery for osteosynthesis (Ringel et al. 2006) and cementoplasty (Taylor, Fritzell, and Taylor 2007; Galibert et al. 1987). Its main advantage is that it keeps the AF intact and it is therefore assumed not to induce IVD degeneration. Additionally, it could be used as a route to inject biologics or biomaterials into the NP. Nevertheless, the route through the vertebra and the endplate remains open after the procedure and needs to be filled to avoid leaking of the intradiscally injected materials or cells. Finally, this procedure seems to induce the transport of bony fragments into the IVD that could later provoke DDD (Le Fournier et al. 2017). This route also loses the advantage of simplicity because it can only be performed by well-trained surgeons.

2.6 POTENTIAL APPLICATIONS OF THE DIFFERENT MODELS

In vivo models are needed to integrate all physiological interactions that cannot be reproduced *ex vivo*. Among all the animal models proposed, it remains difficult to choose the most appropriate. Several criteria have to be considered before judging the relevance of a particular animal model such as cost and ease of breeding, size, anatomy and disc geometry, histology, and biomechanics. Ideally, the perfect animal model for DDD is easily obtained and not expensive, its IVD should be easy to surgically target and be near the human size and anatomy, and the onset of IVD degeneration should occur reliably and quite quickly. Moreover, the histological features of disc degeneration should reproduce the histology observed in humans. Finally, the model must conform to ethical guidelines.

In reality, the ultimate selection will be strongly dependent on the targeted application. Hence, fundamental studies on the pathogenesis of DDD do not necessarily require using the largest species. The influence of mechanical stress on the IVD can be explored with mechanical models such as tail models or bipedal animals that are quite developed in mice and rats. In particular, tail models are easily accessible. Instability models can also be used for this purpose. Lesions of the AF are also relevant in fundamental studies because AF lesions are thought to be a component of spontaneous IVD degeneration as well as altered nutrient supply by endplate lesions. To investigate the effect of a specific protein on the DDD, genetic models can be developed. Nevertheless, one must keep in mind that the presence of notochordal cells can deeply influence the degenerative process in species in which they persist in adulthood, which means the vast majority of animal models (sand rats, mice, rats, rabbits, nonchondrodystrophic dogs, pigs, and monkeys).

To evaluate regenerative therapies, spontaneous models are difficult to use because of both the relative long time of onset and the important variability of the induced degeneration, which could dramatically increase the required number of animals. The same issue arises with the instability model.

For experiments examining cell regenerative therapies, the use of animal models exhibiting the remnance of notochordal cells after skeletal maturity should be avoided to prevent the influence of potential regenerative effects of these cells. The most frequently used surgically induced degeneration models in preclinical trials are annular injuries, which are highly reproducible.

To test implantable medical devices, the need for large-animal models with surgically accessible IVDs has hampered the use of animals smaller than dogs. For that purpose, dogs constitute an attractive model notably because spontaneous DDD exists in pets in which canine clinical trials may be envisioned.

2.7 IMAGING OF IVD DEGENERATION AND PERSPECTIVES

2.7.1 Physiopathology Context

Human disc degeneration, as a complex multifactorial process, alters both the biochemical composition and architecture of IVD. It is mainly characterized by a decreased cellularity with modifications of the ECM that further compromise the structure of the IVD. The ECM exhibits a progressive loss of proteoglycans with a reduced aggrecan synthesis and a decrease in GAG content while the catabolic activity of MMPs, and a disintegrin and metalloproteinase with thrombospondin repeats (ADAMTS) increases (Roberts et al. 2000). In parallel, there is a transition from type II to type I collagen synthesis in NP tissue. ECM begins to calcify and becomes impermeable, thus blocking both the diffusion of nutrients and the removal of metabolites within the NP (Antoniou et al. 1996). The dehydration of the ECM causes a loss of mechanical integrity with a reduction in the pressure in the NP and a decrease in IVD height that leads to subsequent cracks and tears. Because of the drop of intradiscal pressure, the AF deforms and structural defects such as radial cracks appear.

At the histological level, chondrocytes of the endplate become hypertrophic and synthesize type X collagen. Cracks, fractures, and clefts form in the endplate as sclerosis of subchondral bone develops. This sclerosis may be involved in impairment of nutrition. In the NP, one can observe a gradual decrease in the number of notochordal cells that are progressively replaced by nucleopulpocytes, followed by senescence (loss of the capacity of cells to divide) and apoptosis of nucleopulpocytes. This gradual disappearance of notochordal cells seems to be the common denominator in the process leading to IVD degeneration, and it is observed in humans and species such as chondrodystrophic dogs, goats, sheep, and monkeys (Table 2.1).

Numerous techniques and methodologies are developed to assess DDD, among them biochemical analysis as well as histological characterization and imaging. Biochemical and histological alterations of IVD are easily evidenced, but it is important to note that this type of analysis generally requires animal euthanasia and therefore requires an increased number of sacrificed animals. Conversely, modern imaging technologies make it possible to evaluate IVD *in vivo*, and with incremental analyses, longitudinal follow-up can be performed over several months while reducing the numbers of sacrificed animals. Imaging also enables the use of animals as their own control with a statistical benefit. Moreover, imaging gathers various techniques that are all noninvasive and thus are nondestructive for tissues intended for histology. Imaging technologies include X-ray imaging (radiography and CT), nuclear imaging (single photon emission CT, or SPECT), and MRI. Recently, numerous technological breakthroughs have reinforced the interest in MRI in the exploration of IVD.

2.7.2 X-Ray Imaging

The first imaging technology that uses X-rays is radiography. This technique lacks contrast despite good spatial resolution. All soft tissues (including articular cartilage and IVD) are of the same radio-opacity. This therefore gives little direct information about the disc, which cannot be seen except for its height. Nevertheless, radiography remains useful to evaluate bony elements, most particularly the vertebral plates. It can detect major abnormalities such as narrowing of IVD space, osteophytes, sclerosis, and spondylolisthesis or subluxation (displacement of one vertebra in relation to another). IVD height is measured on lateral plain radiographs of the lumbar spine, as illustrated in Figure 2.6. Its excellent spatial resolution allows its use on species of any size: rat (Chai et al. 2016; Han et al. 2008; Lai et al. 2007), rabbit (Fusellier et al. 2016; Kroeber et al. 2002; Lucas et al. 2012; Phillips, Reuben, and Wetzel 2002; Masuda et al. 2004), dog (Fry et al. 1991), pig (Omlor et al. 2009), sheep (Zhou et al. 2007), goat (Gullbrand et al. 2017), and monkey (Lauerman et al. 1992; Nuckley et al. 2008; Zhou et al. 2013b). Moreover, radiography has a number of advantages:

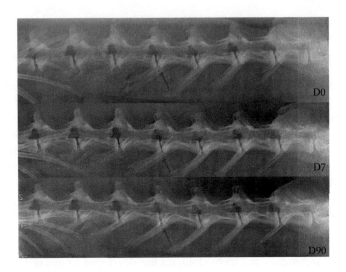

FIGURE 2.6 Lateral radiographs of the lumbar spine of a rabbit showing progressive evolution of disc height after chemonucleolysis at day (D) 0, 7, and 90. The treated IVD is highlighted by the red arrow.

low cost, the possibility of longitudinal follow-up, and the lack of anesthesia needed for most examinations.

However, radiography has several limitations. First, because of the variation in disc height depending on the depth of sedation and muscle relaxation, the same level of anesthesia is required for each examination. Second, because of the cone-shaped X-ray beam, geometric deformation can impact the measurement of IVD height.

The second technique that uses X-rays is tomography. Tomography is an imaging technique providing cross-sectional images of the region explored. The advantage is that the structures lying above or below the level under examination are not superimposed. CT produces transverse images by attenuation of X-rays that pass through the imaged object. The advantages of CT compared with radiography are its great resolution, particularly for micro-CT (under 100 μm), and its better contrast (ability to distinguish different types of soft tissue). Nevertheless, CT remains less suitable than MRI to resolve unmineralized soft tissues such as the IVD. It is important to note that CT, like radiography, delivers radiation doses to the explored tissues, and particularly in longitudinal studies, these doses could be high enough to induce tissue lesions that can interfere with the experiments. To improve the detection threshold of unmineralized soft tissues, contrast-enhanced techniques have been developed and can distinguish between the AF and the NP (Lin et al. 2016). Despite these improvements, the use of CT in animal models of DDD remains moderate, essentially when a CT-guided approach is required to spatially target the IVD during surgery (Mackenzie et al. 2014; Ohnishi et al. 2016; Shi et al. 2016; Wei et al. 2015; Zhou et al. 2013a). CT has also been used for measuring disc height and for the morphologic description of vertebrae and endplates or the lesions of bony structures (Vadalà et al. 2013; Colloca et al. 2012).

Using X-ray, discography can be easily performed. Discography consists in the injection of iodinated contrast agent in IVD followed by radiography, fluoroscopy (real-time X-ray imaging), or CT. It is mainly used to determine the location of a puncture (Vadalà et al. 2013). In the evaluation of DDD, it remains marginally employed in animal models because it provides less information than MRI, and above all, it is suspected of inducing IVD degeneration (Carragee et al. 2009) because of needle puncture to the AF and toxicity of injected anesthetics and iodinated contrast agents (Chee et al. 2014; Lee et al. 2010). Nevertheless, discography may help define tears that extend to the outer part of the AF and annular rupture. These tears seem associated with discogenic pain in humans (Lim et al. 2005). CT discography can depict more lesions than discography (Zhou et al. 2007). While discography has good concordance with MRI in the evaluation of IVD degeneration (Buirski 1992; Gibson et al. 1986; Schneiderman et al. 1987), it remains superior to MRI in depicting annular lesions (Osti and Fraser 1992).

2.7.3 Nuclear Medicine

Nuclear medicine techniques use radiotracers to target *in vivo* physiological or pathological processes. After administration of the tracer, previously labeled with a radio-element that emits gamma radiation (SPECT) or a positron (positron emission tomography [PET]), a camera is used to generate pictures of the radiopharmaceutical agent distribution. Depending on the radiotracer, the target can be an organ, a tumor, a metabolite, or a gene. The major advantages of these techniques are the ability to track injected labeled cells with a high sensitivity (picomolar concentrations). Thus, mesenchymal stem cells labeled with a radioactive tracer were followed for 3 days using PET in pigs (Prologo et al. 2012). The spatial resolution of SPECT or PET is relatively low in comparison with that of CT or MRI, but it can be improved by combination with CT to allow precise anatomic localization. Like X-ray techniques, the exposure to radiation is not insignificant and has to be considered when planning longitudinal studies.

2.7.4 Magnetic Resonance Imaging

MRI produces an image that most often depicts the mobility of hydrogen atoms in their structural environments after application of a strong static magnetic field and radiofrequency pulses. MRI physics is very complex, and the reader is referred to specialized studies on this subject (Heuck and Glaser 2014; Plewes and Kucharczyk 2012). MRI is an excellent noninvasive tool that provides a precise morphological analysis of the IVD and its degeneration in *in vivo* models. Static magnetic field strengths range from 0.2 to 9.7 and even 17 Tesla. The strength of the magnetic field improves the resolution and quality of images. So the lack of spatial resolution of low-field MRI (under 1 Tesla) has led to the replacement by very-high-field MRI when studying small species such as mice or rats (Ohnishi et al. 2016). MRI is now the most widely used imaging modality because of its high spatial resolution without radiation and most particularly for its excellent intrinsic soft tissue contrast, which is very sensitive to changes such as edema. MRI requires the use of contrast agents (gadolinium or iron-based agents) to enhance lesion visibility or to label cells or tissues. Structures that accumulate gadolinium consequently appear bright (hyperintense), with some sequences. Different sequences of radio pulses produce different contrasts on the image: each of them emphasizes different tissue characteristics and thus enables identification of different structures. The most frequently used contrasts are T1 and T2. Classical MRI examination is a qualitative analysis of T1-weighted (T1w) and T2-weighted (T2w) images to identify any signal intensity modification in the discs (Erdem et al. 2005; Sobajima et al. 2005; Vadalà et al. 2012). These evaluations are often completed by measurements of the disc height (Chai et al. 2016; Fusellier et al. 2016; Gebhard et al. 2010; Grunert et al. 2014; Kang et al. 2015; Noury et al. 2008; Omlor et al. 2009).

Healthy NP appears to have a low signal intensity compared to the vertebral bodies on T1w images and exhibits a high signal intensity on T2w because of its high water and proteoglycan content. The AF usually shows lower signal intensities on T1w and T2w images. The endplates are well defined and smooth with an intact covering of cortical bone. With degeneration, all biochemical and structural changes lead to modifications in IVD height and MRI signal (Mwale, Iatridis, and Antoniou 2008). Therefore, in the early stages, a decrease in T2w signal intensity is observed in the NP that parallels the loss of water content and progressive fibrosis. A loss of distinction between the AF and NP is possible (Pfirrmann et al. 2001a). Further degeneration induces fluid-filled fissures visible as linear areas of high signal on the T2w images (Cassar-Pullicino 1998).

A semiquantitative evaluation of IVD degeneration is possible with the Pfirrmann five-level grading sometimes modified by increasing the number of grades from five to eight (Griffith et al. 2007). The Pfirrmann grading scale provides a reliable and reproducible evaluation of DDD on T2w images and is based on MRI signs: disc signal intensity, differentiation of AF and NP, and IVD height that demonstrated sufficient intraobserver and interobserver correlation (Pfirrmann et al. 2001b). In grade I, the disc structure is homogeneous with a net hypersignal of the NP and a normal disc height. In grade II, the disc structure becomes heterogeneous, with a slightly less intense hypersignal of the NP. The distinction between AF and NP is possible with a normal disc height and with or without a horizontal gray band. Lowering the T2 signal is thought to reflect a diminished GAG concentration in the IVD (Johannessen et al. 2006; Weidenbaum et al. 1992). In grade III, the IVD structure is heterogeneous with an intermediate signal. The distinction between AF and NP is not very clear and the IVD height is normal or slightly decreased. In grade IV, the IVD structure is heterogeneous with a marked hypointense signal. There is a complete loss of distinction between AF and NP and normal or decreased IVD height. Finally in grade V, the IVD structure is heterogeneous, with a deep hypointense signal and complete loss of a distinction between AF and NP, and IVD collapse starts to be detectable. This grading system is now largely used in all the studies related to disc degeneration (Bergknut et al. 2012; Clouet et al. 2011; Ohnishi et al. 2016; Omlor et al. 2009; Zhou et al. 2007), sometimes in association with the classification of Modic to better characterize degenerative discopathies. Associated with changes on the disc, modification of the adjacent structures can be observed: Modic described three types of endplate changes on MRI depending on the underlying bone marrow

TABLE 2.3
Modic Changes of the Endplate Signals in MRI

		Modic I	Modic II	Modic III
Bone marrow Signal	T1w image	Low signal intensity	High signal intensity	Low signal intensity
	T2w image	High signal intensity	High signal intensity	Low signal intensity
Clinical significance		Inflammation	Replacement of bone marrow by yellow marrow	Sclerosis

signal (De Roos et al. 1987; Modic et al. 1988). Type I Modic changes lead to T1w hypointense and T2w hyperintense endplate, probably because of an acute reaction of the bone marrow with replacement of the hematopoietic bone marrow by vascularized fibrous tissue (Table 2.3). Type II Modic changes are an increased signal on T1w and isointense to hyperintense signal on T2w because of the presence of fat marrow. Modic type III changes are seen as T1w and T2w hypointense signals because of the presence of dense woven bone and absence of bone marrow.

Such clinical standard MRI provides useful information, but it is subjective and usually not sensitive enough to detect the early onset of DDD (Zobel et al. 2012). New MRI techniques, still being developed, have been proposed to assess the early degenerative process and allow accurate measurement of quantifiable parameters such as the T2 relaxation time, T1ρ relaxation time, and apparent diffusion coefficient (ADC).

2.8 NEW DEVELOPMENTS IN MRI

Apart from conventional imaging techniques such as standard MRI, some newly developed magnetic resonance (MR) techniques hold great promise and make it possible to assess early degenerative alterations and define the degree of degeneration. These techniques include MR relaxation times (T1ρ and T2 mapping), diffusion imaging, cell tracking, and MR spectroscopy (MRS). More unusual techniques such as chemical exchange saturation transfer (CEST)–MRI (CEST-MRI), sodium MRI, and delayed gadolinium-enhanced MRI of cartilage (dGEMRIC) are also being developed.

Quantitative MRI measures relaxation times to evaluate the deterioration of the molecular composition and structural integrity of the IVD (Perry et al. 2006). T2 relaxation time reflects the molecular environment of a tissue. In the IVD, it correlates with water content and, to a lesser extent, with proteoglycan content and (negatively) with collagen (Marinelli et al. 2009; Reitmaier et al. 2013; Weidenbaum et al. 1992). In a rabbit model, T2 relaxation time significantly correlates with histological scores (Chai et al. 2016), and it correlates well with aggrecan and type II collagen gene expression (Sun et al. 2013). In humans, T2 relaxation time in the NP decreases with increasing Pfirrmann grading (Marinelli, Haughton, and Anderson 2010; Marinelli et al. 2009; Stelzeneder et al. 2012; Wang et al. 2013; Watanabe et al. 2007). T2 mapping will probably become a more sensitive tool than classic MRI, notably for the detection of early DDD and the follow-up of innovative therapeutics. The T1ρ MR technique shows slow-motion interactions between macromolecules and bulk water in the ECM. The T1ρ relaxation time may also provide a new and sensitive tool for the diagnosis of early degenerative changes in the IVD (Wei et al. 2014) and the monitoring of new treatments (Lotz et al. 2012). Proteoglycan concentrations in the NP are directly proportional to T1ρ values in the sagittal and axial planes (Johannessen et al. 2006; Mulligan et al. 2014), and early changes in proteoglycan content of the NP can be detected with T1ρ imaging (Johannessen et al. 2006). T1ρ strongly correlates with pressure in the NP and degenerative grades (Blumenkrantz et al. 2006, 2010; Pandit et al. 2016). In rhesus monkeys, T1ρ values rapidly decrease with DDD and then tend to stabilize (Zhou et al. 2013b).

ADC values provide an estimation of free diffusion of unbound water and could be used as a quantitative tool to estimate the IVD. The T2 signal of the disc is significantly correlated with ADC values (Niinimäki et al. 2009), and a decrease in GAGs or water content in the NP results in a direct decrease in the ADC values (Antoniou et al. 2004; Wu et al. 2013; Zhang, Ma, and Wang 2014). Because of a substantial overlap between ADC values of normal and degenerated IVDs, this parameter is difficult to manipulate (Niinimäki et al. 2009).

Through the use of contrast agents, it becomes possible to track cells *in vivo* using MRI. To this end, two main types of contrast agents are used: paramagnetic metal ion agents such as gadolinium, and a superparamagnetic agent such as ferumoxides or superparamagnetic iron oxide (SPIO). Paramagnetic metal ions are positive agents; i.e., they enhance the signal on images by accelerating the relaxation of nearby water protons. Gadolinium can label immune cells for cell tracking studies after cellular internalization by endocytosis or by electroporation. Gadolinium has been used to track different types of stem cells, such as mesenchymal stem cells in small animals (Guenoun et al. 2012). Unfortunately, these agents possess a weak sensitivity and a low uptake by cells, thus limiting their use. Moreover, they exhibit a high degree of toxicity for the labeled cells or even for the kidney. SPIO contrast agents strongly disturb the surrounding magnetic field, thereby inducing a local signal loss, particularly in T2. They are therefore considered as negative contrast agents. Cells can be labeled extracellularly by large iron oxide particles. Nanoparticles or ultrasmall SPIO nanoparticles can be internalized. Iron oxide particles have the advantage of enabling the detection of single cells (Shapiro et al. 2007). SPIO has been successfully used to label mesenchymal stem cells implanted in the disc for tissue regeneration (Saldanha et al. 2008).

MRS is based on the assessment of the resonance frequency of protons, which depends on their chemical bonds and environment. It allows detection of hydrogenate molecule abundance and thus evaluation of DDD. MRS has been used *ex vivo* (Keshari et al. 2005) and *in vivo* (Zuo et al. 2013) to assess proteoglycans and collagen content in IVD. MRS is still under development and will probably improve the investigation of the chemical condition of the IVD.

The following techniques are being developed and need further evaluation before coming into routine use. Biochemical imaging techniques such as CEST, sodium imaging, and dGEMRIC can assess GAG content. In the CEST method, exchangeable protons are saturated, and the saturation is transferred by chemical exchange to the bulk water of their environment, inducing a considerable contrast enhancement of bulk water. GAG CEST has been used to evaluate GAG content and pH changes during DDD *in vitro* (Saar et al. 2013) and *in vivo* (Schleich et al. 2016; Zhou et al. 2016). Sodium MRI consists in quantification of the ^{23}Na concentration by MRI (Ooms et al. 2008). In IVD, the sodium concentration correlates with the GAG concentration (Shapiro et al. 2002) and disc degeneration induces a decrease in IVD ^{23}Na content (Haneder et al. 2014). This technique is feasible on small-animal models such as rabbits (Moon et al. 2012). dGEMRIC was initially developed for examination of articular cartilage after intravenous administration of gadolinium. In IVD, gadolinium diffuses from the capillaries of the EP and can provide an estimation of the *in vivo* GAG content of disc tissue (Vaga et al. 2008).

2.9 CONCLUSION

The perfect animal model of DDD probably does not exist. The most attractive ones are likely the spontaneous models, but one should keep in mind that significant differences exist between animal and human DDD processes, particularly at the cell level. Consequently, there are several limitations to using small-animal models, such as persistence of notochordal cells and small size unsuitable for implementation of surgical techniques clinically transposable to humans. Larger species models seem to have greater relevance when intended to be used for assessing the efficacy of cell-based treatments. It remains important, however, to interpret the results of such studies with caution. Data generated in the context of one animal model remain difficult to transpose to another model or even to humans. The animal model of choice should be (i) perfectly adapted to the issue and (ii) carefully defined at the very beginning of the experiments to avoid any unnecessary animal sacrifices. In the validation of these animal models and their follow-up, imaging techniques and, in particular, MRI have resulted in significant progress. Indeed, the high sensitivity of these techniques allows the longitudinal follow-up of most animal models, thus decreasing the number of sacrificed animals.

REFERENCES

Acosta, F.L., L. Metz, H.D. Adkisson et al. 2011. "Porcine intervertebral disc repair using allogeneic juvenile articular chondrocytes or mesenchymal stem cells." *Tissue Eng Part A* 17 (23–24): 3045–55.

Adler, J.H., M. Schoenbaum, and R. Silberberg. 1983. "Early onset of disk degeneration and spondylosis in sand rats (*Psammomys obesus*)." *Vet Pathol* 20 (1): 13–22.

Aguiar, D.J., S.L. Johnson, and T.R. Oegema. 1999. "Notochordal cells interact with nucleus pulposus cells: Regulation of proteoglycan synthesis." *Exp Cell Res* 246 (1): 129–37.

Alini, M., S.M. Eisenstein, K. Ito et al. 2008. "Are animal models useful for studying human disc disorders/degeneration?" *Eur Spine J* 17 (1): 2–19.

Antoniou, J., C.N. Demers, G. Beaudoin et al. 2004. "Apparent diffusion coefficient of intervertebral discs related to matrix composition and integrity." *Magn Reson Imaging* 22 (7): 963–72.

Antoniou, J., F. Mwale, C.N. Demers et al. 2006. "Quantitative magnetic resonance imaging of enzymatically induced degradation of the nucleus pulposus of intervertebral discs." *Spine* 31 (14): 1547–54.

Antoniou, J., T. Steffen, F. Nelson et al. 1996. "The human lumbar intervertebral disc: Evidence for changes in the biosynthesis and denaturation of the extracellular matrix with growth, maturation, aging, and degeneration." *J Clin Invest* 98 (4): 996–1003.

Atilola, M.A.O., J.P. Morgan, C.S. Bailey, and T. Miyabayashi. 1988. "Canine chemonucleolysis." *Vet Radiol* 29 (4): 168–75.

Bach, F.C., Y. Zhang, A. Miranda-Bedate et al. 2016. "Increased caveolin-1 in intervertebral disc degeneration facilitates repair." *Arthritis Res Ther* 18 (1): 59.

Bailey, A.S., F. Adler, S.M. Lai, and M.A. Asher. 2001. "A comparison between bipedal and quadrupedal rats: Do bipedal rats actually assume an upright posture?" *Spine* 26 (14): 308–13.

Battié, M.C., T. Videman, and E. Parent. 2004. "Lumbar disc degeneration epidemiology and genetic influences." *Spine* 29 (23): 2679–90.

Beckstein, J.C., S. Sen, T.P. Schaer, E.J. Vresilovic, and D.M. Elliott. 2008. "Comparison of animal discs used in disc research to human lumbar disc: Axial compression mechanics and glycosaminoglycan content." *Spine* 33 (6): E166–73.

Benneker, L.M., A. Gisep, J. Krebs et al. 2012. "Development of an *in vivo* experimental model for percutaneous vertebroplasty in sheep." *Vet Comp Orthop Traumatol* 25 (3): 173–77.

Benneker, L.M., P.F. Heini, M. Alini, S.E. Anderson, and K. Ito. 2005. "2004 Young Investigator Award Winner: Vertebral endplate marrow contact channel occlusions and intervertebral disc degeneration." *Spine* 30 (2): 167–73.

Bergknut, N., E. Auriemma, S. Wijsman et al. 2011. "Evaluation of intervertebral disk degeneration in chondrodystrophic and nonchondrodystrophic dogs by use of Pfirrmann Grading of images obtained with low-field magnetic resonance imaging." *Am J Vet Res* 72 (7): 893–8.

Bergknut, N., J.P.H.J. Rutges, H.-J.C. Kranenburg et al. 2012. "The dog as an animal model for intervertebral disc degeneration?" *Spine* 37 (5): 351–58.

Bernick, S., and R. Cailliet. 1982. "Vertebral end-plate changes with aging of human vertebrae." *Spine* 7 (2): 97–102.

Blanco, G. 2001. "The kyphoscoliosis (ky) mouse is deficient in hypertrophic responses and is caused by a mutation in a novel muscle-specific protein." *Hum Mol Genet* 10 (1): 9–16.

Blanco, G., C. Pritchard, P. Underhill et al. 2004. "Molecular phenotyping of the mouse ky mutant reveals UCP1 upregulation at the neuromuscular junctions of dystrophic soleus muscle." *Neuromuscul Disord* 14 (3): 217–28.

Blumenkrantz, G., X. Li, E.T. Han et al. 2006. "A feasibility study of *in vivo* T1rho imaging of the intervertebral disc." *Magn Reson Imaging* 24 (8): 1001–7.

Blumenkrantz, G., J. Zuo, X. Li et al. 2010. "*In vivo* 3.0-Tesla magnetic resonance T1rho and T2 relaxation mapping in subjects with intervertebral disc degeneration and clinical symptoms." *Magn Reson Med* 63 (5): 1193–1200.

Bowles, R. D., H.H. Gebhard, J.P. Dyke et al. 2012. "Image-based tissue engineering of a total intervertebral disc implant for restoration of function to the rat lumbar spine." *NMR Biomed* 25 (3): 443–51.

Boxberger, J.I., J.D. Auerbach, S. Sen, and D.M. Elliott. 2008. "An *in vivo* model of reduced nucleus pulposus glycosaminoglycan content in the rat lumbar intervertebral disc." *Spine (Phila Pa 1976)* 33 (2): 146–54.

Boyd, L.M., W.J. Richardson, K.D. Allen et al. 2008. "Early-onset degeneration of the intervertebral disc and vertebral end plate in mice deficient in type IX collagen." *Arthritis Rheum* 58 (1): 164–71.

Bradford, D.S., K.M. Cooper, and T.R. Oegema Jr. 1983. "Chymopapain, chemonucleolysis, and nucleus pulposus regeneration." *J Bone Joint Surg [Am]* 65 (9): 1220–315.

Brisson, B.A. 2010. "Intervertebral disc disease in dogs." *Vet Clin N Am Small Anim Pract* 40 (5): 829–58.

Bromley, J.W., J.W. Hirst, M. Osman et al. 1980. "Collagenase: An experimental study of intervertebral disc dissolution." *Spine* 5 (2): 126–32.

Buirski, G. 1992. "Magnetic resonance signal patterns of lumbar discs in patients with low back pain. a prospective study with discographic correlation." *Spine* 17 (10): 1199–1204.

Bushell, G.R., P. Ghosh, T.K.F. Taylor, J.M. Sutherland, and K.G. Braund. 1978. "The effect of spinal fusion on the collagen and proteoglycans of the canine intervertebral disc." *J Surg Res* 25 (1): 61–69.

Cao, Y., S. Liao, H. Zeng, S. Ni et al. 2017. "3D characterization of morphological changes in the intervertebral disc and endplate during aging: A propagation phase contrast synchrotron micro-tomography study." *Sci Rep* 7: 1–12.

Carragee, E.J., A.S. Don, E.L. Hurwitz et al. 2009. "2009 ISSLS prize winner: Does discography cause accelerated progression of degeneration changes in the lumbar disc: A ten-year matched cohort study." *Spine* 34 (21): 2338–45.

Cassar-Pullicino, V.N. 1998. "MRI of the aging and herniating intervertebral disc." *Eur J Radiol* 27 (3): 214–28.

Cassidy, J.D., K. Yong-Hing, W.H. Kirkaldy-Willis, and A.A. Wilkinson. 1988. "A study of the effects of bipedism and upright posture on the lumbosacral spine and paravertebral muscles of the Wistar rat." *Spine* 13 (3): 301–8.

Chai, J.W., H.S. Kang, J.W. Lee, S.J. Kim, and S. Hwan Hong. 2016. "Quantitative analysis of disc degeneration using axial T2 mapping in a percutaneous annular puncture model in rabbits." *Korean J Radiol* 17 (1): 103–10.

Chan, S.C.W., A. Bürki, H.M. Bonél, L.M. Benneker, and B. Gantenbein-Ritter. 2013. "Papain-induced *in vitro* disc degeneration model for the study of injectable nucleus pulposus therapy." *Spine J* 13 (3): 273–83.

Chee, A.V., J. Ren, B.A. Lenart et al. 2014. "Cytotoxicity of local anesthetics and nonionic contrast agents on bovine intervertebral disc cells cultured in a three-dimensional culture system." *Spine J* 14 (3): 491–98.

Chen, C., Z. Jia, Z. Han et al. 2015. "Quantitative T2 relaxation time and magnetic transfer ratio predict endplate biochemical content of intervertebral disc degeneration in a canine model." *BMC Musculoskelet Disord* 16 (6):157.

Cherrone, K.L., C.W. Dewey, J.R. Coates, and R.L. Bergman. 2004. "A retrospective comparison of cervical intervertebral disk disease in nonchondrodystrophic large dogs versus small dogs." *J Am Anim Hosp Assoc* 40 (4): 316–20.

Chicheportiche, V., C. Parlier-Cuau, P. Champsaur, and J.-D. Laredo. 1997. "Lumbar chymopapain chemonucleolysis." *Semin Musculoskelet Radiol* 1 (2): 197–206.

Ching, C.T.S., D.H.K. Chow, F.Y.D. Yao, and A.D. Holmes. 2003. "The effect of cyclic compression on the mechanical properties of the inter-vertebral disc: An *in vivo study* in a rat tail model." *Clin Biomech* 18 (3): 182–89.

Ching, C.T.S., D.H.K. Chow, F.Y.D. Yao, and A.D. Holmes. 2004. "Changes in nuclear composition following cyclic compression of the intervertebral disc in an *in vivo* rat-tail model." *Med Eng Phys* 26 (7): 587–94.

Cinotti, G., C.D. Rocca, S. Romeo, F. Vittur, R. Toffanin, and G. Trasimeni. 2005. "Degenerative changes of porcine intervertebral disc induced by vertebral endplate injuries." *Spine* 30 (2): 174–80.

Clouet, J., M. Pot-Vaucel, G. Grimandi et al. 2011. "Characterization of the age-dependent intervertebral disc changes in rabbit by correlation between MRI, histology and gene expression." *BMC Musculoskelet Disord* 12 (January): 147.

Colloca, C.J., R. Gunzburg, B.J. Freeman et al. 2012. "Biomechancial quantification of pathologic manipulable spinal lesions: An *in vivo* ovine model of spondylolysis and intervertebral disc degeneration." *J Manipulative Physiol Ther* 35 (5): 354–66.

Colombier, P., A. Camus, L. Lescaudron, J. Clouet, and J. Guicheux. 2014. "Intervertebral disc regeneration: A great challenge for tissue engineers." *Trends Biotechnol* 32 (9): 433–35.

Cotterill, P.C., J.P. Kostuik, G. D'Angelo, G.R. Fernie, and B.E. Maki. 1986. "An anatomical comparison of the human and bovine thoracolumbar spine." *J Orthop Res* 4 (3): 298–303.

Court, C., O.K. Colliou, J.R. Chin et al. 2001. "The effect of static *in vivo* bending on the murine intervertebral disc." *Spine J* 1 (4): 239–45.

Daly, C., P. Ghosh, G. Jenkin, D. Oehme, and T. Goldschlager. 2016. "A review of animal models of intervertebral disc degeneration: Pathophysiology, regeneration, and translation to the clinic." *BioMed Res Int* 2016: 5952165.

De Roos, A., H. Kressel, C. Spritzer, and M. Dalinka. 1987. "MR imaging of marrow changes adjacent to end plates in degenerative lumbar disk disease." *Am J Roentgenol* 149 (3): 531–34.

Dickey, D.T., K.E. Bartels, G.A. Henry et al. 1996. "Use of the holmium yttrium aluminum garnet laser for percutaneous thoracolumbar intervertebral disk ablation in dogs." *J Am Vet Med Assoc* 208 (8): 1263–67.

Elliott, D.M., and J.J. Sarver. 2004. "Young Investigator Award Winner: Validation of the mouse and rat disc as mechanical models of the human lumbar disc." *Spine* 29 (7): 713–22.

Elliott, D.M., C.S. Yerramalli, J.C. Beckstein, J.I. Boxberger, W. Johannessen, and E.J. Vresilovic. 2008. "The effect of relative needle diameter in puncture and sham injection animal models of degeneration." *Spine* 33 (6): 588–96.

Erdem, L.O., C. Erdem, B. Acikgoz, and S. Gundogdu. 2005. "Degenerative disc disease of the lumbar spine: A prospective comparison of fast T1-weighted fluid-attenuated inversion recovery and T1-weighted turbo spin echo MR Imaging." *Eur J Radiol* 55 (2): 277–82.

Faller, K., J. Penderis, C. Stalin et al. 2014. "The effect of kyphoscoliosis on intervertebral disc degeneration in dogs." *Vet J* 200 (3): 449–51.

Freeman, B.J.C., J.S. Kuliwaba, C.F. Jones et al. 2016. "Allogeneic mesenchymal stem cells promote healing in postero-lateral annular lesions and improve indices of lumbar intervertebral disc degeneration in an ovine model." *Spine* 41 (March): 1.

Fry, T.R., J.C. Eurell, L. Johnson et al. 1991. "Radiographic and histologic effects of chondroitinase ABC on normal canine lumbar intervertebral disc." *Spine* 16: 816–19.

Fusellier, M., P. Colombier, J. Lesoeur et al. 2016. "Longitudinal comparison of enzyme- and laser-treated intervertebral disc by MRI, X-Ray, and histological analyses reveals discrepancies in the progression of disc degeneration: A rabbit study." *BioMed Res Int* 2016:5498271.

Galibert, P., H. Deramond, P. Rosat, and D. Le Gars. 1987. "[Preliminary note on the treatment of vertebral angioma by percutaneous acrylic vertebroplasty]." *Neuro-Chirurgie* 33 (2): 166–68.

Ganey, T., W.C. Hutton, T. Moseley, M. Hedrick, and H.-J. Meisel. 2009. "Intervertebral disc repair using adipose tissue-derived stem and regenerative cells: Experiments in a canine model." *Spine* 34 (21): 2297–304.

Gebhard, H., R. Bowles, J. Dyke et al. 2010. "Total disc replacement using a tissue-engineered intervertebral disc *in vivo*: New animal model and initial results." *Evid Based Spine Care J* 1 (2): 62–6.

Gibson, M.J., J. Buckley, R. Mawhinney, R.C. Mulholland, and B.S. Worthington. 1986. "Magnetic resonance imaging and discography in the diagnosis of disc degeneration. a comparative study of 50 discs." *J Bone Joint Surg* 68 (3): 374–8.

Griffin, J.F., J. Levine, S. Kerwin, and R. Cole. 2009. "Canine thoracolumbar invertebral disk disease: Diagnosis, prognosis, and treatment." *Compendium (Yardley, PA)* 31 (3): E3.

Griffith, J.F., Y.-X.J. Wang, G.E. Antonio et al. 2007. "Modified Pfirrmann grading system for lumbar intervertebral disc degeneration." *Spine* 32 (24): E708–12.

Gruber, H.E., J.A. Ingram, and E.N. Hanley. 2006. "Immunolocalization of thrombospondin in the human and sand rat intervertebral disc." *Spine* 31 (22): 2556–61.

Gruber, H.E., T.L. Johnson, K. Leslie et al. 2002. "Autologous intervertebral disc cell implantation: A model using *Psammomys obesus*, the sand rat." *Spine* 27 (15): 1626–33.

Grunert, P., H.H. Gebhard, R.D. Bowles et al. 2014. "Tissue-engineered intervertebral discs: MRI results and histology in the rodent spine." *J Neurosurg Spine* 20 (4): 443–51.

Guder, E., S. Hill, F. Kandziora, and K.J. Schnake. 2009. "Partial nucleotomy of the ovine disc as an *in vivo* model for disc degeneration." *Z Orthop Unfall* 147 (1): 52–58.

Guehring, T., F. Unglaub, H. Lorenz et al. 2006. "Intradiscal pressure measurements in normal discs, compressed discs and compressed discs treated with axial posterior disc distraction: An experimental study on the rabbit lumbar spine model." *Eur Spine J* 15 (5): 597–604.

Guenoun, J., G.A. Koning, G. Doeswijk et al. 2012. "Cationic Gd-DTPA liposomes for highly efficient labeling of mesenchymal stem cells and cell tracking with MRI." *Cell Transplant* 21 (1): 191–205.

Gullbrand, S.E., B.G. Ashinsky, J.T. Martin et al. 2016. "Correlations between quantitative T2 and T1ρ MRI, mechanical properties and biochemical composition in a rabbit lumbar intervertebral disc degeneration model." *J Orthop Res* 34 (April 2016): 1–25.

Gullbrand, S.E.E., N.R.R. Malhotra, T.P.P. Schaer et al. 2017. "A large animal model that recapitulates the spectrum of human intervertebral disc degeneration." *Osteoarthritis Cartilage* 25 (1): 146–56.

Hadjipavlou, A.G., J.W. Simmons, J.P. Yang et al. 1998. "Torsional injury resulting in disc degeneration: I. an *in vivo* rabbit model." *J Spinal Disord* 11 (4): 312–17.

Hammer, R.E., S.D. Maika, J.A. Richardson, J.P. Tang, and J.D. Taurog. 1990. "Spontaneous inflammatory disease in transgenic rats expressing HLA-B27 and human β2m: An animal model of HLA-b27-associated human disorders." *Cell* 63 (5): 1099–112.

Hamrick, M.W., C. Pennington, and C.D. Byron. 2003. "Bone architecture and disc degeneration in the lumbar spine of mice lacking GDF-8 (myostatin)." *J Orthop Res* 21 (6): 1025–32.

Han, B., K. Zhu, F.-C.C. Li et al. 2008. "A simple disc degeneration model induced by percutaneous needle puncture in the rat tail." *Spine (Phila Pa 1976)* 33 (18): 1925–34.

Haneder, S., M.M.L. Ong, J.M. Budjan et al. 2014. "23Na-magnetic resonance imaging of the human lumbar vertebral discs: *In vivo* measurements at 3.0 T in healthy volunteers and patients with low back pain." *Spine J* 14 (7): 1343–50.

Henderson, N., Vi. Stanescu, and J. Cauchoix. 1991. "Nucleolysis of the rabbit Intervertebral disc using chondroitinase ABC." *Spine* 16 (2): 203–8.

Heuck, A., and C. Glaser. 2014. "Basic aspects in MR imaging of degenerative lumbar disk disease." *Semin Musculoskelet Radiol* 18 (3): 228–39.

Higuchi, M., K. Abe, and K. Kaneda. 1983. "Changes in the nucleus pulposus of the intervertebral disc in bipedal mice." *Clin Orthop Relat Res* 175: 251–7.

Holm, S., A. Kaigle Holm, L. Ekström, A. Karladani, and T. Hansson. 2004. "Experimental disc degeneration due to endplate injury." *J Spinal Disord Tech* 17 (1): 64–71.

Hoogendoorn, R.J.W., M.N. Helder, R.J. Kroeze et al. 2008. "Reproducible long-term disc degeneration in a large animal model." *Spine* 33 (9): 949–54.

Hoogendoorn, R.J., P.I. Wuisman, T.H. Smit, V.E. Everts, and M.N. Helder. 2007. "Experimental intervertebral disc degeneration induced by chondroitinase ABC in the goat." *Spine* 32 (17): 1816–25.

Hoy, D., L. March, P. Brooks et al. 2014. "The global burden of low back pain: Estimates from the Global Burden of Disease 2010 Study." *Ann Rheum Dis* 73 (6): 968–74.

Hsieh, A.H., D. Hwang, D.A. Ryan, A.K. Freeman, and H. Kim. 2009. "Degenerative anular changes induced by puncture are associated with insufficiency of disc biomechanical function." *Spine* 34 (10): 998–1005.

Hunter, C.J., J.R. Matyas, and N.A. Duncan. 2003. "The three-dimensional architecture of the notochordal nucleus pulposus: Novel observations on cell structures in the canine intervertebral disc." *J Anat* 202: 279–91.

Hunter, C.J., J.R. Matyas, and N.A. Duncan. 2004. "Cytomorphology of notochordal and chondrocytic cells from the nucleus pulposus: A species comparison." *J Anat* 205 (5): 357–62.

Iatridis, J.C., P.L. Mente, I.A. Stokes, D.D. Aronsson, and M. Alini. 1999. "Compression-induced changes in intervertebral disc properties in a rat tail model." *Spine* 24 (10): 996–1002.

Issy, A.C., V. Castania, M. Castania et al. 2013. "Experimental model of intervertebral disc degeneration by needle puncture in Wistar rats." *Braz J Med Biol Res* 46 (3).

Jaumard, N.V., J. Leung, A.J. Gokhale et al. 2015. "Relevant anatomic and morphological measurements of the rat spine: Considerations for rodent models of human spine trauma." *Spine* 40 (20): E1084–92.

Johannessen, W., J.D. Auerbach, A.J. Wheaton et al. 2006. "Assessment of human disc degeneration and proteoclycan content using T1rho-weighted magnetic resonance imaging." *Spine* 31 (11): 1253–57.

Johnson, J.A., R.C. Da Costa, and M.J. Allen. 2010. "Micromorphometry and cellular characteristics of the canine cervical intervertebral discs." *J Vet Intern Med* 24 (6): 1343–49.

Juneja, S.C., A. Vonica, C. Zeiss et al. 2014. "Deletion of Mecom in mouse results in early-onset spinal deformity and osteopenia." *Bone* 60: 148–61.

Kääpä, E., S. Holm, X. Han, T. Takala, V. Kovanen, and H. Vanharanta. 1994. "Collagens in the injured porcine intervertebral disc." *J Orthop Res* 12 (1): 93–102.

Kalaf, G., E.A. Scott, A. Sell, and J.G. Bledsoe. 2014. "Developing a mechanical and chemical model of degeneration in young bovine lumbar intervertebral disks and reversing loss in mechanical function." *J Spinal Disord Tech* 27 (5): E168–75.

Kang, R., H. Li, K. Rickers, S. Ringgaard, L. Xie, and C. Bünger. 2015. "Intervertebral disc degenerative changes after intradiscal injection of TNF-α in a porcine model." *Eur Spine J* 24 (9):2010–6.

Kasch, R., B. Mensel, F. Schmidt et al. 2012. "Percutaneous disc decompression with nucleoplasty–volymetry of the nucleus pulposus using ultrahigh-field MRI." *PLoS One* 7 (7): 1–8.

Keshari, K.R., J.C. Lotz, J. Kurhanewicz, and S. Majumdar. 2005. "Correlation of HR-MAS spectroscopy derived metabolite concentrations with collagen and proteoglycan levels and Thompson grade in the degenerative disc." *Spine* 30 (23): 2683–8.

Kiester, D.P., J.M. Williams, G.B. Andersson, E.J. Thonar, and T.W. McNeill. 1994. "The dose-related effect of intradiscal chymopapain on rabbit intervertebral discs." *Spine* 19 (7): 747–51.

Kim, D.W., H.-J. Chun, and S.-K. Lee. 2015. "Percutaneous needle puncture technique to create a rabbit model of traumatic degenerative disc disease." *World Neurosurg* 84 (2): 1–8.

Kim, K.S., S.T. Yoon, J. Li, J.S. Park, and W.C. Hutton. 2005. "Disc degeneration in the rabbit: A biochemical and radiological comparison between four disc injury models." *Spine* 30 (1): 33–37.

Kimura, T., K. Nakata, N. Tsumaki et al. 1996. "Progressive degeneration of articular cartilage and intervertebral discs. An experimental study in transgenic mice bearing a type IX collagen mutation." *Int Orthop* 20 (3): 177–81.

Kroeber, M., F. Unglaub, T. Guehring et al. 2005. "Effects of controlled dynamic disc distraction on degenerated intervertebral discs: An *in vivo* study on the rabbit lumbar spine model." *Spine* 30 (2): 181–7.

Kroeber, M.W., F. Unglaub, H. Wang et al. 2002. "New *in vivo* animal model to create intervertebral disc degeneration and to investigate the effects of therapeutic strategies to stimulate disc regeneration." *Spine* 27 (23): 2684–90.

Kudo, T., A. Sumi, and A. Hashimoto. 1993. "Experimental chemonucleolysis with chymopapain in canine intervertebral Disks." *J Vet Med Sci* 55 (2): 211–15.

Lai, A., D.H.K. Chow, W.S. Siu, A.D. Holmes, and F.H. Tang. 2007. "Reliability of radiographic intervertebral disc height measurement for *in vivo* rat-tail model." *Med Eng Phys* 29 (7): 814–9.

Lauerman, W.C., R.C. Platenberg, J.E. Cain, and V.F. Deeney. 1992. "Age-related disk degeneration: Preliminary report of a naturally occurring baboon model." *J Spinal Disord* 5 (2): 170–5.

Le Fournier, L., M. Fusellier, B. Halgand et al. 2017. "The transpedicular surgical approach for the development of intervertebral disc-targeting regenerative strategies in an ovine model." *Eur Spine J* 26 (8): 2072–2083.

Lee, H., G. Sowa, N. Vo et al. 2010. "Effect of bupivacaine on intervertebral disc cell viability." *Spine J* 10 (2): 159–66.

Leung, V.Y.L., S.C. Hung, L.C. Li et al. 2008. "Age-related degeneration of lumbar intervertebral discs in rabbits revealed by deuterium oxide-assisted MRI." *Osteoarthritis Cartilage* 16 (11): 1312–8.

Lim, C.-H., W.-H. Jee, B.C. Son et al. 2005. "Discogenic lumbar pain: Association with MR imaging and CT discography." *Eur J Radiol* 54 (3): 431–37.

Lin, K.H., Q. Wu, D.J. Leib, and S.Y. Tang. 2016. "A novel technique for the contrast-enhanced microCT imaging of murine intervertebral discs." *J Mech Behav Biomed Mater* 63: 66–74.

Lindblom, K. 1957. "Intervertebral disc degeneration considered as a pressure atrophy." *J Bone Joint Surg* 39 (4): 933–45.

Lipson, S.J., and H. Muir. 1980. "Vertebral osteophyte formation in experimental disc degeneration: Morphologic and proteoglycan changes over time." *Arthritis Rheum* 23 (3): 319–24.

Lipson, S.J., and H. Muir. 1981. "Experimental intervertebral disc degeneration: Morphologic and proteoglycan changes over time." *Arthritis Rheum* 24 (1): 12–21.

Lotz, J.C., O.K. Colliou, J.R. Chin, N.A. Duncan, and E. Liebenberg. 1998. "Compression-induced degeneration of the intervertebral disc: An *in vivo* mouse model and finite-element study." *Spine* 23 (23): 2493–2506.

Lotz, J.C., V. Haughton, S.D. Boden et al. 2012. "New treatments and imaging strategies in degenerative disease of the intervertebral disks." *Radiology* 264 (1): 6–19.

Lü, D.S., Y. Shono, I. Oda, K. Abumi, and K. Kaneda. 1997. "Effects of chondroitinase ABC and chymopapain on spinal motion segment biomechanics. An *in vivo* biomechanical, radiologic, and histologic canine study." *Spine* 22(16): 1828–35.

Lucas, O., O. Hamel, A. Blanchais et al. 2012. "Laser-treated nucleus pulposus as an innovative model of intervertebral disc degeneration." *Exp Biol Med (Maywood, N.J.)* 237 (11): 1359–67.

Mackenzie, S.D., J.L. Caswell, B.A. Brisson, L. Gaitero, and H. J. Chalmers. 2014. "Comparison between computed tomography, fluoroscopy, and ultrasonography for guiding percutaneous injection of the canine intervertebral disc." *Vet Radiol Ultrasound* 55 (5): 571–81.

Mageed, M., D. Berner, H. Jülke et al. 2013a. "Is sheep lumbar spine a suitable alternative model for human spinal researches? Morphometrical comparison study." *Lab Anim Res* 29 (4): 183–89.

Mageed, M., D. Berner, H. Jülke et al. 2013b. "Morphometrical dimensions of the sheep thoracolumbar vertebrae as seen on digitised CT images." *Lab Anim Res* 29 (3): 138–47.

Marinelli, N.L., V.M. Haughton, and P.A. Anderson. 2010. "T2 relaxation times correlated with stage of lumbar intervertebral disk degeneration and patient age." *Am J Neuroradiol* 31 (7): 1278–82.

Marinelli, N.L., V.M. Haughton, A. Muñoz, and P.A. Anderson. 2009. "T2 relaxation times of intervertebral disc tissue correlated with water content and proteoglycan content." *Spine* 34 (5): 520–24.

Martin, J.T., D.J. Gorth, E.E. Beattie et al. 2013. "Needle puncture injury causes acute and long-term mechanical deficiency in a mouse model of intervertebral disc degeneration." *J Orthop Res* 31 (8): 1276–82.

Mason, R.M., and A.J. Palfrey. 1984. "Intervertebral disc degeneration in adult mice with hereditary kyphoscoliosis." *J Orthop Res* 2 (4): 333–8.

Masuda, K., Y. Aota, C. Muehleman et al. 2004. "A novel rabbit model of mild, reproducible disc degeneration by an anulus needle puncture: Correlation between the degree of disc injury and radiological and histological appearances of disc degeneration." *Spine* 30 (1): 5–14.

Melrose, J., S. Roberts, S. Smith, J. Menage, and P. Ghosh. 2002. "Increased nerve and blood vessel ingrowth associated with proteoglycan depletion in an ovine anular lesion model of experimental disc degeneration." *Spine* 27 (12): 1278–85.

Melrose, J., T.K. Taylor, P. Ghosh et al. 1996. "Intervertebral disc reconstitution after chemonucleolysis with chymopapain is dependent on dosage." *Spine* 21 (1): 9–17.

Miao, J., S. Wang, Z. Wan et al. 2013. "Motion characteristics of the vertebral segments with lumbar degenerative spondylolisthesis in elderly patients." *Eur Spine J* 22 (2): 425–31.

Michalek, A.J., K.L. Funabashi, and J.C. Iatridis. 2010. "Needle puncture injury of the rat intervertebral disc affects torsional and compressive biomechanics differently." *Eur Spine J* 19 (12): 2110–6.

Miyamoto, S., K. Yonenobu, and K. Ono. 1991. "Experimental cervical spondylosis in the mouse." *Spine* 16 (10 Suppl): S495–500.

Modic, M.T., P.M. Steinberg, J.S. Ross, T.J. Masaryk, and J.R. Carter. 1988. "Degenerative disk disease: Assessment of changes in vertebral body marrow with MR imaging." *Radiology* 166 (1): 193–99.

Monaco, L.A., S.J. Dewitte-Orr, and D.E. Gregory. 2016. "A comparison between porcine, ovine, and bovine intervertebral disc anatomy and single lamella annulus fibrosus tensile properties." *J Morphol* 277 (2): 244–51.

Moon, C.H., J.-H. Kim, L. Jacobs et al. 2012. "Part 1: Dual-tuned proton/sodium magnetic resonance imaging of the lumbar spine in a rabbit model." *Spine (Phila Pa 1976)* 37 (18): E1106–12.

Moon, S.M., J.H. Yoder, A.C. Wright et al. 2013. "Evaluation of intervertebral disc cartilaginous endplate structure using magnetic resonance imaging." *Eur Spine J* 22 (8): 1820–28.

Moskowitz, R.W., I. Ziv, C.W. Denko, B. Boja, P.K. Jones, and J.H. Adler. 1990. "Spondylosis in sand rats: A model of intervertebral disc degeneration and hyperostosis." *J Orthop Res* 8 (3): 401–11.

Moss, I.L., Y. Zhang, P. Shi, A. Chee, M.J. Piel, and H.S. An. 2013. "Retroperitoneal approach to the intervertebral disc for the annular puncture model of intervertebral disc degeneration in the rabbit." *Spine J* 13 (3): 229–34.

Mulligan, K.R., C.E. Ferland, R. Gawri et al. 2014. "Axial T1ρ MRI as a diagnostic imaging modality to quantify proteoglycan concentration in degenerative disc disease." *Eur Spine J* 24 (11): 2395–2401.

Mwale, F., C.N. Demers, A.J. Michalek et al. 2008. "Evaluation of quantitative magnetic resonance imaging, biochemical and mechanical properties of trypsin-treated intervertebral discs under physiological compression loading." *J Magn Reson Imaging* 27 (3): 563–73.

Mwale, F., J.C. Iatridis, and J. Antoniou. 2008. "Quantitative MRI as a diagnostic tool of intervertebral disc matrix composition and integrity." *Eur Spine J* 17 (Suppl 4): 432–40.

Niinimäki, J., A. Korkiakoski, O. Ojala et al. 2009. "Association between visual degeneration of intervertebral discs and the apparent diffusion coefficient." *Magn Reson Imaging* 27 (5): 641–47.

Nisolle, J.-F., B. Bihin, N. Kirschvink et al. 2016. "Prevalence of age-related changes in ovine lumbar intervertebral discs during computed tomography and magnetic resonance imaging." *Comp Med* 66 (4): 300–7.

Norcross, J.P., G.E. Lester, P. Weinhold, and L.E. Dahners. 2003. "An *in vivo* model of degenerative disc disease." *J Orthop Res* 21 (1): 183–88.

Noury, F., J. Mispelter, F. Szeremeta, S. Même, B.T. Doan, and J.C. Beloeil. 2008. "MRI methodological development of intervertebral disc degeneration: A rabbit *in vivo* study at 9.4 T." *Magn Reson Imaging* 26 (10): 1421–32.

Nuckley, D.J., P.A. Kramer, A. Del Rosario et al. 2008. "Intervertebral disc degeneration in a naturally occurring primate model: Radiographic and biomechanical evidence." *J Orthop Res* 26 (9): 1283–88.

O'Connell, G.D., E.J. Vresilovic, and D. M Elliott. 2007. "Comparison of animals used in disc research to human lumbar disc geometry." *Spine* 32 (3): 328–33.

Oegema, T.R., S. Swedenberg, S.L. Johnson, M. Madison, and D.S. Bradford. 1992. "Residual chymopapain activity after chemonucleolysis in normal intervertebral discs in dogs." *J Bone Joint Surg* 74 (6): 831–38.

Oehme, D., T. Goldschlager, P. Ghosh, J.V. Rosenfeld, and G. Jenkin. 2015. "Cell-based therapies used to treat lumbar degenerative disc disease: A systematic review of animal studies and human clinical trials." *Stem Cells Int* 2015: 946031.

Ohnishi, T., H. Sudo, K. Iwasaki et al. 2016. "*In vivo* mouse intervertebral disc degeneration model based on a new histological classification." *PLoS One* 11 (8).

Olsewski, J.M., M.J. Schendel, L.J. Wallace, J.W. Ogilvie, and C.R. Gundry. 1996. "Magnetic resonance imaging and biological changes in injured intervertebral discs under normal and increased mechanical demands." *Spine* 21 (17): 1945–51.

Omlor, G.W., A.G. Nerlich, H.-J. Wilke et al. 2009. "A new porcine in vivo animal model of disc degeneration: Response of annulus fibrosus cells, chondrocyte-like nucleus pulposus cells, and notochordal nucleus pulposus cells to partial nucleotomy." *Spine* 34 (25): 2730–9.

Ooms, K.J., M. Cannella, A.J. Vega, M. Marcolongo, and T. Polenova. 2008. "23Na TQF NMR imaging for the study of spinal disc tissue." *J Magn Reson* 195 (1): 112–5.

Oshima, H., H. Ishihara, J.P.G. Urban, and H. Tsuji. 1993. "The use of coccygeal discs to study intervertebral disc metabolism." *J Orthop Res* 11 (3): 332–8.

Osti, O.L., and R.D. Fraser. 1992. "MRI and discography of annular tears and intervertebral disc degeneration. A prospective clinical comparison." *J Bone Joint Surg Br* 74 (3): 431–5.

Osti, O.L., B. Vernon-Roberts, and R.D. Fraser. 1990. "1990 Volvo Award in Experimental Studies: Anulus tears and intervertebral disc degeneration: An experimental study using an animal model." *Spine* 15 (8): 762–67.

Paassilta, P., J. Lohiniva, H.H. Goring et al. 2001. "Identification of a novel common genetic risk factor for lumbar disk disease." *JAMA* 285 (14): 1843–9.

Pandit, P., J.F. Talbott, V. Pedoia, W. Dillon, and S. Majumdar. 2016. "$T_{1\rho}$ and T_2-based characterization of regional variations in intervertebral discs to detect early degenerative changes." *J Orthop Res* May: 1–9.

Paul, C.P.L., M. De Graaf, A. Bisschop et al. 2017. "Static axial overloading primes lumbar caprine intervertebral discs for posterior herniation." *Plos One* 12 (4): 1–23.

Perry, J., V. Haughton, P.A. Anderson et al. 2006. "The value of T2 relaxation times to characterize lumbar intervertebral disks: Preliminary results." *Am J Neuroradiol* 27 (2): 337–42.

Pfeiffer, M., P. Griss, P. Franke et al. 1994. "Degeneration model of the porcine lumbar motion segment: Effects of various intradiscal procedures." *Eur Spine J* 3 (1): 8–16.

Pfirrmann, C.W., A. Metzdorf, M. Zanetti, J. Hodler, and N. Boos. 2001a. "Magnetic resonance classification of lumbar intervertebral disc degeneration." *Spine* 26 (17): 1873–78.

Pfirrmann, C.W., A. Metzdorf, M. Zanetti, J. Hodler, and N. Boos. 2001b. "Magnetic resonance classification of lumbar intervertebral disc degeneration." *Spine* 26 (17): 1873–8.

Phillips, F.M., J. Reuben, and F.T. Wetzel. 2002. "Intervertebral disc degeneration adjacent to a lumbar fusion. an experimental rabbit model." *J Bone Joint Surg* 84 (2): 289–94.

Platenberg, R.C., G.B. Hubbard, W.J. Ehler, and C.J. Hixson. 2001. "Spontaneous disc degeneration in the baboon model: Magnetic resonance imaging and histopathologic correlation." *J Med Primatol* 30: 268–72.

Plewes, D.B., and W. Kucharczyk. 2012. "Physics of MRI: A primer." *J Magn Reson Imaging* 35 (5): 1038–54.

Pokharna, H.K., B. Boja, V. Monnier, and R.W. Moskowitz. 1994. "Effect of age on pyridinoline and pentosidine matrix cross-links in the desert sand rat intervertebral disc." *Glycosylation Dis* 1 (3): 185–90.

Priester, W.A. 1976. "Canine intervertebral disc disease–occurrence by age, breed, and sex among 8.117 cases." *Theriogenology* 6 (2–3): 293–303.

Prologo, J.D., A. Pirasteh, N. Tenley et al. 2012. "Percutaneous image-guided delivery for the transplantation of mesenchymal stem cells in the setting of degenerated intervertebral discs." *J Vasc Interv Radiol* 23 (8).

Puustjärvi, K., M. Lammi, H. Helminen, R. Inkinen, and M. Tammi. 1994. "Proteoglycans in the intervertebral disc of young dogs following strenuous running exercise." *Connect Tissue Res* 30: 225–40.

Puustjärvi, K., M. Lammi, I. Kiviranta, H.J. Helminen, and M. Tammi. 1993. "Proteoglycan synthesis in canine intervertebral discs after long-distance running training." *J Orthop Res* 11 (5): 738–46.

Reid, J.E., J.R. Meakin, S.P. Robins, J.M.S. Skakle, and D.W.L. Hukins. 2002. "Sheep lumbar intervertebral discs as models for human discs." *Clin Biomech* 17 (4): 312–4.

Reitmaier, S., H. Schmidt, R. Ihler et al. 2013. "Preliminary investigations on intradiscal pressures during daily activities: An *in vivo* study using the merino sheep." *PLoS One* 8 (7): 1–10.

Reitmaier, S., D. Volkheimer, N. Berger-Roscher, H.-J. Wilke, and A. Ignatius. 2014. "Increase or decrease in stability after nucleotomy? Conflicting *in vitro* and *in vivo* results in the sheep model." *J R Soc. Interf* 11 (100): 20140650.

Rhodes, S.D., W. Zhang, D. Yang et al. 2015. "Dystrophic spinal deformities in a neurofibromatosis type 1 murine model." *PLoS One* 10 (3).

Ringel, F., M. Stoffel, C. Stüer, and B. Meyer. 2006. "Minimally invasive transmuscular pedicle screw fixation of the thoracic and lumbar spine." *Neurosurgery* 59 (4 Suppl 2): ONS361-367.

Roberts, S., B. Caterson, J. Menage et al. 2000. "Matrix metalloproteinases and aggrecanase: Their role in disorders of the human intervertebral disc." *Spine* 25 (23): 3005–13.

Roberts, S., H. Evans, J. Trivedi, J. Menage. 2006. "Histology and pathology of the human intervertebral disc." *J Bone Joint Surg* 88 Suppl 2 (suppl 2): 10–14.

Roberts, S., J. Menage, S. Sivan, and J.P.G. Urban. 2008. "Bovine explant model of degeneration of the intervertebral disc." *BMC Musculoskelet Disord* 9: 24.

Rousseau, M.-A., J.A. Ulrich, E.C. Bass et al. 2007. "Stab incision for inducing intervertebral disc degeneration in the rat." *Spine* 32 (1): 17–24.

Russo, F., R.A. Hartman, K.M. Bell et al. 2016. "Biomechanical evaluation of transpedicular nucleotomy with intact annulus fibrosus." *Spine (Phila Pa 1976)* 42 (4): 193–201.

Saamanen, A.M., K. Puustjarvi, K. Ilves et al. 1993. "Effect of running exercise on proteoglycans and collagen content in the intervertebral disc of young dogs." *Int J Sports Med* 14 (1): 48–51.

Saar, G., Zhang B., W. Ling et al. 2013. "Assessment of GAG concentration changes in the intervertebral disc via CEST." *NMR Biomed* 25 (2): 255–61.

Sahlman, J., R. Inkinen, T. Hirvonen et al. 2001. "Premature Vertebral endplate ossification and mild disc degeneration in mice after inactivation of one allele belonging to the Col2a1 gene for type II collagen." *Spine* 26 (23): 2558–65.

Sakai, D., J. Mochida, T. Iwashina et al. 2005. "Differentiation of mesenchymal stem cells transplanted to a rabbit degenerative disc model: Potential and limitations for stem cell therapy in disc regeneration." *Spine* 30 (21): 2379–87.

Saldanha, K.J., S.L. Piper, K.M. Ainslie, T.A. Desai, H.T. Kim, S. Majumdar. 2008. "Magnetic resonance imaging of iron oxide labelled stem cells: Applications to tissue engineering based regeneration of the intervertebral disc." *Eur Cells Mater* 16: 17–25.

Sasaki, M., T. Takahashi., K. Miyahara, and aT. Hirose. 2001. "Effects of chondroitinase ABC on intradiscal pressure in sheep: An *in vivo* study." *Spine* 26 (5): 463–8.

Sato, M., T. Asazuma, M. Ishihara, M. Ishihara, T. Kikuchi, M. Kikuchi, and K. Fujikawa. 2003. "An experimental study of the regeneration of the intervertebral disc with an allograft of cultured annulus fibrosus cells using a tissue-engineering method." *Spine* 28 (6): 548–53.

Schleich, C., A. Müller-Lutz, L. Zimmermann et al. 2016. "Biochemical imaging of cervical intervertebral discs with glycosaminoglycan chemical exchange saturation transfer magnetic resonance imaging: Feasibility and initial results." *Skelet Radiol* 45 (1): 79–85.

Schmidt, H., and S. Reitmaier. 2012. "Is the ovine intervertebral disc a small human one? A finite element model study." *J Mech Behav Biomed Mater* 17: 229–41.

Schneiderman, G., B. Flannigan, S. Kingston et al. 1987. "Magnetic resonance imaging in the diagnosis of disc degeneration: Correlation with discography." *Spine* 12 (3): 276–81.

Schollum, M.L., R.C. Appleyard, C.B. Little, and J. Melrose. 2010. "A detailed microscopic examination of alterations in normal anular structure induced by mechanical destabilization in an ovine model of disc degeneration." *Spine* 35 (22): 1965–73.

Serigano, K., D. Sakai, A. Hiyama et al. 2010. "Effect of cell number on mesenchymal stem cell transplantation in a canine disc degeneration model." *J Orthop Res* 28 (10): 1267–75.

Shapiro, E.M., A. Borthakur, A. Gougoutas, and R. Reddy. 2002. "23Na MRI accurately measures fixed charge density in articular cartilage." *Magn Reson Med* 47 (2): 284–91.

Shapiro, E.M., L.N. Medford-Davis, T.M. Fahmy, C.E. Dunbar, and A.P. Koretsky. 2007. "Antibody-mediated cell labeling of peripheral T Cells with micron-sized iron oxide particles (MPIOs) allows single cell detection by MRIy." *Contrast Media Mol Imaging* 2 (5): 147–53.

Sheng, S.R., X.Y. Wang, H.Z. Xu, G.Q. Zhu, and Y.F. Zhou. 2010. "Anatomy of large animal spines and its comparison to the human spine: A systematic review." *Eur Spine J* 19 (1): 46–56.

Shi, Z.Y., T. Gu, C. Zhang et al. 2016. "Computed tomography-guided nucleus pulposus biopsy for canine intervertebral disc degeneration preparation: A radiology and histology study." *Spine J* 16 (2): 252–8.

Silberberg, R. 1988a. "Histologic and morphometric observations on vertebral bone of aging sand rats." *Spine (Phila Pa 1976)* 13 (2): 202–8.

Silberberg, R. 1988b. "The vertebral column of diabetic sand rats (*Psammomys obesus*)." *Exp Cell Biol* 56 (4): 217–20.

Sobajima, S., J.F. Kompel, J.S. Kim et al. 2005. "A slowly progressive and reproducible animal model of intervertebral disc degeneration characterized by MRI, X-ray, and histology." *Spine* 30 (1): 15–245.

Spencer, D.L., J.A. Miller, and A.B. Schultz. 1985. "The effects of chemonucleolysis on the mechanical properties of the canine lumbar disc." *Spine (Phila Pa 1976)* 10 (6): 555–61.

Stelzeneder, D., G.H. Welsch, B.K. Kovács et al. 2012. "Quantitative T2 evaluation at 3.0T compared to morphological grading of the lumbar intervertebral disc: A standardized evaluation approach in patients with low back pain." *Eur J Radiol* 81 (2): 324–30.

Stern, W.E., and W.F. Coulson. 1976. "Effects of collagenase upon the intervertebral disc in monkeys." *J Neurosurg* 44 (1): 32–44.

Stevens, J.W., G.L. Kurriger, A.S. Carter, and J.A. Maynard. 2000. "CD44 expression in the developing and growing rat intervertebral disc." *Dev Dyn* 219 (3): 381–90.

Stokes, I.A.F., and J.C. Iatridis. 2004. "Mechanical conditions that accelerate intervertebral disc degeneration: Overload versus immobilization." *Spine* 29 (23): 2724–32.

Sugimura, T., F. Kato, K. Mimatsu, O. Takenaka, and H. Iwata. 1996. "Experimental chemonucleolysis with chondroitinase ABC in monkeys." *Spine* 21 (2): 161–5.

Sullivan, J.D., H.F. Farfan, and D.S. Kahn. 1971. "Pathologic changes with intervertebral joint rotational instability in the rabbit." *Can J Surg* 14 (1): 71–9.

Sun, W., K. Zhang, C.-Q. Zhao et al. 2013. "Quantitative T2 mapping to characterize the process of intervertebral disc degeneration in a rabbit model." *BMC Musculoskelet Disord* 14: 357.

Sweet, H.O., and M.C. Green. 1981. "Progressive ankylosis, a new skeletal mutation in the mouse." *J Hered* 72: 87–93.

Takahashi, T., H. Kurihara, S. Nakajima et al. 1996. "Chemonucleolytic effects of chondroitinase ABC on normal rabbit intervertebral discs. Course of action up to 10 days postinjection and minimum effective dose." *Spine* 21 (21): 2405–11.

Tapp, H., R. Deepe, J. Ingram, M. Kuremsky, E.N. Hanley, and H.E. Gruber. 2008. "Adipose-derived mesenchymal stem cells from the sand rat: Transforming growth factor beta and 3D co-culture with human disc cells stimulate proteoglycan and collagen type i rich extracellular matrix." *Arthritis Res Ther* 10 (4): R89.

Taurog, J.D., S.D. Maika, W.A. Simmons, M. Breban, and R.E. Hammer. 1993. "Susceptibility to inflammatory disease in hla-b27 transgenic rat lines correlates with the level of B27 expression." *J Immunol* 150 (9): 4168–78.

Taylor, R.S., P. Fritzell, and R.J. Taylor. 2007. "Balloon kyphoplasty in the management of vertebral compression fractures: An updated systematic review and meta-analysis." *Eur Spine J* 16 (8): 1085–1100.

Taylor, T.K.F., P. Ghosh, K.G. Braund, J.M. Sutherland, and A.A. Sherwood. 1976. "The effect of spinal fusion on intervertebral disc composition: An experimental study." *J Surg Res* 21 (2): 91–104.

Vadalà, G., F. De Strobel, M. Bernardini et al. 2013. "The transpedicular approach for the study of intervertebral disc regeneration strategies: *In vivo* characterization." *Eur Spine J* 22 (Suppl. 6): 972–78.

Vadalà, G., F. Russo, G. Pattappa et al. 2015. "A nucleotomy model with intact annulus fibrosus to test intervertebral disc regeneration strategies." *Tissue Eng Part C Methods*, 1–29.

Vadalà, G., F. Russo, G. Pattappa, and D. Schiuma. 2013. "The transpedicular approach as an alternative route for intervertebral disc regeneration." *Spine (Phila, Pa 1978)* 38 (6): 319–24.

Vadalà, G., G. Sowa, M. Hubert et al. 2012. "Mesenchymal stem cells injection in degenerated intervertebral disc: Cell leakage may induce osteophyte formation." *J Tissue Eng Regener Med* 6 (5): 348–55.

Vaga, S., M.T. Raimondi, E.G. Caiani et al. 2008. "Quantitative assessment of intervertebral disc glycosaminoglycan distribution by gadolinium-enhanced MRI in orthopedic patients." *Magn Reson Med* 59 (1): 85–95.

Venn, G., and R.M. Mason. 1986. "Changes in mouse intervertebral-disc proteoglycan synthesis with age. Hereditary kyphoscoliosis is associated with elevated synthesis." *Biochem J* 234 (2): 475–9.

Vo, N., H.-Y. Seo, A. Robinson et al. 2012. "Accelerated aging of intervertebral discs in a mouse model of progeria." *J Orthop Res* 28 (12): 1600–7.

Wakano, K., R. Kasman, E.Y. Chao, D.S. Bradford, and T.R. Oegema. 1983. "Biomechanical analysis of canine intervertebral discs after chymopapain injection. A preliminary report." *Spine (Phila Pa 1976)* 8 (1): 59–68.

Walsh, A.J.L., and Jeffrey C. Lotz. 2004. "Biological response of the intervertebral disc to dynamic loading." *J Biomech* 37 (3): 329–37.

Wang, W., J.S. Nyman, K. Ono, D.A. Stevenson, X. Yang, and F. Elefteriou. 2011. "Mice lacking Nf1 in osteochondroprogenitor cells display skeletal dysplasia similar to patients with neurofibromatosis Type I." *Hum Mol Genet* 20 (20): 3910–24.

Wang, Y.X.J., F. Zhao, J.F. Griffith et al. 2013. "T1rho and T2 relaxation times for lumbar disc degeneration: An *in vivo* comparative study at 3.0-Tesla MRI." *Eur Radiol* 23 (1): 228–34.

Watanabe, A., L.M. Benneker, C. Boesch, T. Watanabe, T. Obata, and S.E. Anderson. 2007. "Classification of intervertebral disk degeneration with axial T2 mapping." *Am J Roentgenol* 189 (4): 936–42.

Watanabe, H., and Y. Yamada. 2002. "Chondrodysplasia of gene knockout mice for aggrecan and link protein." *Glycoconj J* 19(4–5):269–73.

Wei, F., R. Zhong, X. Pan et al. 2015. "Computed tomography guided subendplate injection of pingyangmycin for a novel rabbit model of slowly progressive disc degeneration." *Spine J* 7: S1529–9430.

Wei, F., R. Zhong, Z. Zhou et al. 2014. "*In vivo* experimental intervertebral disc degeneration induced by bleomycin in the Rhesus monkey." *BMC Musculoskelet Disord* 15 (January): 340.

Weidenbaum, M., R.J. Foster, B.A. Best et al. 1992. "Correlating magnetic resonance imaging with the biochemical content of the normal human intervertebral disc." *J Orthop Res 10* (4): 552–61.

Wilke, H.J., A. Kettler, K.H. Wenger, and L.E. Claes. 1997. "Anatomy of the sheep spine and its comparison to the human spine." *Anat Record* 247 (4): 542–55.

Wilke, H.J., P. Neef, M. Caimi, T. Hoogland, and L.E. Claes. 1999. "New *in vivo* measurements of pressures in the intervertebral disc in daily life." *Spine* 24 (8): 755–62.

Wilke, H.J., A. Rohlmann, S. Neller et al. 2003. "ISSLS prize winner: A novel approach to determine trunk muscle forces during flexion and extension." *Spine* 28 (23): 2585–93.

Wu, N., H. Liu, J. Chen, L. et al. 2013. "Comparison of apparent diffusion coefficient and T2 relaxation time variation patterns in assessment of age and disc level related intervertebral disc changes." *Plos One* 8 (7): 1–7.

Xi, Y., J. Kong, Y. Liu, Z. Wang, S. Ren, Z. Diao, and Y. Hu. 2013. "Minimally invasive induction of an early lumbar disc degeneration model in rhesus monkeys." *Spine* 38 (10): E579–86.

Yovich, J.V., B.E. Powers, and T.S. Stashak. 1985. "Morphologic features of the cervical intervertebral disks and adjacent vertebral bodies of horses." *Am J Vet Res* 46 (11): 2372–7.

Yurube, T., K. Nishida, T. Suzuki et al. 2010. "Matrix metalloproteinase (MMP)-3 gene upregulation in a rat tail compression loading-induced disc degeneration model." *J Orthop Res* 28 (8): 1026–32.

Zhang, W., X. Ma, and Y. Wang. 2014. "Assessment of apparent diffusion coefficient in lumbar intervertebral disc degeneration." *Eur Spine J* 23 (9): 1830–6.

Zhang, Y., S. Drapeau, H.S. An et al. 2011. "Histological features of the degenerating intervertebral disc in a goat disc-injury model." *Spine* 36 (19): 1519–27.

Zhou, H.W., S.X. Hou, W.L. Shang et al. 2007. "A new *in vivo* animal model to create intervertebral disc degeneration characterized by MRI, radiography, CT/discogram, biochemistry, and histology." *Spine* 32 (8): 864–725.

Zhou, R.P., Z.M. Zhang, L. Wang et al. 2013a. "Establishing a disc degeneration model using computed tomography-guided percutaneous puncture technique in the rabbit." *J Surg Res* 181 (2): e65–74.

Zhou, Z., M. Bez, W. Tawackoli et al. 2016. "Quantitative chemical exchange saturation transfer MRI of intervertebral disc in a porcine model." *Magn Reson Med* 76 (6): 1677–83.

Zhou, Z.Z., B. Jiang, Z.Z. Zhou et al. 2013b. "Intervertebral disk degeneration: T1ρ MR imaging of human and animal models." *Radiology* 268 (2): 492–500.

Ziran, B.H., S. Pineda, H. Pokharna et al. 1994. "Biomechanical, radiologic, and histopathologic correlations in the pathogenesis of experimental intervertebral disc disease." *Spine* 19 (19): 2159–63.

Ziv, I., R.W. Moskowitz, I. Kraise, J.H. Adler, and A. Maroudas. 1992. "Physicochemical properties of the aging and diabetic sand rat intervertebral disc." *J Orthop Res* 10 (2): 205–10.

Zobel, B.B., G. Vadalà, R. Del Vescovo et al. 2012. "T1ρ magnetic resonance imaging quantification of early lumbar intervertebral disc degeneration in healthy young adults." *Spine* 37 (14): 1224–30.

Zook, B.C., and A.I. Kobrine. 1986. "Effects of collagenase and chymopapain on spinal nerves and intervertebral discs of cynomolgus monkeys." *J Neurosurg* 64 (3): 474–83.

Zuo, J., G.B. Joseph, X. Li et al. 2013. "*In-vivo* intervertebral disc characterization using magnetic resonance spectroscopy and T1ρ imaging: Association with discography and Oswestry Disability Index and SF-36." *Spine (Phila Pa 1976)* 37 (3): 214–21.

3 Intervertebral Disc Whole Organ Cultures
How to Choose the Appropriate Model

Sebastian Wangler, Zhen Li, Sibylle Grad, and Marianna Peroglio

CONTENTS

3.1 Introduction 67
3.2 Changes in Human IVD Degeneration 69
    3.2.1 Proteoglycans 70
    3.2.2 Collagen 70
    3.2.3 Cytokines 71
    3.2.4 Enzymes 72
    3.2.5 Mechanobiology 72
3.3 Established Whole-Organ Culture Models 73
3.4 Monitoring Degenerative Changes in *Ex Vivo* IVD Organ Culture 85
    3.4.1 Proteoglycan Analysis 85
    3.4.2 Collagen Analysis 86
    3.4.3 Cytokine Analysis 87
    3.4.4 Enzyme Analysis 89
    3.4.5 Cell Viability 89
    3.4.6 Macroscopic Evaluation 91
    3.4.7 Recent Developments in Mechano-Biological Assessment of IVDs ...92
3.5 Conclusion 93
References 93

3.1 INTRODUCTION

"*It started about 3 days ago. A sharp, burning pain and stiffness in my lower back like I had never felt before. The pain was not only radiating to the right calf and the right sole of my foot but also associated with numbness and a tingling sensation. Changing my position did not lead to a significant relief.*" This is the story of Peter, a 46-year-old patient who is presenting with the described typical symptoms for the first time, at the outpatient department of the local hospital. To find the reason for the described symptoms, a clinician has two major diagnostic tools:

1. Physical examination
2. Imaging: magnetic resonance imaging (MRI), computed tomography (CT), or radiography (X-ray)

While the physical examination describes the functional loss, e.g., missing reflexes, changes in muscle tonus, or sensory loss, imaging (especially the MRI) provides information about the morphological changes of the investigated structure. The physical examination of this case can typically be described as follows: "Palpable muscle spasm in lower back. Positive straight leg raising test and femoral stretch test right. Sensory loss in the right lateral sole and the right calf according to the S1 dermatome. Power plantar flexion: right grade III, left grade IV with diminished ankle jerk reflex right." This status leads to the working diagnosis of nerve root compression, to be confirmed by the result of an MRI. The MRI uses the content of water to describe the grade of degeneration in intervertebral discs (IVDs). The most popular classification to characterize the phenotype of degenerated IVDs based on routine T2-weighted MRI images has been introduced by Pfirrmann et al. (2001). The classification uses four different criteria such as disc structure (homogenous/inhomogeneous, white and black signal), distinction between annulus fibrosus (AF) and nucleus pulposus (NP) (clear/lost), signal intensity of the disc (hyperintense/hypointense) and disc height (normal/collapsed) to describe the degenerative grade of a human IVD. Figure 3.1 shows the MRI of the earlier described clinical case.

The black arrow points to the disc herniation in the lower lumbar discs at the level L5-S1 and the white arrow indicates the herniated NP, causing a compression of the left S1 traversing nerve root and leading to the described symptoms. Our patient's medical history together with the MRI findings (Figure 3.1) leads to the diagnosis of radiculopathy caused by a lumbar disc herniation (LDH). Radiculopathy is one of the most disabling types of pain emanating from the lower back. Lifetime prevalence for radiculopathies varies between 12% and 34%, and one of the main causes is LDH (Konstantinou and Dunn 2008).

LDH is defined as a displacement of disc material (NP or AF) beyond the IVD space and affects the spine in young and middle-aged patients, with a male-to-female ratio of 2:1. Current treatment options involve conservative (physiotherapy and pain management by drugs) or invasive (surgery with nerve root decompression) strategies, which lead to an enormous healthcare burden and uncertain long-term outcome (Anderson, McCormick, and Angevine 2008; Jordan, Konstantinou, and O'Dowd 2009; Schoenfeld and Weiner 2010). Researchers from all over the world working in the field of tissue engineering and regenerative medicine take this as a motivation to develop new treatment options for degenerative disc diseases. To test these new techniques, researchers must rely on models, which mimic the changes observed in human degenerated IVDs.

The aim of this chapter is to provide an overview of (i) the main morphological changes observed in human degenerated IVDs, (ii) the established *ex vivo* organ

Intervertebral Disc Whole Organ Cultures

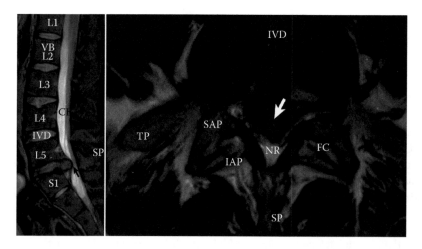

FIGURE 3.1 MRI lumbar spine, axial (left) and sagittal (right). Disc degeneration (b/w arrow) at level L5/S1 with right paracentral disc extrusion and superior migration of herniated NP causing compression of the right S1 traversing nerve root. NR, nerve root; CE, caudal equine; VB, vertebral body; TP, transverse process; SP, spinous process; SAP, superior articular process; IAP, inferior articular process; FC, facet joint.

culture models aiming to simulate these changes, and (iii) the methods used to monitor these simulated changes. The latter part of the chapter is based on recent review articles and will focus on the methodology of measured output parameters in degenerative organ cultures with the aim to provide guidance on the choice of the most appropriate model for future experiments (Gantenbein et al. 2015; Iatridis et al. 2013; Peroglio et al. 2016).

3.2 CHANGES IN HUMAN IVD DEGENERATION

Two main mechanisms have been described to initiate degenerative changes in IVDs. The first expects degeneration to be started through mechanical overloading, which leads to microfissures in the tissue and eventually to matrix breakdown, caused by the dysregulation of tissue homeostasis (Adams and Dolan 1997; Adams et al. 2000). The second hypothesis describes a lack of disc-cell nutrition over lifetime, resulting in a continuous cell loss, as the nutrient diffusion through the cartilaginous endplate (CEP) decreases with aging. Besides, metabolic waste products accumulate in the disc, leading to a loss of homeostasis and resulting, similar to the first scenario, in the breakdown of the extracellular matrix (ECM) (Grunhagen et al. 2011; Urban, Smith, and Fairbank 2004).

The main function of the ECM is to provide a stiff tissue that can absorb mechanical stress while enabling a wide range of motion. This is achieved with the help of a high osmotic pressure, which results from the ECM capability to bind and capture water. Therefore, clinicians use the water content of an IVD (measured by MRI) as an indicator for its grade of degeneration (Pfirrmann et al. 2001).

The following subsections provide an overview of the components of the ECM, the mediators and signaling molecules, and the resulting biomechanical changes observed during degeneration of human IVDs.

3.2.1 PROTEOGLYCANS

Proteoglycans are built of a core protein onto which glycosaminoglycan (GAG) chains are bound. Depending on the size and the repeating disaccharide of the GAG chain, proteoglycans can be divided in two groups, large- and small-chain proteoglycans (Iozzo and Schaefer 2015). The most frequent proteoglycan found in human IVDs is aggrecan, with 220 kDa, representing one of the large proteoglycans (Urban and Roberts 2003). Due to their polarity, the GAG chains possess the capability to bind high amounts of water. The highest amount of GAG in a human IVD is found in the NP. The high water-binding capacity of the NP, and its confinement by the AF and endplates contribute to a high intradiscal pressure. The functionality is comparable to the mechanisms of a water bed (NP = water, AF/endplate = envelope). During the process of IVD degeneration, the IVD's proteoglycans, especially the aggrecan molecule, become degraded and the smaller fragments diffuse out of the IVD. Furthermore, cleaved aggrecan fractions lose the ability to bind water. Therefore, this degradation process and the washout effect are responsible for a decreased hydration and a loss of osmotic pressure, which leads then to mechanical malfunction (Vergroesen et al. 2015).

Protective mechanisms were described, slowing down the washout effect by entrapping the small, degraded proteoglycan chains through the AF tissue and the cartilage endplates of the vertebrae (Antoniou et al. 1996; Gan et al. 2003; Urban and Roberts 2003). Further, there is evidence that the amounts of small-chain proteoglycans decorin and biglycan (both acting as linking proteins between collagen and large proteoglycans) are elevated in degenerate human discs, which might increase the entrapment of GAG fragments by providing an increased ECM density (Inkinen et al. 1998). On the other hand, an increased crosslinking slows down matrix turnover and also captures damaged macromolecules, possibly leading to reduced tissue strength (Roughley 2004).

Another member of the proteoglycans, fibronectin, was also observed to accumulate and fragment during the degeneration process. *In vitro*, fibronectin fragments have been shown to down-regulate the synthesis of aggrecan and stimulate the production of matrix degrading metalloproteinases (MMPs) (Oegema et al. 2000).

3.2.2 COLLAGEN

Several types of collagen (I, II, III, V, VI, IX, XI, XII, and XIV) have been found in the human IVD. Different from the observed quantitative loss of proteoglycans, it is rather a change in the type and distribution of collagen that occurs in the process of IVD degeneration. These changes are induced by enzymatic activity through denaturation and crosslinking (Eyre, Matsui, and Wu 2002; Sivan et al. 2008). For example, increases in collagen IX, denatured collagen II, and the mostly pericellular concentrated collagen III were observed in degenerated human IVDs (Antoniou et al.

1996; Hollander et al. 1996). In the NP, a shift from collagen type II to collagen type I was described, leading to more fibrous tissue-like properties with lower hydration capacities (Antoniou et al. 1996).

Furthermore, an increase of crosslinks between collagen and glucose can lead to advanced glycation end-products, limiting as well the swelling properties of the NP (DeGroot et al. 2004). The gradual change in color to a brown appearance observed in old human IVDs can be explained by this accumulation of carbohydrate-derived adducts (Hormel and Eyre 1991).

3.2.3 Cytokines

Degeneration of the IVD involves an imbalance between anabolic and catabolic responses. Cytokines appear as key modulatory players in these degenerative processes, triggering matrix degradation, production of chemokines, cell phenotype transformation, and attraction of cells from the systemic immune system (Risbud and Shapiro 2014). Elevated levels of tumor necrosis factor (TNF)-α, interleukin (IL)-1 α/β, IL-6, and IL-17 have been observed in degenerated human IVDs (Andrade et al. 2013; Gruber et al. 2013; Le Maitre, Hoyland, and Freemont 2007a). These inflammatory cytokines are released by the disc cells triggered through smoking, infection, abnormal biomechanical loading, decreased nutrient transport across the endplate, and biological aging (Adams et al. 2000; Battie et al. 2009; Cheung et al. 2009; Mayer et al. 2013; Wang et al. 2012). Risbud et al. defined three distinct, but overlapping, phases in the process of disc degeneration resulting in neovascularization, nerve ingrowth, and finally back and radicular pain (Risbud and Shapiro 2014) (Table 3.1).

TABLE 3.1
Phases in the Process of IVD Degeneration

Phase	Trigger	Target	Reaction	Effect
Phase 1	Mechanical trauma, genetics, infection, smoking	NP cell	Production: TNF-α, IL-1b and IL-6, ADAMTS 4/5, MMPs, SDC4, neurotrophins	ACAN, collagen type II degradation, and stimulation of the dorsal root ganglion
Phase 2	Chemokines	Immunocytes	Production: TNF-α, IL-1b, IL-6, NGF, BDNF; migration	Stimulation of the dorsal root ganglion
Phase 3	Chemokines	Dorsal root ganglion	Production: ASIC3, Trpv1	Back and radicular pain

Note: ADAMTS, a disintegrin and MMP with thrombospondin motifs; SDC4, syndecan 4; NGF, nerve growth factor; BDNF brain-derived neutrophic factors; ASIC3, Acid-sensing ion channel 3; Trpv1, transient receptor potential cation channel subfamily V member 1, also known as the capsaicin receptor and the vanilloid receptor 1.

TABLE 3.2
MMPs Expressed in Human IVDs

Group	MMP	Target
Collagenases	MMP-1, -8, and -13	Fibrillary collagen
Gelatinases	MMP-2 and -9	Denatured collagens, gelatins, and laminin
Stromelysins	MMP-3 and -10	Proteoglycans, gelatins, collagens, and pro-MMPs
Matrilysins	MMP-7	Aggrecan, factors, and cytokines
Other	MMP-28	Casein

3.2.4 ENZYMES

While cytokines "orchestrate" the degenerative process, the fragmentation of collagen, proteoglycan, and fibronectin is mainly caused by catabolic enzymes, capable of breaking down these ECM molecules. They can be classified in two major groups: MMPs (matrix MMPs) and a disintegrin and MMP with thrombospondin motifs (ADAMTS).

Goupille et al. (1998) were one of the first ones to systematically describe the concept of unbalanced MMP activities as a possible cause of IVD degeneration. In total, 24 MMPs have been defined in humans, of which nine were reported to be present in human IVDs. They can be categorized within five traditionally described activity groups: collagenases, gelatinases, stromelysins, matrilysins, and "others" (Table 3.2). With the exception of MMP-28, all MMPs found in human IVDs are reported to show an increased expression in degenerated IVDs (Vo et al. 2013).

ADAMTS account for degradation of aggrecan and versican (Kuno et al. 1997). Five of the known 20 ADAMTS are reported to be active in human IVDs. They all belong to the group of hyalectanases, possessing aggrecanolytic functions (Murphy and Nagase 2008; Pasternak and Aspenberg 2009). Four of the five, ADAMTS-1, -4, -5, and -15, are described to be up-regulated in degenerated human IVDs (Vo et al. 2013).

Both MMPs and ADAMTS are secreted in an inactive form and their catabolic activity is controlled by regulative molecules, in particular by tissue inhibitors of MMPs (TIMPs), which specifically bind and inhibit the MMPs (Visse and Nagase 2003). TIMP-1, -2, and -3 are expressed in human IVDs, and higher expressions of TIMP-1 and -2 were found in degenerated human IVDs (Vo et al. 2013).

3.2.5 MECHANOBIOLOGY

The mechanical environment has a great influence on IVD metabolism, especially on its matrix turnover and cell viability (Lotz and Chin 2000; Lotz et al. 2002). Unphysiological loading, like static compressive loading or absence of loading, can result in a decreased proteoglycan content of the IVD (Hutton et al. 1998, 2002). In a healthy disc, the intradiscal pressure tensions the AF fibers, and preserves the disc height and stiffness when applying axial pressure. In degenerated discs, the loss of proteoglycans and the resulting decrease in hydration and intradiscal pressure lead to a disc height loss and an increased radial bulge (Brinckmann and Grootenboer 1991; Masuoka et al. 2007; Vergroesen et al. 2014). Furthermore, NP GAG content is related to the neutral zone properties, whereby a GAG loss leads to hypermobility and, through that, to the development of hoop stresses in the AF. Therefore, an increased shear stress can be observed in the AF and NP of degenerated discs (Hwang et al. 2012; Inoue and Espinoza Orias 2011; Urban 2002).

3.3 ESTABLISHED WHOLE-ORGAN CULTURE MODELS

Different cell and organ culture methods are used to mimic and investigate the mechanisms involved in human IVD degeneration. The best controllable method is the two-dimensional (2D) cell culture system. It provides an easy culture system for the investigated cell population and allows understanding of specific processes in a controlled manner. However, as the cells in an IVD reside in a 3D architecture, results found in the 2D cannot directly be compared to the interactions between cells and tissues found in an organ. 3D *in vitro* culture systems aim to overcome the problem of the missing orientation in space (Gruber and Hanley 2000). With this method, harvested disc cells or mesenchymal stem cells (MSCs) are cultured within a gel or a biomaterial scaffold, leading to a disc-like 3D cell arrangement. By mimicking the natural 3D arrangement, the phenotype of disc cells is better preserved.

Whole-organ cultures using IVDs from mammalians provide a junction between *in vitro* systems like 2D or 3D systems and *in vivo* studies. Their function is to provide an environment that helps to better predict the potential of a new treatment strategy by allowing the study of cell–matrix interactions in their native milieu, including the complexity of self-regulating mechanisms. Isolated human IVDs have been cultured under free swelling conditions or under applied loading in a bioreactor system (Gawri et al. 2011; Haglund et al. 2011; Walter et al. 2014). However, whole human IVDs are scarcely available, and therefore, research is focusing on the development of alternative organ culture models. Different species have been used to mimic the biomechanical and biochemical changes observed in human IVD degeneration, and every species has its advantages and disadvantages. In respect of the organ size, a range from 2 mm (mouse) up to 25 mm (bovine tail) disc diameter has been used for *ex vivo* organ cultures. Large animals have the benefit that they possess transport and metabolic rates similar to human IVDs. The bovine coccygeal disc is the model most often used, as it approximates human IVDs with respect to size, diffusion and biomechanical properties. Furthermore, the easy accessibility of coccygeal bovine IVDs allows researchers to perform trials with high experimental numbers (N) without the need of ethical approval, as the material is obtained from and regarded as waste by local abattoirs. Moreover, organ culture systems reduce the amount of *in vivo* experiments since only promising methods are further investigated within small or large animal models.

Several methods have been established to induce degeneration of harvested whole IVDs, which can be categorized in the following groups: static loading, unphysiological dynamic loading, injection of degrading enzymes, single impact strike load and disc stabbing (Figure 3.2). While static loading requires only a device capable of applying a

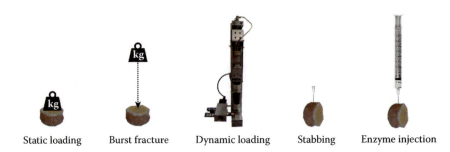

FIGURE 3.2 Methods used to induce degeneration in isolated IVDs.

constant load on the disc, dynamic loading regimes demand the usage of a bioreactor system. The advantage of a force-controlled bioreactor system is the ability to apply both degenerative and physiological loading regimes as found in human IVDs (Illien-Junger et al., 2010). Dynamic loading is important to maintain the exchange of nutrients through the endplates and enable extended culture periods. As for the injection of catabolic enzymes, their role is to mimic the ECM degradation observed in human IVDs, although with a much shorter time scale. Indeed, it has to be considered that the degeneration of human IVD is a process that takes decades. Therefore, induced degenerative IVD organ culture systems aim to mimic a certain time point or stage of degeneration. Whereas degeneration by enzyme injection and ECM breakdown mimics features of later stages of IVD degeneration, dynamic/static loading provides a platform to investigate triggers, diagnostic markers, and treatment strategies for early stages of IVD degeneration.

However, it is important to consider that every model has its strengths and limitations. Taking an organ out of its physiological environment allows on one hand an easy accessibility, but on the other hand excludes systemic players like the immune system or the surrounding structures. As the immune response is the first reaction after introducing a new material or cell population *in vivo*, this represents one of the biggest limitations even for an organ considered immune-privileged like the IVD. Besides the absence of systemic effectors, time needs to be considered as a key player when simulating degenerative changes found in human IVDs. Degeneration is a process taking place over decades, whereby organ culture models can mimic a certain time point in the degenerative cascade.

Table 3.3 provides an overview of established whole organ culture models. Each organ culture model is designed to represent a specific set of changes and to answer questions connected to these mimicked changes. When planning a new experiment, we therefore suggest the following workflow:

1. Define your research question (RQ).
2. Based on your RQ, search Table 3.3 for suitable organ culture models, e.g., type of degeneration, with/without loading. Based on flow charts (Figure 3.9), select methods described to assess the relevant outcome parameters (e.g., dimethylmethylene blue [DMMB]). All models listed in Table 3.3 are also cited on the flow charts.

TABLE 3.3
List of Established Organ Culture Systems Including Loading Regime, Used Species, Loading Time, RQ, and Main Finding

No.	Reference	Loading	Species	Time	RQ/Aim	Results
		Free Swelling Culture				
1	Gawri et al. 2011	Free swelling culture	Human	28 days	Development of a long-term organ culture system for intact human IVDs with potential to study biologic repair of disc degeneration	Retained cell viability without matrix degeneration after 4 weeks
2	Haschtmann et al. 2006	Free swelling culture	Rabbit	49 days	Establishment of a whole-organ IVD culture model to study disc degeneration *in vitro*	4 weeks: Viability preserved, ↓ proteoglycans ↓ collagen I/II, ACAN mRNA; 1/3 metabolic activity left
3	Grant et al. 2016	Free swelling culture	Bovine coccygeal	5 months	Development of a long-term IVD organ culture model that retains vertebral bone and is easy to prepare	PrimeGrowth culture medium allowed IVD culture for 5 months without loss of viability
4	Pereira et al. 2016	Enucleation through EP cavity	Bovine coccygeal	21 days	Investigation of MSC migration toward degenerated IVD as a strategy to promote tissue repair/regeneration	MSC treatment: ↑ collagen II, aggrecan in IVD; production of FGF-6/7, PDGF-R, GM-CSF, IGF-1sR
		Osmotic Stress				
5	Risbud et al. 2003	Hyperosmotic: 410 mOsm/kg	Rat, lumbar	14 days	Development of a model to maintain IVD tissue architecture and metabolic function in organ culture	Hyperosmotic medium + TGF-β: cells remained viable for at least 1 week

(*Continued*)

TABLE 3.3 (CONTINUED)
List of Established Organ Culture Systems Including Loading Regime, Used Species, Loading Time, RQ, and Main Finding

No.	Reference	Loading	Species	Time	RQ/Aim	Results
6	Risbud et al. 2006	Hyperosmotic: 410 mOsm/kg	Rat, lumbar	7 days	Development of an *in vitro* organ culture of rat IVD and effects of TGF-β3 on disc cell function	TGF-β3: ↑ matrix genes, ↑ [35S] incorporation. Inhibition of ERK activity in the presence TGF-β3 resulted in suppression of collagen type II, ACAN, TGF-RI, TGF-RII, and TGF-RIII gene expression
7	van Dijk, Potier, and Ito 2013	Isotonic (NP): 430 mOsm/kg Hypertonic: 570 mOsm/kg	Bovine coccygeal	42 days	Development of a physiological *in vitro* explant model that incorporates the native environment of IVDs	Both hypertonic NP culture and artificial AF maintained tissue matrix composition for 42 days
8	van Dijk, Potier, and Ito 2011	Hypotonic: 323 mOsm/kg Hypertonic: 570 mOsm/kg	Bovine coccygeal	21 days	Balancing the inherent tissue osmolarity to prevent swelling and thus maintain NP tissue in a native state	Hypertonic medium maintained the NP tissue specific matrix composition
9	Haschtmann, Stoyanov, and Ferguson 2006	Isoosmotic/hyperosmotic media (485 mosmol/kg)	Rabbit	28 days	Effects of cyclic osmotic loading (28 days) on metabolism and matrix gene expression of IVDs	Diurnal osmotic stimulation: no effect; hyperosmolarity: ↑ cell death, ↓ collagen I expression

(*Continued*)

TABLE 3.3 (CONTINUED)
List of Established Organ Culture Systems Including Loading Regime, Used Species, Loading Time, RQ, and Main Finding

No.	Reference	Loading	Species	Time	RQ/Aim	Results
				Static Loading		
10	Alkhatib et al. 2014	IVD compression: noninjured (5% of disc height), injured (30% of disc height)	Human (Th11–L5)	14 days	Investigation of biochemical changes and ECM disruption following acute mechanical injury	Injure disc tissue: ↑ aggrecan fragmentation. Inj. medium: ↑ neurite sprouting in PC12 cells, ↑ IL-5, IL-6, IL-7, IL-8, MCP-2, GROα, MIG; ELISA: ↑ nerve growth factor
11	Lee et al. 2006	Static: 5 kg, 0.25 MPa	Bovine coccygeal	7 days	Effect of EPs on biochemical/cellular stability of large discs	+EPs: better viability/activity 0–24 hours, ↓ cell viability after 1 week −EPs: ↓ biosynthetic activity 0–2 days, sensitive to loading
12	Walter et al. 2011	Static: 0.2 MPa	Bovine	7 days	Effect of low magnitudes of axial compression loading applied asymmetrically on IVD	Asymmetric compression: ↑ cell death, ↑ caspase-3, ↓ aggrecan, ↑ MMP-1, ADAMTS4, IL-1β, IL-6 mRNA
13	Chan et al. 2011	Static: 20 N/0.1 Hz with 0°, 2°, 5°, and 10° torsion	Bovine	4 days	IVD response to repetitive cyclic torsion of varying magnitudes at a physiological frequency	2°: ↑ cell viability in iAF. 5°: ↑ apoptotic activity in NP

(Continued)

TABLE 3.3 (CONTINUED)
List of Established Organ Culture Systems Including Loading Regime, Used Species, Loading Time, RQ, and Main Finding

No.	Reference	Loading	Species	Time	RQ/Aim	Results
14	Korecki, MacLean, and Iatridis 2007	Static: 0.2 MPa Diurnal: 0.1/0.3 MPa (change every 12 hours)	Bovine	8 days	Refinement of variables to monitor IVDs culture response and develop a technique to measure cell viability	Disc height/water content ↓ after culture. Cell viability ↓ with culture duration in iAF/NP; 8 days: GAG content stable
			Burst Fracture			
15	Haschtmann et al. 2008	Burst endplate fracture	Rabbit	9 days	Establishment of disc/endplate trauma (fracture) culture model to investigate concurrent disc changes *in vitro*	↑ LDH levels (1–3 days). Gene expression AF/NP: ↑ caspase 3. NP: ↑ FasL, TNF-α. AF: ↑ MMP-1/-13
16	Dudli, Haschtmann, and Ferguson 2012	Endplate trauma (0.76 J impact) to provoke an EP fracture	Rabbit	28 days	Effect of significant endplate damage or impact loading alone on the initiation of DD	Fracture: ↑ LDH, ↑ caspase-3/7 activity, ↓ GAG, ↓ ACAN mRNA, ↑ collagen II, MMP-1/-3/-13, TNF-α, IL-6, IL-8, MCP-1, FASL, caspase 3 mRNA
			Burst Fracture + Dynamic Loading			
17	Dudli, Ferguson, and Haschtmann 2014	Burst fracture/EP puncturing Dynamic: 1 MPa/1 Hz	Rabbit	28 days	Comparison of pathogenesis of posttraumatic disc degeneration in different injury models	Structural perturbation of EP/IVD, but not the loading or nuclear depressurization, promotes degeneration

(*Continued*)

TABLE 3.3 (CONTINUED)
List of Established Organ Culture Systems Including Loading Regime, Used Species, Loading Time, RQ, and Main Finding

No.	Reference	Loading	Species	Time	RQ/Aim	Results
18	Dudli, Haschtmann, and Ferguson 2015	Burst fracture Phys: dynamic, 1 MPa/1 Hz	Rabbit	28 days	Establishment of *in vitro* burst fracture model, including posttraumatic physiological loading, to investigate DD	Burst fracture + phys day 28: 65%↓ GAG/DNA, ↑ MMP-1/-3, TNF-α, FASL, IL-1/6, iNOS expression in NP
				Dynamic Loading		
19	Rosenzweig et al. 2017	Dynamic; low: 0.1–0.3, medium: 0.1–0.6, high: 0.1–1.2 MPa	Human	14 days	Development of a bioreactor for intact human discs involving dynamic loading	Low/medium: cell viability >80%. High: cell viability ~60%–70%. GAG (~50 μg sGAG/mg), CHAD and newly synthesized collagen II protein are stable in all conditions
20	Walter et al. 2014	Human diurnal: 0.1 MPa/ 0.2 MPa (12/12 hours); static: 0.2 MPa Bovine diurnal 0.2/0.6 MPa (3/16 hours) +0.6 ± 0.2 MPa/ 0.1 Hz (5 hours)	Human, bovine coccygeal	21 days	Establishment of a method for isolating human IVDs with intact vertebral endplates; development and validation of an organ culture loading system for human and bovine IVDs	Human IVDs obtained within at least 48 hours postmortem: viable in culture up to 21 days Bovine IVDs maintained their mechanical behavior and retained >85% viable cells on day 21.
21	Haglund et al. 2011	Static: 0.1 MPa (48 hours) Dynamic: 0.1 Hz, 0.1–0.3 MPa (2 hours)	Bovine	28 days	Development of a bioreactor for intact large discs in a controlled dynamically loaded environment	Physiological and pathological loading. Monitoring of disc height. 4 weeks: viability/disc height stable under physiological cyclic load

(Continued)

TABLE 3.3 (CONTINUED)
List of Established Organ Culture Systems Including Loading Regime, Used Species, Loading Time, RQ, and Main Finding

No.	Reference	Loading	Species	Time	RQ/Aim	Results
22	Gawri et al. 2014	Trypsin injection (48 hours), static: 0.1 MPa (48 hours), cyclic dynamic: 0.1–0.3 MPa (4 hours/day)	Bovine	14 days	Effect of physiological loading on matrix homeostasis after trypsin-induced degeneration	Dynamic physiological load: repair of the ECM; matrix depletion typical of early disc degeneration
23	Illien-Junger et al. 2012	Phys: 0.6 MPa ± 0.2/0.2 Hz (4.5 g/L glucose) Degen: needle punch 0.6 MPa ± 0.2/10 Hz (1/4.5 g/L glucose)	Bovine coccygeal	14 days	Investigation whether metabolic and mechanical challenges can induce MSC recruitment into the IVD	Degen: ↓ cell viability, ↑ matrix degrading enzymes, ↑ nitric oxide production, ↑ MSC homing IGF-1-transduced MSCs: ↑ IGF-1 secretion, ↑ GAG synthesis rate
24	Li et al. 2016	Phys: 0.06 ± 0.02 MPa, 0.1 Hz (6 hours/day)	Bovine coccygeal	5 days	Development of an *ex vivo* cavity model to study repair strategies in loaded IVDs	Dynamic compressive stiffness ↓, disc height loss ↑
25	Furtwangler et al. 2013	MMP-3, ADAMTS-4, HTRA1 injection, diurnal load: 0.4 MPa, 16 hours (d1–d4), followed by free swelling (d4–d8)	Bovine coccygeal	8 days	Development of an *in vitro* model of enzyme-mediated IVD degeneration to mimic the clinical outcome in humans for investigation of therapeutic treatment options	MMP-3, ADAMTS-4, HTRA1: ↓ cell activity, no matrix degradation, no change in gene expression. HTRA1: ↓ disc height, positively correlated with decreased GAG/DNA
26	Paul et al. 2013	Phys: 0.1 MPa/1 Hz (8 hours/day) + 0.1–0.6 MPa/1 Hz (16 hours/day) Degen: 0.1 MPa/1 Hz (8 hours/day) +0.4–0.6–0.8 MPa/1 Hz (16 hours)	Caprine	21 days	Effect of dynamic vs. static overloading regarding mechanically induced IVD degeneration	21 days: both loading with pathological changes, GAG/water ↓ Dynamic: cell death in all IVD regions; Static: oAF

(*Continued*)

TABLE 3.3 (CONTINUED)
List of Established Organ Culture Systems Including Loading Regime, Used Species, Loading Time, RQ, and Main Finding

No.	Reference	Loading	Species	Time	RQ/Aim	Results
27	Vergroesen et al. 2014	Low load: ±10 N/1 Hz High load: ±20 N/1 Hz	Caprine	4.5 hours	Effect of prolonged dynamic loading on intradiscal pressure, disc height, and compressive stiffness	Intradiscal pressure is influenced by recent loading due to fluid flow
28	Castro et al. 2014	*In vivo*: Chondroitinase-ABC inj. *In vitro*: 16 hours/day loading	Caprine	22 days	Comparison of FE-model creep calculation and bioreactor model data	FE model reproduces the generic behavior (control and injected). Discrepancies: recovery periods
29	Chan et al. 2013b	Low dynamic load (LDL) 0.1–0.2 MPa/1 Hz Phys: 0.1–0.6 MPa/1 Hz (16 h/d) + 0.1–0.2 MPa (8 hours/day)	Caprine, lumbar	21 days	Effect of physiological load to optimally preserve IVD properties	21 days Phys: cell viability, cell density, and gene expression were preserved up to 21 days. Unloaded/LDL: ↓ cell viability, ↓ cell density, change in gene expression. ECM content stable in all groups
30	Gantenbein et al. 2006	Uniaxial diurnal: 0.2 MPa (8 hours), 0.8 MPa (16 hours)	Ovine	7 days	Establishment of an organ culture system with native 3D environment under load	Diffusion under diurnal loading. 7 days: cell viability/GAG production stable, catabolic genes ↑, anabolic genes ↓
31	Paul et al. 2012b	LDL: 0.1–0.2 MPa/1 Hz Diurnal phys: 0.1–0.6 MPa, 1 Hz (16 hours), 0.1–0.2 MPa, 1 Hz (8 hours)	Ovine	21 days	Simulated physiological load will preserve IVD properties after 21 days.	Phys loading: cell viability, density and gene expression preserved. No changes in ECM and IVD biomechanical properties.

(Continued)

TABLE 3.3 (CONTINUED)
List of Established Organ Culture Systems Including Loading Regime, Used Species, Loading Time, RQ, and Main Finding

No.	Reference	Loading	Species	Time	RQ/Aim	Results
32	Junger et al. 2009	Diurnal axial: 0.2/0.6 MPa (8/16 hours) cyclic during 0.6 MPa active phase: 0.2 MPa/0.3 Hz (2/4 hours)	Ovine	7/21 days	Effect of limited nutrition on disc cells in short-term and midterm whole organ disc culture	21 days, sufficient glucose: cell viability maintained MMP-13, MMP-7 (active/inactive forms) and newly formed chondroitin sulfate do not rely on nutrition
33	Illien-Junger et al. 2010	Diurnal axial: 0.2/0.6 MPa (8/16 hours) cyclic during 0.6 MPa Phys: 0.2 MPa/ 0.2 Hz Degen: 0.2 MPa/10 Hz	Ovine	7 days	Additive effects of load and nutrition on cell survival, gene expression, and cell activity on day 7	↓ glucose + Hf load: ↓↓ AF/NP cell viability cell ↑ MMP-13
34	Paul et al. 2012a	*Ex vivo* chondroitinase ABC injection followed by physiological loading	Caprine	21 days	Changes in biomechanical, histological, and quantitative MRI parameters after NP cABC injection	cABC injection: ↓ disc height d21 vs d0: change in T2 and T1-rho sequences in NP

Needle Puncture

No.	Reference	Loading	Species	Time	RQ/Aim	Results
35	Korecki, Costi, and Iatridis 2008	Needle puncture (25G/14G) dynamic: 0.2–1 MPa/1 Hz (1 hour/day)	Bovine coccygeal	6 days	Effect of thin/thick needle puncture on disc structure, mechanics, and cellular response after 1 week	No difference between small or big needles observed. Puncture injury: ↓ dynamic modulus, ↑ increase in creep during 1-hour loading.
36	Michalek, Funabashi, and Iatridis 2010	Needle-puncture (30G, 25G, 21G), compression/torsion test	Rat	—	Acute effects of AF injury on the biomechanics of rat caudal IVDs.	Compression: ↓ 20% elastic stiffness (all needle sizes) Torsion: affected proportionally with needle size

(Continued)

TABLE 3.3 (CONTINUED)
List of Established Organ Culture Systems Including Loading Regime, Used Species, Loading Time, RQ, and Main Finding

No.	Reference	Loading	Species	Time	RQ/Aim	Results
				Enzyme Injection		
37	Malonzo et al. 2015	Papain inj., 9 days culture, 7 days static: 0.1 MPa	Bovine coccygeal	16 days	Evaluation of thermo-reversible hydrogel (+/− MSCs, NPCs) for load bearing properties/cyto compatibility	NP: volume/disc height ↓ under static loading. Cell viability: 86% in 3D control, 72% in organ culture
38	Chan et al. 2013a	PBS or Papain inj. (3, 15, 30, 60, and 150 U/mL)	Bovine coccygeal	10 days	Characterization of an *in vitro* disc degeneration model to determine the initial response of injected MSCs	>30 U/mL papain: ↓ GAG/water content 30, 60, and 150 U/mL: MRI Grade II, III, IV degeneration
39	Bucher et al. 2013	Papain inj. (60 i.U)	Bovine coccygeal	14 days	Effect of MSCs stimulated by nonviral gene delivery of growth and differentiation factor 5 (GDF5)	GDF5 transfected MSCs injected into papain degen. IVD: partial recovery of the GAG/DNA ratio on day 7
40	Jim et al. 2011	Trypsin injection	Bovine	10 days	Establishment of a degenerative organ culture with long-term maintenance of cell viability to study repair strategies	Trypsin injection resulted in 60% loss of aggrecan after 7 days without affecting cell viability
41	Roberts et al. 2008	Trypsin or papain inj.	Bovine coccygeal	21 days	Mimicking of disc degeneration in an explant bovine model via enzymatic digestion	Papain and trypsin inj.: center macro- and microscopically fragmented with ↓ GAG
42	Mwale et al. 2014	Trypsin inj.	Bovine coccygeal	14 days	Effect of MSCs and Link N with regard to tissue repair in the degenerate disc	MSCs and Link N can restore GAG content in degenerate discs and increase collagen II expression

(*Continued*)

TABLE 3.3 (CONTINUED)
List of Established Organ Culture Systems Including Loading Regime, Used Species, Loading Time, RQ, and Main Finding

No.	Reference	Loading	Species	Time	RQ/Aim	Results
43	Chiba et al. 2007	Chondroitinase-ABC or chymopapain in culture medium	Rabbit	28 days	Investigation of repair capacity of alginate-embedded NP/AF cells	IVD cells exposed to chondroitinase ABC produce more GAG than cells exposed to chymopapain
44	Yerramalli et al. 2007	Chondroitinase-ABC, Genipin (Gen) inj.	Rat	12 hours	Effect of decreased GAG (Chondroitinase-ABC) and increased crosslinking (Gen) in NP	↓ NP GAG: alters disc axial mechanics Crosslinking: unlikely to contribute to degen

Proinflammatory Cytokine Treatment

No.	Reference	Loading	Species	Time	RQ/Aim	Results
45	Purmessur et al. 2013	TNF-α in culture medium	Bovine coccygeal	21 days	Effect of TNF-α on whole IVDs in organ culture and its association with IVD degeneration	Day 7: ↓ aggrecan, ↑catabolism, ↑ pro-inflammatory cytokines, ↑ nerve growth factor expression
46	Antunes et al. 2017	Needle-puncture, culture in IL-1β (100 ng/mL)	Bovine coccygeal	8 days	Effect of intradiscal injection of Df-NPs (diclofenac nanoparticles) in degenerated IVDs	Df-NP treatment: ↓ IL-6/8, MMP-1/3, ↓ PGE₂ production. ↑ Collagen II and aggrecan proteins in IVD

Note: EP, endplate; FGF-6/7, fibroblast growth factor 6/7; PDGF-R, platelet-derived growth factor receptors; GM-CSF, granulocyte macrophage colony-stimulating factor; IGF-1sR, insulin-like growth factor 1s receptor; ERK, Extracellular Signal-regulated Kinase-1; MCP-2, monocyte chemoattractant protein 2 (CCL8); GROα, (C-X-C motif) ligand 1; MIG, monokine induced by gamma interferon (CXCL9); IAF, inner AF; FasL, Fas ligand; DD, disc degeneration; iNOS, inducible nitric oxide synthases; CHAD, chondroadherin; oAF, outer AF; HTRA1, High-Temperature Requirement A Serine Peptidase 1; FE, finite element; Hf, high frequency; GDF5, Growth differentiation factor 5; cABC, Chondroitinase ABC; bNPc, bovine NP cells; PBS, phosphate buffered saline; PGE, Prostaglandin E synthase.

Intervertebral Disc Whole Organ Cultures

3.4 MONITORING DEGENERATIVE CHANGES IN *EX VIVO* IVD ORGAN CULTURE

Besides the choice of the appropriate organ culture model, different outcome parameters need to be considered according to the study aim. The following sections provide an overview of the measured output parameters found in common organ culture systems (Gantenbein et al. 2015). All included organ culture systems describe degeneration strategies achieved by static or degenerative dynamic loading, single impact striking, needle stabbing, or the injection of catabolic enzymes. The most frequently used methods to monitor changes in the process of IVD degeneration and regeneration are outlined, and examples of applications based on recent research involving gene and cell delivery for IVD regeneration are given.

3.4.1 Proteoglycan Analysis

Detection of GAGs is the most common method used to monitor the state of degeneration and/or the regenerative effect of an applied treatment in an organ culture system (Figure 3.3). In order to describe the GAG content of IVDs, both the total GAG content and the GAG synthesis rate have to be considered. The total GAG content of the analyzed tissue represents the amount of accumulated GAG over time. The GAG synthesis rate represents the newly synthesized GAG in a defined timeframe and therefore reflects the biological activity of the cells in the analyzed tissue at the time point of the measurement. As GAGs found in IVDs belong to the sulfated GAG family, applied radio-traceable Sulphur 35 is incorporated to monitor the synthesis rate in the analyzed tissue (Hansen and Ullberg 1960).

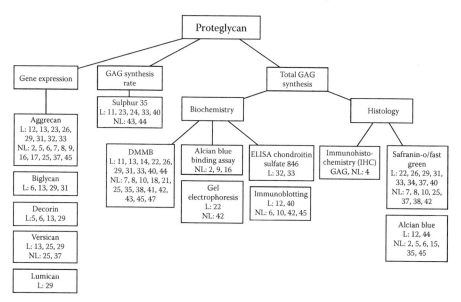

FIGURE 3.3 Methods used to analyze changes in proteoglycan content of IVD organ cultures.

The most common method used to determine the total GAG content or deposition in a tissue sample is the DMMB assay. The hereby used dye (1,9-dimethylmethylene) blue is a thiazine chromo-trope agent that binds to sulfated GAGs and results in a change in the dye's absorption spectrum (Chandrasekhar, Esterman, and Hoffman 1987). As the DMMB reaction can be measured within seconds, this assay allows a rapid quantification of the total amount of GAG. GAG content is usually normalized to the DNA content or the weight of the analyzed tissue sample (Brunk, Jones, and James 1979).

The most common histological method used to display a tissue's GAG content is the Safranin-O staining. It is usually applied in combination with the dye "fast-green," which stains collagens. The Safranin-O dye is bound to the tissue GAG stoichiometrically and can therefore be used for semiquantitative determination if all samples are stained in the same batch (Rosenberg 1971). Immunohistochemistry (IHC) represents a more specific staining as it uses antibodies to identify tissue components of interest (Gotz et al. 1997; Pereira et al. 2016).

Besides the synthesis rate or the total amount of deposed GAG, the phenotype of the IVD cells can be assessed using gene expression (Doege et al. 1991). The most frequently analyzed gene ACAN encodes for the aggrecan core protein, which provides a large extended region for GAG chain attachment between the G2 and G3 domain (Kiani et al. 2002).

Example: Mwale et al. used the DMMB reaction and the Safranin-O/Fast green staining to investigate the effect of injected MSCs and/or Link N on GAG restoration in a trypsin-induced degenerated bovine IVD model. They found a significantly higher GAG content (DMMB) in discs that were treated with MSCs, Link N, or a combination of both Link N and MSCs compared to the untreated degenerated control. Histological analysis (Safranin-O/Fast green) revealed that the newly synthesized proteoglycan was able to diffuse throughout the ECM and restore tissue content even in areas remote from the injected cells (Mwale et al. 2014).

3.4.2 Collagen Analysis

The most popular method to determine the total collagen content of an IVD tissue sample is the hydroxyproline assay. It is a quantitative measurement of the amino acid hydroxyproline, representing a major component of all triple-helical collagen proteins (type I, II, III, V, XI, XXIV, and XXVII) and functioning as an enhancer of collagen stability (Nemethy and Scheraga 1986).

Fast-green staining allows a histological evaluation of collagen in IVD tissue sample. It is about to replace the Masson's trichrome staining, as its color is more brilliant and less likely to fade, representing a semiquantitative method to determine the total amount of collagen. Picrosirius Red is an alternative histological stain targeting the cationic collagen fibers through its sulfonated groups. As the stain has a linear structure, picrosirius red enables the identification of fibrillar collagen networks in tissue sections (Junqueira, Bignolas, and Brentani 1979).

An alternative to the historical dye-based staining methods is IHC using specific collagen-binding antibodies to detect the collagen proteins. This method allows a staining for selected collagen types and therefore provides a more differentiated picture of the IVD's collagen composition (Chelberg et al. 1995).

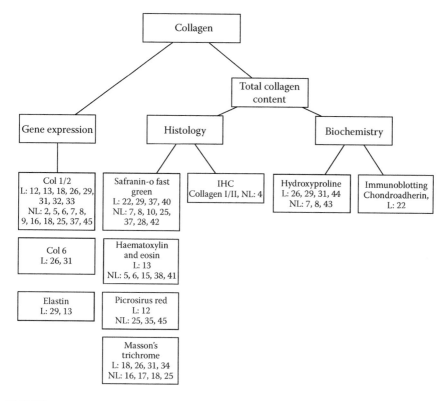

FIGURE 3.4 Methods used to analyze changes in collagen content and structure of IVD organ cultures.

In addition to the semiquantitative morphological assessment, real-time polymerase chain reaction (PCR) represents a quantitative analytical method to describe the phenotype of the investigated cells (Figure 3.4).

Example: Pereira et al. investigated the long-term response (3 weeks) of bovine nucleotomized IVDs to MSCs seeded on CEPs. Collagen type I and II expression and distribution in situ were investigated in the IVD tissue using immunofluorescence staining and quantified by fluorescence intensity. MSCs induced a higher expression of type II collagen in the NP, measured by immunofluorescence intensity, compared to the untreated, nucleotomized controls. An up-regulation of type I collagen was found in the NP of all discs undergoing nucleotomy; however, the presence of MSCs did not affect the expression of type I collagen.

3.4.3 Cytokine Analysis

The most common method used to describe the cytokine profile of an investigated tissue sample is gene expression. Phillips et al. performed a low-density array to investigate the expression of 91 cytokine and chemokine associated genes in NP cells from degenerate human IVDs. Compared to nondegenerated IVDs, increased levels of IL-16, CCL2, CCL7, and CXCL8 were found, and these results were confirmed histologically (Phillips et al. 2013).

Inflammatory cytokines, such as TNF-α and IL-1β among others, have been described to be up-regulated in degenerated IVDs, especially at the herniation site. Recent results suggest that TNF-α acts as a coordinator of macrophage infiltration in order to absorb the herniated disc tissue (Wang et al. 2013).

As the biological activity of a transcribed gene starts with its protein expression, IHC assays can be used to confirm the presence and localization of the targeted cytokines. Furthermore, signaling molecules like ERK1/2 (extracellular signal–regulated kinases) or growth factors like transforming growth factor (TGF) can be detected by enzyme-linked immunosorbent assay (ELISA) or immunoblotting (Western blot) of digested tissue or collected media (Figure 3.5).

Example: Teixeira et al. used a proinflammatory bovine IVD organ culture model (static loading followed by needle puncture and IL-1β stimulation) to investigate the effect of injected nanoparticles carrying the anti-inflammatory drug diclofenac (Df-NPs). Gene expression analyses showed that the injection of Df-NPs prevented the up-regulation of several proinflammatory markers such as PGE$_2$, IL-6, IL-8, MMP-1, and MMP-3. Interestingly, Df-NP injection into degenerative bovine IVDs (induced

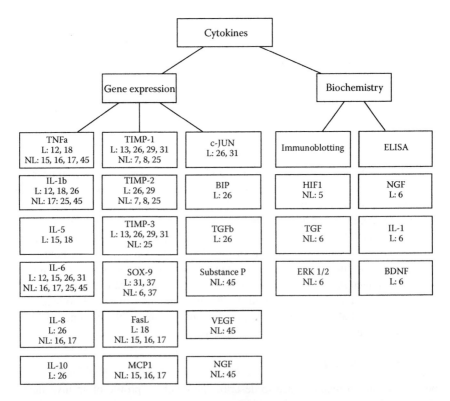

FIGURE 3.5 Methods used to describe the cytokine profile of IVD organ cultures. SOX-9, Sex Determining Region Y)-Box9; c-JUN, Jun Proto-Oncogene; BIP, Binding immunoglobulin protein; VEGF, vascular endothelial growth factor; HIF1, Hypoxia-inducible factor 1.

by needle puncture, static loading, and IL-1β) promoted an up-regulation of ECM proteins (by IHC), namely collagen type II and aggrecan, compared to untreated controls (Antunes et al. 2017; Teixeira et al. 2016). The authors explain this finding by a local control of inflammation, and the chondrogenic effect of poly-γ-glutamic acid, which is a component of the NPs (Antunes et al. 2015).

3.4.4 Enzyme Analysis

Real-time PCR or microarrays are the most frequent methods used to assess the phenotype of the analyzed cell population. To assess the concentration of an enzyme at the very moment of analysis, techniques based on antigen–antibody (or protein–ligand)-specific reactions can be used, such as ELISA, immunoblotting, and extraction zymography (Magi and Liberatori 2005). A limitation of these methods is that the enzymatic activity depends on spatial distribution, natural inhibitors, pro-enzyme activation, and other factors. Therefore, the observed concentration of an analyzed enzyme is not equal to its biological activity. A different approach, in situ zymography (ISZ), overcomes this problem by analyzing only active, uninhibited enzymes and therefore providing a better picture of the actual state of the investigated sample (Vandooren et al. 2013) (Figure 3.6).

Example: Le Maitre et al. used ISZ and IHC to examine the effects of IL-1 and IL-1Ra (IL-1 receptor antagonist) on matrix degradation and metal-dependent protease expression in explants of nondegenerated and degenerated human IVDs. They found that IL-1 is a key cytokine driving matrix degradation in the degenerated human IVD. Furthermore, IL-1Ra delivered directly or by gene therapy into disc cells effectively inhibited IVD matrix degradation (Le Maitre, Hoyland, and Freemont 2007b).

3.4.5 Cell Viability

Several methods have been used to assess cell viability or cell metabolic activity (Figure 3.7). The so-called "live/dead" assay is the most commonly used method to assess the ratio between live and dead cells in an unfixed tissue sample. This assay is based on the calcein/ethidium homodimer staining: calcein (calcein-AM = green) indicates intracellular esterase activity and therefore marks all living cells, while the ethidium homodimer (ethidium homodimer-1 = red) binds to the nucleic acid of cells that lost the integrity of their plasma membrane and therefore marks the dead cells (Somodi and Guthoff 1995). Another method is the lactate dehydrogenase (LDH)/ethidium homodimer staining of unfixed tissue sections, which has the advantage of not being affected by incubation steps that can lead to artifacts due to tissue swelling (Li et al. 2016). Metabolic assays can be performed by measuring the fluorescence of a cell metabolized dye, e.g., Alamar blue (Nakayama et al. 1997), or indirectly by measuring the LDH released from damaged cells (Korzeniewski and Callewaert 1983).

If information about the number of cells undergoing programmed cell death (apoptosis) is required, terminal deoxynucleotidyl transferase dUTP nick end labeling (TUNEL) or the caspase 3/7 assay is chosen. TUNEL enables the quantification of

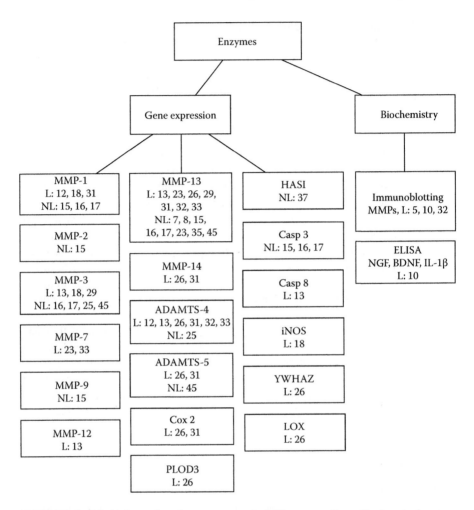

FIGURE 3.6 Methods used to detect enzymes in IVD organ culture. Cox2, cytochrome c oxidase II (also known as prostaglandin-endoperoxide synthase); PLOD3, procollagen-lysine, 2-oxoglutarate 5-dioxygenase 3; HAS1, hyaluronan synthase 1; iNOS, inducible nitric oxide synthase; YWHAZ, tyrosine 3-monooxygenase/tryptophan 5-monooxygenase activation protein zeta; LOX, lysyl oxidase.

apoptotic cells by detecting their DNA fragmentation, while the caspase assay detects cleaved caspase 3/7 fragments occurring during cell apoptosis (Gavrieli, Sherman, and Ben-Sasson 1992).

Example: Grant et al. used calcein/ethidium homodimer staining to validate a new long-term organ culture system for isolated coccygeal bovine IVDs. Using newly developed isolation and culture media, a culture time up to 5 months was possible without changes in cell viability and cellularity (Grant et al. 2016). This long-term organ culture model might therefore provide a novel platform to investigating the long-term effects of loading paradigms or applied treatments on disc degeneration and repair.

Intervertebral Disc Whole Organ Cultures

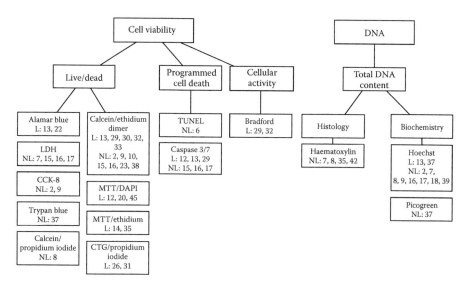

FIGURE 3.7 Methods used to describe cell number and cell viability in IVD organ cultures. CCK-8, cell counting kit-8; MTT, 3-(45-dimethylthiazol-2-yl)-2,5-diphenyltetrazolium bromide; DAPI, 4′,6-diamidin-2-phenylindol; CTG, CellTiter-Glo®.

3.4.6 Macroscopic Evaluation

In 1990, Thompson et al. proposed a grading system for isolated human IVDs in which every part of a spinal segment—NP, AF, endplate, and the vertebral body—was analyzed separately for defined morphological markers (Thompson et al. 1990).

In clinics, MRI is used for morphological characterization purposes. The relaxation time of T1, T2, or T1ρ can be described qualitatively with the help of the Pfirrmann score (Pfirrmann et al. 2001). Other software-based, quantitative methods exist but are not very common as handling is still quite demanding (Mwale, Iatridis, and Antoniou 2008) (Figure 3.8).

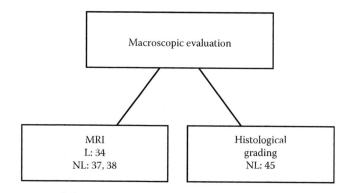

FIGURE 3.8 Methods used to describe macroscopic changes in IVD organ cultures.

Example: Rosenzweig et al. used T1 and T2 MRI signals to evaluate the effect of cell/hydrogel implantation on human isolated IVDs. The small animal MRI (Bruker, 7.5 Tesla) allowed them to obtain well-defined images of isolated discs, with details of tissue integrity and proteoglycan content, and therefore assess the repair potential of the therapies by comparing MRI scans before and after therapy (Rosenzweig et al. 2017).

3.4.7 Recent Developments in Mechano-Biological Assessment of IVDs

If an IVD organ culture system involves loading, the changes in dimensions before and after loading are generally measured with calipers. In this way, disc height loss (and increase in diameter) following loading and the disc height (and diameter) recovery after loading can be monitored. IVD stiffness can be evaluated from the recorded loading and displacement data (Figure 3.9).

More complex bioreactors allow the simultaneous application of compressive and rotational forces (Chan et al. 2011). However, a full characterization of IVD mechanics in all degrees of freedom can so far be performed only on embedded spine motion segments. Developments are underway to perform similar studies also in whole organ culture (Amin et al. 2016).

A novel noninvasive imaging technique is the magnetic resonance (MR) elastography. Differently from the standard MRI sequences, it tracks propagating strain waves as they move through soft tissue. This allows a relative assessment of the IVD shear stiffness and therefore reflects a functional measurement of tissue quality. As the matrix breakdown that occurs throughout degeneration leads to a progressive change in the mechanical behavior and material properties of the IVD, MR elastography might provide an objective biomarker for early stages of degeneration. However, to be clinically applicable, this method has to be highly reproducible. In particular, the mechanical behavior of the IVD is recognized to be time dependent and influenced by the degree of tissue hydration, which can vary by up to 20% throughout the day.

Example: Walter et al. investigated the repeatability of MR elastography throughout the day (morning/afternoon) in patients with and without IVD degeneration. They could demonstrate that shear stiffness measurements derived from MR elastography are highly repeatable, poorly correlate with age, and increase with advancing IVD degeneration (Walter et al. 2017).

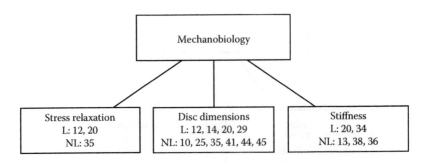

FIGURE 3.9 Methods used to describe mechano-biological properties of IVD organ cultures.

3.5 CONCLUSION

The patient introduced in our clinical case is suffering from a symptomatic disc herniation. In such a case, discectomy is the most common surgical treatment strategy, as the removal of the herniated tissue allows a decompression of the irritated nerve root, resulting in a relief of pain and numbness. Clinical studies commonly report a beneficial outcome for most of the patients undergoing this type of surgery (Jansson et al. 2005; Stromqvist et al. 2013; Weinstein et al. 2006). However, standard surgical approaches do not include any primary repair of the AF defect, and if the endogenous healing fails, recurrent re-herniation can occur. With an incidence of 0.5% to 24%, recurrent disc herniation is reported to be the most common cause for a second surgical procedure within 1 or 2 years after the primary operation (Ahsan et al. 2012; Aizawa et al. 2012; Ambrossi et al. 2009; Cheng et al. 2013; Fritzell et al. 2015). Ongoing research is therefore focusing on developing new strategies for AF repair to prevent secondary herniation. Two main strategies have been proposed: primary closure of the AF defect with anchor systems or the application of glue-like/adhesive biomaterials and secondary closure through growth factor and/or cell application. Whole IVD organ culture models have recently been applied to evaluate such new biological approaches.

Example: Pirvu et al. used a dynamically loaded whole organ culture AF defect model to evaluate a new repair strategy by closing the AF defect with a MSC-loaded poly-trimethylene carbonate (PTMC) scaffold (Pirvu et al. 2015). The applied scaffolds restored disc height of annulotomized discs and prevented herniation of NP tissue under dynamic loading. The implanted MSCs showed an up-regulated gene expression of type V collagen, a potential AF marker, indicating in situ differentiation capability. Furthermore, MSCs delivered within PTMC scaffolds induced an up-regulation of anabolic gene expression and down-regulation of catabolic gene expression in adjacent native disc tissue.

In summary, IVD organ cultures provide a tool for research that can be used to predict the potential of new treatment strategies and help to better understand the biological and biomechanical processes associated with disc degeneration and regeneration. However, it is important to take in to account that *ex vivo* IVD models are designed to answer specific RQs. Therefore, the proper model has to be chosen with care in order to investigate new strategies in the treatment of IVDs. Our flowcharts and Table 3.3 shall support researchers, especially young investigators, to propose valid IVD organ cultures for their studies.

REFERENCES

Adams, M.A., and P. Dolan. 1997. "Could sudden increases in physical activity cause degeneration of intervertebral discs?" *Lancet* 350 (9079):734–5. doi: 10.1016/s0140-6736(97)03021-3.

Adams, M.A., B.J. Freeman, H.P. Morrison, I.W. Nelson, and P. Dolan. 2000. "Mechanical initiation of intervertebral disc degeneration." *Spine (Phila Pa 1976)* 25 (13):1625–36.

Ahsan, K., Najmus-Sakeb, A. Hossain, S.I. Khan, and M.A. Awwal. 2012. "Discectomy for primary and recurrent prolapse of lumbar intervertebral discs." *J Orthop Surg (Hong Kong)* 20 (1):7–10. doi: 10.1177/230949901202000102.

Aizawa, T., H. Ozawa, T. Kusakabe et al. 2012. "Reoperation for recurrent lumbar disc herniation: A study over a 20-year period in a Japanese population." *J Orthop Sci* 17 (2):107–13. doi: 10.1007/s00776-011-0184-6.

Alkhatib, B., D.H. Rosenzweig, E. Krock et al. 2014. "Acute mechanical injury of the human intervertebral disc: Link to degeneration and pain." *Eur Cell Mater* 28:98–110; discussion 110–1.

Ambrossi, G.L., M.J. McGirt, D.M. Sciubba et al. 2009. "Recurrent lumbar disc herniation after single-level lumbar discectomy: Incidence and health care cost analysis." *Neurosurgery* 65 (3):574–8; discussion 578. doi: 10.1227/01.neu.0000350224.36213.f9.

Amin, D.B., D. Sommerfeld, I.M. Lawless et al. 2016. "Effect of degeneration on the six degree of freedom mechanical properties of human lumbar spine segments." *J Orthop Res* 34 (8):1399–409. doi: 10.1002/jor.23334.

Anderson, P.A., P.C. McCormick, and P.D. Angevine. 2008. "Randomized controlled trials of the treatment of lumbar disk herniation: 1983–2007." *J Am Acad Orthop Surg* 16 (10):566–73.

Andrade, P., G. Hoogland, M.A. Garcia et al. 2013. "Elevated IL-1beta and IL-6 levels in lumbar herniated discs in patients with sciatic pain." *Eur Spine J* 22 (4):714–20. doi: 10.1007/s00586-012-2502-x.

Antoniou, J., T. Steffen, F. Nelson et al. 1996. "The human lumbar intervertebral disc: Evidence for changes in the biosynthesis and denaturation of the extracellular matrix with growth, maturation, aging, and degeneration." *J Clin Invest* 98 (4):996–1003. doi: 10.1172/jci118884.

Antunes, J.C., C.L. Pereira, G.Q. Teixeira et al. 2017. "Poly(gamma-glutamic acid) and poly (gamma-glutamic acid)-based nanocomplexes enhance type II collagen production in intervertebral disc." *J Mater Sci Mater Med* 28 (1):6. doi: 10.1007/s10856-016-5787-1.

Antunes, J.C., R. Tsaryk, R.M. Goncalves et al. 2015. "Poly(gamma-glutamic acid) as an exogenous promoter of chondrogenic differentiation of human mesenchymal stem/stromal cells." *Tissue Eng Part A* 21 (11–12):1869–85. doi: 10.1089/ten.TEA.2014.0386.

Battie, M.C., T. Videman, J. Kaprio et al. 2009. "The Twin Spine Study: Contributions to a changing view of disc degeneration." *Spine J* 9 (1):47–59. doi: 10.1016/j.spinee.2008.11.011.

Brinckmann, P., and H. Grootenboer. 1991. "Change of disc height, radial disc bulge, and intradiscal pressure from discectomy. An *in vitro* investigation on human lumbar discs." *Spine (Phila Pa 1976)* 16 (6):641–6.

Brunk, C.F., K.C. Jones, and T.W. James. 1979. "Assay for nanogram quantities of DNA in cellular homogenates." *Anal Biochem* 92 (2):497–500.

Bucher, C., A. Gazdhar, L.M. Benneker, T. Geiser, and B. Gantenbein-Ritter. 2013. "Nonviral gene delivery of growth and differentiation factor 5 to human mesenchymal stem cells injected into a 3D bovine intervertebral disc organ culture system." *Stem Cells Int* 2013:326828. doi: 10.1155/2013/326828.

Castro, A.P., C.P. Paul, S.E. Detiger et al. 2014. "Long-term creep behavior of the intervertebral disk: Comparison between bioreactor data and numerical results." *Front Bioeng Biotechnol* 2:56. doi: 10.3389/fbioe.2014.00056.

Chan, S.C., A. Burki, H.M. Bonel, L.M. Benneker, and B. Gantenbein-Ritter. 2013a. "Papain-induced *in vitro* disc degeneration model for the study of injectable nucleus pulposus therapy." *Spine J* 13 (3):273–83. doi: 10.1016/j.spinee.2012.12.007.

Chan, S.C., S.J. Ferguson, K. Wuertz, and B. Gantenbein-Ritter. 2011. "Biological response of the intervertebral disc to repetitive short-term cyclic torsion." *Spine (Phila Pa 1976)* 36 (24):2021–30. doi: 10.1097/BRS.0b013e318203aea5.

Chan, S.C., J. Walser, P. Kappeli et al. 2013a. "Region specific response of intervertebral disc cells to complex dynamic loading: An organ culture study using a dynamic torsion-compression bioreactor." *PLoS One* 8 (8):e72489. doi: 10.1371/journal.pone.0072489.

Chandrasekhar, S., M.A. Esterman, and H.A. Hoffman. 1987. "Microdetermination of proteoglycans and glycosaminoglycans in the presence of guanidine hydrochloride." *Anal Biochem* 161 (1):103–8.

Chelberg, M.K., G.M. Banks, D.F. Geiger, and T.R. Oegema. 1995. "Identification of heterogeneous cell populations in normal human intervertebral disc." *J Anat* 186 (Pt 1):43–53.

Cheng, J., H. Wang, W. Zheng et al. 2013. "Reoperation after lumbar disc surgery in two hundred and seven patients." *Int Orthop* 37 (8):1511–7. doi: 10.1007/s00264-013-1925-2.

Cheung, K.M., J. Karppinen, D. Chan et al. 2009. "Prevalence and pattern of lumbar magnetic resonance imaging changes in a population study of one thousand and forty-three individuals." *Spine (Phila Pa 1976)* 34 (9):934–40. doi: 10.1097/BRS.0b013e3181a01b3f.

Chiba, K., K. Masuda, G.B. Andersson, S. Momohara, and E.J. Thonar. 2007. "Matrix replenishment by intervertebral disc cells after chemonucleolysis *in vitro* with chondroitinase ABC and chymopapain." *Spine J* 7 (6):694–700. doi: 10.1016/j.spinee.2006.09.005.

DeGroot, J., N. Verzijl, M.J. Wenting-van Wijk et al. 2004. "Accumulation of advanced glycation end products as a molecular mechanism for aging as a risk factor in osteoarthritis." *Arthritis Rheum* 50 (4):1207–15. doi: 10.1002/art.20170.

Doege, K.J., M. Sasaki, T. Kimura, and Y. Yamada. 1991. "Complete coding sequence and deduced primary structure of the human cartilage large aggregating proteoglycan, aggrecan. Human-specific repeats, and additional alternatively spliced forms." *J Biol Chem* 266 (2):894–902.

Dudli, S., S.J. Ferguson, and D. Haschtmann. 2014. "Severity and pattern of post-traumatic intervertebral disc degeneration depend on the type of injury." *Spine J* 14 (7):1256–64. doi: 10.1016/j.spinee.2013.07.488.

Dudli, S., D. Haschtmann, and S.J. Ferguson. 2012. "Fracture of the vertebral endplates, but not equienergetic impact load, promotes disc degeneration *in vitro*." *J Orthop Res* 30 (5):809–16. doi: 10.1002/jor.21573.

Dudli, S., D. Haschtmann, and S.J. Ferguson. 2015. "Persistent degenerative changes in the intervertebral disc after burst fracture in an *in vitro* model mimicking physiological post-traumatic conditions." *Eur Spine J* 24 (9):1901–8. doi: 10.1007/s00586-014-3301-3.

Eyre, D.R., Y. Matsui, and J.J. Wu. 2002. "Collagen polymorphisms of the intervertebral disc." *Biochem Soc Trans* 30 (Pt 6):844–8. doi: 10.1042/.

Fritzell, P., B. Knutsson, B. Sanden, B. Strömqvist, and O. Hägg. 2015. "Recurrent versus primary lumbar disc herniation surgery: Patient-reported outcomes in the Swedish Spine Register Swespine." *Clin Orthop Relat Res* 473 (6):1978–84. doi: 10.1007/s11999-014-3596-8.

Furtwangler, T., S.C. Chan, G. Bahrenberg, P.J. Richards, and B. Gantenbein-Ritter. 2013. "Assessment of the matrix degenerative effects of MMP-3, ADAMTS-4, and HTRA1, injected into a bovine intervertebral disc organ culture model." *Spine (Phila Pa 1976)* 38 (22):E1377–87. doi: 10.1097/BRS.0b013e31829ffde8.

Gan, J.C., P. Ducheyne, E.J. Vresilovic, W. Swaim, and I.M. Shapiro. 2003. "Intervertebral disc tissue engineering I: Characterization of the nucleus pulposus." *Clin Orthop Relat Res* (411):305–14. doi: 10.1097/01.blo.0000063796.98363.9a.

Gantenbein, B., T. Grunhagen, C.R. Lee et al. 2006. "An *in vitro* organ culturing system for intervertebral disc explants with vertebral endplates: A feasibility study with ovine caudal discs." *Spine (Phila Pa 1976)* 31 (23):2665–73. doi: 10.1097/01.brs.0000244620.15386.df.

Gantenbein, B., S. Illien-Junger, S.C. Chan et al. 2015. "Organ culture bioreactors – Platforms to study human intervertebral disc degeneration and regenerative therapy." *Curr Stem Cell Res Ther* 10 (4):339–52.

Gavrieli, Y., Y. Sherman, and S.A. Ben-Sasson. 1992. "Identification of programmed cell death in situ via specific labeling of nuclear DNA fragmentation." *J Cell Biol* 119 (3):493–501.

Gawri, R., J. Moir, J. Ouellet et al. 2014. "Physiological loading can restore the proteoglycan content in a model of early IVD degeneration." *PLoS One* 9 (7):e101233. doi: 10.1371/journal.pone.0101233.

Gawri, R., F. Mwale, J. Ouellet et al. 2011. "Development of an organ culture system for long-term survival of the intact human intervertebral disc." *Spine (Phila Pa 1976)* 36 (22):1835–42. doi: 10.1097/BRS.0b013e3181f81314.

Gotz, W., S. Barnert, R. Bertagnoli et al. 1997. "Immunohistochemical localization of the small proteoglycans decorin and biglycan in human intervertebral discs." *Cell Tissue Res* 289 (1):185–90.

Goupille, P., M.I. Jayson, J.P. Valat, and A.J. Freemont. 1998. "Matrix metalloproteinases: The clue to intervertebral disc degeneration?" *Spine (Phila Pa 1976)* 23 (14):1612–26.

Grant, M., L.M. Epure, O. Salem et al. 2016. "Development of a large animal long-term intervertebral disc organ culture model that includes the bony vertebrae for *ex vivo* studies." *Tissue Eng Part C Methods* 22 (7):636–43. doi: 10.1089/ten.TEC.2016.0049.

Gruber, H.E., and E.N. Hanley. 2000. "Human disc cells in monolayer vs 3D culture: Cell shape, division and matrix formation." *BMC Musculoskelet Disord* 1:1. doi: 10.1186/1471-2474-1-1.

Gruber, H.E., G.L. Hoelscher, J.A. Ingram, H.J. Norton, and E.N. Hanley, Jr. 2013. "Increased IL-17 expression in degenerated human discs and increased production in cultured annulus cells exposed to IL-1ss and TNF-alpha." *Biotech Histochem* 88 (6):302–10. doi: 10.3109/10520295.2013.783235.

Grunhagen, T., A. Shirazi-Adl, J.C. Fairbank, and J.P. Urban. 2011. "Intervertebral disk nutrition: A review of factors influencing concentrations of nutrients and metabolites." *Orthop Clin North Am* 42 (4):465–77, vii. doi: 10.1016/j.ocl.2011.07.010.

Haglund, L., J. Moir, L. Beckman et al. 2011. "Development of a bioreactor for axially loaded intervertebral disc organ culture." *Tissue Eng Part C Methods* 17 (10):1011–9. doi: 10.1089/ten.TEC.2011.0025.

Hansen, H.J., and S. Ullberg. 1960. "Uptake of S35 in the intervertebral discs after injection of S35-sulphate. An autoradiographic study." *Acta Orthop Scand* 30:84–90.

Haschtmann, D., J.V. Stoyanov, L. Ettinger, L.P. Nolte, and S.J. Ferguson. 2006. "Establishment of a novel intervertebral disc/endplate culture model: Analysis of an *ex vivo in vitro* whole-organ rabbit culture system." *Spine (Phila Pa 1976)* 31 (25):2918–25. doi: 10.1097/01.brs.0000247954.69438.ae.

Haschtmann, D., J.V. Stoyanov, and S.J. Ferguson. 2006. "Influence of diurnal hyperosmotic loading on the metabolism and matrix gene expression of a whole-organ intervertebral disc model." *J Orthop Res* 24 (10):1957–66. doi: 10.1002/jor.20243.

Haschtmann, D., J.V. Stoyanov, P. Gedet, and S.J. Ferguson. 2008. "Vertebral endplate trauma induces disc cell apoptosis and promotes organ degeneration *in vitro*." *Eur Spine J* 17 (2):289–99. doi: 10.1007/s00586-007-0509-5.

Hollander, A.P., T.F. Heathfield, J.J. Liu et al. 1996. "Enhanced denaturation of the alpha (II) chains of type-II collagen in normal adult human intervertebral discs compared with femoral articular cartilage." *J Orthop Res* 14 (1):61–6. doi: 10.1002/jor.1100140111.

Hormel, S.E., and D.R. Eyre. 1991. "Collagen in the ageing human intervertebral disc: An increase in covalently bound fluorophores and chromophores." *Biochim Biophys Acta* 1078 (2):243–50.

Hutton, W.C., Y. Toribatake, W.A. Elmer et al. 1998. "The effect of compressive force applied to the intervertebral disc *in vivo*. A study of proteoglycans and collagen." *Spine (Phila Pa 1976)* 23 (23):2524–37.

Hutton, W.C., S.T. Yoon, W.A. Elmer et al. 2002. "Effect of tail suspension (or simulated weightlessness) on the lumbar intervertebral disc: Study of proteoglycans and collagen." *Spine (Phila Pa 1976)* 27 (12):1286–90.

Hwang, D., A.S. Gabai, M. Yu, A.G. Yew, and A.H. Hsieh. 2012. "Role of load history in intervertebral disc mechanics and intradiscal pressure generation." *Biomech Model Mechanobiol* 11 (1–2):95–106. doi: 10.1007/s10237-011-0295-1.

Iatridis, J.C., S.B. Nicoll, A.J. Michalek, B.A. Walter, and M.S. Gupta. 2013. "Role of biomechanics in intervertebral disc degeneration and regenerative therapies: What needs repairing in the disc and what are promising biomaterials for its repair?" *Spine J* 13 (3):243–62. doi: 10.1016/j.spinee.2012.12.002.

Illien-Junger, S., B. Gantenbein-Ritter, S. Grad et al. 2010. "The combined effects of limited nutrition and high-frequency loading on intervertebral discs with endplates." *Spine (Phila Pa 1976)* 35 (19):1744–52. doi: 10.1097/BRS.0b013e3181c48019.

Illien-Junger, S., G. Pattappa, M. Peroglio et al. 2012. "Homing of mesenchymal stem cells in induced degenerative intervertebral discs in a whole organ culture system." *Spine (Phila Pa 1976)* 37 (22):1865–73. doi: 10.1097/BRS.0b013e3182544a8a.

Inkinen, R.I., M.J. Lammi, S. Lehmonen et al. 1998. "Relative increase of biglycan and decorin and altered chondroitin sulfate epitopes in the degenerating human intervertebral disc." *J Rheumatol* 25 (3):506–14.

Inoue, N., and A.A. Espinoza Orias. 2011. "Biomechanics of intervertebral disk degeneration." *Orthop Clin North Am* 42 (4):487–99, vii. doi: 10.1016/j.ocl.2011.07.001.

Iozzo, R.V., and L. Schaefer. 2015. "Proteoglycan form and function: A comprehensive nomenclature of proteoglycans." *Matrix Biol* 42:11–55. doi: 10.1016/j.matbio.2015.02.003.

Jansson, K.A., G. Nemeth, F. Granath, B. Jonsson, and P. Blomqvist. 2005. "Health-related quality of life in patients before and after surgery for a herniated lumbar disc." *J Bone Joint Surg Br* 87 (7):959–64. doi: 10.1302/0301-620x.87b7.16240.

Jim, B., T. Steffen, J. Moir, P. Roughley, and L. Haglund. 2011. "Development of an intact intervertebral disc organ culture system in which degeneration can be induced as a prelude to studying repair potential." *Eur Spine J* 20 (8):1244–54. doi: 10.1007/s00586-011-1721-x.

Jordan, J., K. Konstantinou, and J. O'Dowd. 2009. "Herniated lumbar disc." *BMJ Clin Evid* 2009.

Junger, S., B. Gantenbein-Ritter, P. Lezuo et al. 2009. "Effect of limited nutrition on in situ intervertebral disc cells under simulated-physiological loading." *Spine (Phila Pa 1976)* 34 (12):1264–71. doi: 10.1097/BRS.0b013e3181a0193d.

Junqueira, L.C., G. Bignolas, and R.R. Brentani. 1979. "Picrosirius staining plus polarization microscopy, a specific method for collagen detection in tissue sections." *Histochem J* 11 (4):447–55.

Kiani, C., L. Chen, Y.J. Wu, A.J. Yee, and B.B. Yang. 2002. "Structure and function of aggrecan." *Cell Res* 12 (1):19–32. doi: 10.1038/sj.cr.7290106.

Konstantinou, K., and K.M. Dunn. 2008. "Sciatica: Review of epidemiological studies and prevalence estimates." *Spine (Phila Pa 1976)* x33 (22):2464–72. doi: 10.1097/BRS.0b013e318183a4a2.

Korecki, C.L., J.J. Costi, and J.C. Iatridis. 2008. "Needle puncture injury affects intervertebral disc mechanics and biology in an organ culture model." *Spine (Phila Pa 1976)* 33 (3):235–41. doi: 10.1097/BRS.0b013e3181624504.

Korecki, C.L., J.J. MacLean, and J.C. Iatridis. 2007. "Characterization of an *in vitro* intervertebral disc organ culture system." *Eur Spine J* 16 (7):1029–37. doi: 10.1007/s00586-007-0327-9.

Korzeniewski, C., and D.M. Callewaert. 1983. "An enzyme-release assay for natural cytotoxicity." *J Immunol Methods* 64 (3):313–20.

Kuno, K., N. Kanada, E. Nakashima et al. 1997. "Molecular cloning of a gene encoding a new type of metalloproteinase-disintegrin family protein with thrombospondin motifs as an inflammation associated gene." *J Biol Chem* 272 (1):556–62.

Le Maitre, C.L., J.A. Hoyland, and A.J. Freemont. 2007a. "Catabolic cytokine expression in degenerate and herniated human intervertebral discs: IL-1beta and TNFalpha expression profile." *Arthritis Res Ther* 9 (4):R77. doi: 10.1186/ar2275.

Le Maitre, C.L., J.A. Hoyland, and A.J. Freemont. 2007b. "Interleukin-1 receptor antagonist delivered directly and by gene therapy inhibits matrix degradation in the intact degenerate human intervertebral disc: An in situ zymographic and gene therapy study." *Arthritis Res Ther* 9 (4):R83. doi: 10.1186/ar2282.

Lee, C.R., J.C. Iatridis, L. Poveda, and M. Alini. 2006. "*In vitro* organ culture of the bovine intervertebral disc: Effects of vertebral endplate and potential for mechanobiology studies." *Spine (Phila Pa 1976)* 31 (5):515–22. doi: 10.1097/01.brs.0000201302.59050.72.

Li, Z., P. Lezuo, G. Pattappa et al. 2016. "Development of an *ex vivo* cavity model to study repair strategies in loaded intervertebral discs." *Eur Spine J* 25 (9):2898–908. doi: 10.1007/s00586-016-4542-0.

Lotz, J.C., and J.R. Chin. 2000. "Intervertebral disc cell death is dependent on the magnitude and duration of spinal loading." *Spine (Phila Pa 1976)* 25 (12):1477–83.

Lotz, J.C., A.H. Hsieh, A.L. Walsh, E.I. Palmer, and J.R. Chin. 2002. "Mechanobiology of the intervertebral disc." *Biochem Soc Trans* 30 (Pt 6):853–8. doi: 10.1042/.

Magi, B., and S. Liberatori. 2005. "Immunoblotting techniques." *Methods Mol Biol* 295: 227–54.

Malonzo, C., S.C. Chan, A. Kabiri et al. 2015. "A papain-induced disc degeneration model for the assessment of thermo-reversible hydrogel-cells therapeutic approach." *J Tissue Eng Regen Med* 9 (12):E167–76. doi: 10.1002/term.1667.

Masuoka, K., A.J. Michalek, J.J. MacLean, I.A. Stokes, and J.C. Iatridis. 2007. "Different effects of static versus cyclic compressive loading on rat intervertebral disc height and water loss *in vitro*." *Spine (Phila Pa 1976)* 32 (18):1974–9. doi: 10.1097/BRS.0b013e318133d591.

Mayer, J.E., J.C. Iatridis, D. Chan et al. 2013. "Genetic polymorphisms associated with intervertebral disc degeneration." *Spine J* 13 (3):299–317. doi: 10.1016/j.spinee.2013.01.041.

Michalek, A.J., K.L. Funabashi, and J.C. Iatridis. 2010. "Needle puncture injury of the rat intervertebral disc affects torsional and compressive biomechanics differently." *Eur Spine J* 19 (12):2110–6. doi: 10.1007/s00586-010-1473-z.

Murphy, G., and H. Nagase. 2008. "Reappraising metalloproteinases in rheumatoid arthritis and osteoarthritis: Destruction or repair?" *Nat Clin Pract Rheumatol* 4 (3):128–35. doi: 10.1038/ncprheum0727.

Mwale, F., J.C. Iatridis, and J. Antoniou. 2008. "Quantitative MRI as a diagnostic tool of intervertebral disc matrix composition and integrity." *Eur Spine J* 17 (Suppl 4):432–40. doi: 10.1007/s00586-008-0744-4.

Mwale, F., H.T. Wang, P. Roughley, J. Antoniou, and L. Haglund. 2014. "Link N and mesenchymal stem cells can induce regeneration of the early degenerate intervertebral disc." *Tissue Eng Part A* 20 (21–22):2942–9. doi: 10.1089/ten.TEA.2013.0749.

Nakayama, G.R., M.C. Caton, M.P. Nova, and Z. Parandoosh. 1997. "Assessment of the Alamar Blue assay for cellular growth and viability *in vitro*." *J Immunol Methods* 204 (2):205–8.

Nemethy, G., and H.A. Scheraga. 1986. "Stabilization of collagen fibrils by hydroxyproline." *Biochemistry* 25 (11):3184–8.

Oegema, T.R., Jr., S.L. Johnson, D.J. Aguiar, and J.W. Ogilvie. 2000. "Fibronectin and its fragments increase with degeneration in the human intervertebral disc." *Spine (Phila Pa 1976)* 25 (21):2742–7.

Pasternak, B., and P. Aspenberg. 2009. "Metalloproteinases and their inhibitors—Diagnostic and therapeutic opportunities in orthopedics." *Acta Orthop* 80 (6):693–703. doi: 10.3109/17453670903448257.

Paul, C.P., T. Schoorl, H.A. Zuiderbaan et al. 2013. "Dynamic and static overloading induce early degenerative processes in caprine lumbar intervertebral discs." *PLoS One* 8 (4): e62411. doi: 10.1371/journal.pone.0062411.

Paul, C.P., G.J. Strijkers, M. de Graaf et al. 2012a. "Changes in biomechanical, histological and quantitative MRI parameters in lumbar caprine intervertebral discs subjected to chondroitinase-induced degeneration." *Global Spine J* 02-P86. doi: 10.1055/s-0032-1319960.

Paul, C.P., H.A. Zuiderbaan, B. Zandieh Doulabi et al. 2012b. "Simulated-physiological loading conditions preserve biological and mechanical properties of caprine lumbar intervertebral discs in *ex vivo* culture." *PLoS One* 7 (3):e33147. doi: 10.1371/journal.pone.0033147.

Pereira, C.L., G.Q. Teixeira, C. Ribeiro-Machado et al. 2016. "Mesenchymal stem/stromal cells seeded on cartilaginous endplates promote intervertebral disc regeneration through extracellular matrix remodeling." *Sci Rep* 6. doi: 10.1038/srep33836.

Peroglio, M., Z. Li, L.M. Benneker, M. Alini, and S. Grad. 2016. "Intervertebral disc whole organ cultures." In *Biological Approaches to Spinal Disc Repair and Regeneration for Clinicians*, ed. R. Härtl and L.J. Bonassar, 66–74. New York, Stuttgart: Thieme—Medical Publishers.

Pfirrmann, C.W., A. Metzdorf, M. Zanetti, J. Hodler, and N. Boos. 2001. "Magnetic resonance classification of lumbar intervertebral disc degeneration." *Spine (Phila Pa 1976)* 26 (17):1873–8.

Phillips, K.L., N. Chiverton, A.L. Michael et al. 2013. "The cytokine and chemokine expression profile of nucleus pulposus cells: Implications for degeneration and regeneration of the intervertebral disc." *Arthritis Res Ther* 15 (6):R213. doi: 10.1186/ar4408.

Pirvu, T., S.B. Blanquer, L.M. Benneker et al. 2015. "A combined biomaterial and cellular approach for annulus fibrosus rupture repair." *Biomaterials* 42:11–9. doi: 10.1016/j.biomaterials.2014.11.049.

Purmessur, D., B.A. Walter, P.J. Roughley et al. 2013. "A role for TNF – alpha, in intervertebral disc degeneration: A non-recoverable catabolic shift." *Biochem Biophys Res Commun* 433 (1):151–6. doi: 10.1016/j.bbrc.2013.02.034.

Risbud, M.V., A. Di Martino, A. Guttapalli et al. 2006. "Toward an optimum system for intervertebral disc organ culture: TGF-beta 3 enhances nucleus pulposus and anulus fibrosus survival and function through modulation of TGF-beta-R expression and ERK signaling." *Spine (Phila Pa 1976)* 31 (8):884–90. doi: 10.1097/01.brs.0000209335.57767.b5.

Risbud, M.V., M.W. Izzo, C.S. Adams et al. 2003. "An organ culture system for the study of the nucleus pulposus: Description of the system and evaluation of the cells." *Spine (Phila Pa 1976)* x28 (24):2652–8; discussion 2658–9. doi: 10.1097/01.brs.0000099384.58981.c6.

Risbud, M.V., and I.M. Shapiro. 2014. "Role of cytokines in intervertebral disc degeneration: Pain and disc content." *Nat Rev Rheumatol* 10 (1):44–56. doi: 10.1038/nrrheum.2013.160.

Roberts, S., J. Menage, S. Sivan, and J.P. Urban. 2008. "Bovine explant model of degeneration of the intervertebral disc." *BMC Musculoskelet Disord* 9:24. doi: 10.1186/1471-2474-9-24.

Rosenberg, L. 1971. "Chemical basis for the histological use of safranin O in the study of articular cartilage." *J Bone Joint Surg Am* 53 (1):69–82.

Rosenzweig, D.H., P. Gawri, J. Moir et al. 2017. "Evaluation of tissue repair using MRI imaging of human intervertebral discs cultured in a bioreactor." *Global Spine J* 6 (1 suppl):s-0036-1582621–s-0036-1582621.

Roughley, P.J. 2004. "Biology of intervertebral disc aging and degeneration: Involvement of the extracellular matrix." *Spine (Phila Pa 1976)* x29 (23):2691–9.

Schoenfeld, A.J., and B.K. Weiner. 2010. "Treatment of lumbar disc herniation: Evidence-based practice." *Int J Gen Med* 3:209–14.

Sivan, S.S., E. Wachtel, E. Tsitron et al. 2008. "Collagen turnover in normal and degenerate human intervertebral discs as determined by the racemization of aspartic acid." *J Biol Chem* 283 (14):8796–801. doi: 10.1074/jbc.M709885200.

Somodi, S., and R. Guthoff. 1995. "Visualization of keratocytes in the human cornea with fluorescence microscopy." *Ophthalmologe* 92 (4):452–7.

Stromqvist, B., P. Fritzell, O. Hagg, B. Jonsson, and B. Sanden. 2013. "Swespine: The Swedish spine register: The 2012 report." *Eur Spine J* 22 (4):953–74. doi: 10.1007/s00586-013-2758-9.

Teixeira, G.Q., A. Boldt, I. Nagl et al. 2016. "A degenerative/proinflammatory intervertebral disc organ culture: An *ex vivo* model for anti-inflammatory drug and cell therapy." *Tissue Eng Part C Methods* 22 (1):8–19. doi: 10.1089/ten.tec.2015.0195.

Thompson, J.P., R.H. Pearce, M.T. Schechter et al. 1990. "Preliminary evaluation of a scheme for grading the gross morphology of the human intervertebral disc." *Spine (Phila Pa 1976)* 15 (5):411–5.

Urban, J.P. 2002. "The role of the physicochemical environment in determining disc cell behaviour." *Biochem Soc Trans* 30 (Pt 6):858–64. doi: 10.1042/.

Urban, J.P., and S. Roberts. 2003. "Degeneration of the intervertebral disc." *Arthritis Res Ther* 5 (3):120–30.

Urban, J.P., S. Smith, and J.C. Fairbank. 2004. "Nutrition of the intervertebral disc." *Spine (Phila Pa 1976)* 29 (23):2700–9.

van Dijk, B., E. Potier, and K. Ito. 2011. "Culturing bovine nucleus pulposus explants by balancing medium osmolarity." *Tissue Eng Part C Methods* 17 (11):1089–96. doi: 10.1089/ten.TEC.2011.0215.

van Dijk, B.G., E. Potier, and K. Ito. 2013. "Long-term culture of bovine nucleus pulposus explants in a native environment." *Spine J* 13 (4):454–63. doi: 10.1016/j.spinee.2012.12.006.

Vandooren, J., N. Geurts, E. Martens, P.E. Van den Steen, and G. Opdenakker. 2013. "Zymography methods for visualizing hydrolytic enzymes." *Nat Methods* 10 (3):211–20. doi: 10.1038/nmeth.2371.

Vergroesen, P.P., I. Kingma, K.S. Emanuel et al. 2015. "Mechanics and biology in intervertebral disc degeneration: A vicious circle." *Osteoarthritis Cartilage* 23 (7):1057–70. doi: 10.1016/j.joca.2015.03.028.

Vergroesen, P.P., A.J. van der Veen, B.J. van Royen, I. Kingma, and T.H. Smit. 2014. "Intradiscal pressure depends on recent loading and correlates with disc height and compressive stiffness." *Eur Spine J* 23 (11):2359–68. doi: 10.1007/s00586-014-3450-4.

Visse, R., and H. Nagase. 2003. "Matrix metalloproteinases and tissue inhibitors of metalloproteinases: Structure, function, and biochemistry." *Circ Res* 92 (8):827–39. doi: 10.1161/01.res.0000070112.80711.3d.

Vo, N.V., R.A. Hartman, T. Yurube et al. 2013. "Expression and regulation of metalloproteinases and their inhibitors in intervertebral disc aging and degeneration." *Spine J* 13 (3):331–41. doi: 10.1016/j.spinee.2012.02.027.

Walter, B.A., S. Illien-Junger, P.R. Nasser, A.C. Hecht, and J.C. Iatridis. 2014. "Development and validation of a bioreactor system for dynamic loading and mechanical characterization of whole human intervertebral discs in organ culture." *J Biomech* 47 (9):2095–101. doi: 10.1016/j.jbiomech.2014.03.015.

Walter, B.A., C.L. Korecki, D. Purmessur, P.J. Roughley, A.J. Michalek, and J.C. Iatridis. 2011. "Complex loading affects intervertebral disc mechanics and biology." *Osteoarthritis Cartilage* 19 (8):1011–8. doi: 10.1016/j.joca.2011.04.005.

Walter, B.A., P. Mageswaran, X. Mo et al. 2017. "MR elastography-derived stiffness: A biomarker for intervertebral disc degeneration." *Radiology* 162287. doi: 10.1148/radiol.2017162287.

Wang, D., L.A. Nasto, P. Roughley et al. 2012. "Spine degeneration in a murine model of chronic human tobacco smokers." *Osteoarthritis Cartilage* 20 (8):896–905. doi: 10.1016/j.joca.2012.04.010.

Wang, J., Y. Tian, K.L. Phillips et al. 2013. "Tumor necrosis factor alpha- and interleukin-1beta-dependent induction of CCL3 expression by nucleus pulposus cells promotes macrophage migration through CCR1." *Arthritis Rheum* 65 (3):832–42. doi: 10.1002/art.37819.

Weinstein, J.N., T.D. Tosteson, J.D. Lurie et al. 2006. "Surgical vs nonoperative treatment for lumbar disk herniation: The Spine Patient Outcomes Research Trial (SPORT): A randomized trial." *JAMA* 296 (20):2441–50. doi: 10.1001/jama.296.20.2441.

Yerramalli, C.S., A.I. Chou, G.J. Miller et al. 2007. "The effect of nucleus pulposus cross-linking and glycosaminoglycan degradation on disc mechanical function." *Biomech Model Mechanobiol* 6 (1–2):13–20. doi: 10.1007/s10237-006-0043-0.

4 Adult Stem Cells for Intervertebral Disc Repair

Esther Potier and Delphine Logeart-Avramoglou

CONTENTS

4.1 Intervertebral Disc and Its Degeneration 103
4.2 Committed Cell-Based Therapies 105
4.3 Adult Stem Cells 106
4.4 Adult Stem Cell-Based Therapies 108
4.5 Clinical Studies 110
4.6 Challenges for Adult Stem Cell Therapies for IVD Repair 117
    4.6.1 Do Transplanted Adult Stem Cells Survive in the IVD? 117
    4.6.2 What Is the Mechanism of Action of Transplanted Adult Stem Cells? 120
    4.6.3 Should the Adult Stem Cells Be Predifferentiated? 121
    4.6.4 What Is the Best Adult Stem Cell Source? 122
    4.6.5 What Is the Optimal Dose of Adult Stem Cells? 123
    4.6.6 How Should Adult Stem Cells Be Delivered? 124
    4.6.7 What Is the Target Patient Profile? 125
4.7 Concluding Remarks 125
References 126

4.1 INTERVERTEBRAL DISC AND ITS DEGENERATION

Intervertebral discs (IVDs) are the structures lying between the vertebral bodies, and which function is to transmit the mechanical loads exerted on the spine during motion to assure stable mobility. The stress redistribution is achieved by the nucleus pulposus (NP), a water-rich, compressible core, which is surrounded by the annulus fibrosus (AF), a firm, fibrous outer layer. Both AF and NP interface with two end plates that are thin, horizontal layers of hyaline cartilage located at the top and bottom portions of the adjacent vertebral bodies. The IVD structure is presented in Figure 4.1a. The AF consists of concentric lamellae with alternating orientation, creating an angle-ply structure depicted in Figure 4.1b, and is composed of highly organized collagen fiber bundles, rich in type I collagen. These specific composition and structure provide to

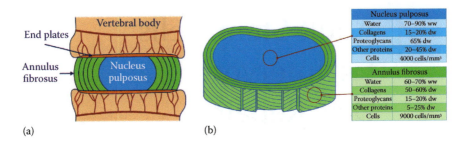

FIGURE 4.1 (a) Schematic representation of the IVD structure. (b) Schematic representation of NP and AF composition. (Adapted from Le Maitre, C.L. et al., *Biochem Soc Trans* 35 (Pt 4): 652–5, 2007b; Maroudas, A. et al., *J Anat* 120 (Pt 1):113–30, 1975.)

the AF high resistance to circumferential, longitudinal, and torsional stresses that are created by intradiscal pressure and by spine motion. The NP is a gelatinous matrix made of proteoglycans immobilized in a fibrous collagen network, comprised mainly of aggrecan and type II collagen as indicated in Figure 4.1b. This peculiar extracellular matrix creates a highly osmotic environment that attracts and retains water within the tissue, enabling the NP to sustain high compressive loading. The IVD is also the largest avascular structure in the body. Supply of nutrients is solely insured by diffusion from blood vessels at the disc's margins, mainly through the end plates (Ogata and Whiteside 1981; Shirazi-Adl, Taheri, and Urban 2010).

Another unique feature of IVDs is their low inherent cell density. The AF contains elongated, thin, fibroblast-like cells aligned parallel to the collagen fibers at an approximate cell density of 9000 cells/mm^3 (Maroudas et al. 1975), while the NP contains small, chondrocyte-like cells (NP cells) dispersed in the NP matrix at approximately 4000/mm^3 (Maroudas et al. 1975). During embryonic development and young age, the NP also contains large, vacuolated, notochordal cells that are residual cells from the embryonic notochord. While notochordal cells persist throughout most of adult life in some species (such as rats or cats) (Hunter, Matyas, and Duncan 2003), they progressively disappear during growth in other species, including humans (Trout, Buckwalter, and Moore 1982). It is still debated whether the notochordal cell disappearance is due to their terminal differentiation into NP cells or to their cell death.

The first signs of degeneration appear in the NP, where type II collagen and proteoglycans are replaced by type I collagen, causing a drop in water content (Antoniou et al. 1996; Le Maitre et al. 2007b) and, subsequently, a decline in the height and the load-bearing capacity of the IVD. The ensuing stress redistribution spreads to adjacent AF, end-plates, and vertebral bodies, which may ultimately lead to crack formation, rupture, and/or structural deterioration in the IVD. These tissue modifications can, in turn, result in painful symptoms and neurological deficits due to nerve root compression, facet joint arthrosis, etc. Changes in the NP composition are the result of increased production of both proinflammatory cytokines, and enzymes degrading the collagens and proteoglycans (Kepler et al. 2013; Le Maitre, Freemont, and Hoyland 2004), combined with the inability of the NP cell population to maintain and/or produce the NP-specific matrix due to their declining number and altered phenotype (Freemont 2009; Raj 2008). Contrary to most of the other musculoskeletal tissues, the IVD starts to degenerate spontaneously as early as in the second decade of life (Boos et al. 2002). This process is the natural aging of the IVD, as shown by the high incidence of IVD degenerative changes in asymptomatic patients, but it could be accelerated and become pathological because of genetic, environmental, or biomechanical factors (Battié and Videman 2006).

IVD degeneration is now well accepted as one of the primary causes of chronic back pain, involved in at least 39% of cases (Zhang et al. 2009). A large majority of patients with back pain will obtain satisfactory pain relief with conservative pharmacologic and physiotherapeutic treatments. Once conservative measures have failed, surgical options are explored. Current surgical treatments aim at immobilizing (e.g., by spinal fusion) or replacing (e.g., using disc prosthesis) the degenerated IVD. These treatments, however, are only palliative in the sense that they offer symptom relief rather than targeting the underlying pathogenesis of IVD degeneration. Moreover, the clinical results of IVD arthroplasty and spinal fusion remain suboptimal (Freeman and Davenport 2006) and these techniques are suspected to accelerate the degenerative process in adjacent segments by altering further the biomechanics of the spine (so-called adjacent segment disease) (Ghiselli et al. 2004; Gruber et al. 2009). In this context, several therapeutical strategies have been proposed to repair the IVD. One of them consists in the introduction of injectable biomaterials with mechanical and swelling properties suitable to restore disc height and to ensure IVD stability (for a recent review, see Bowles and Setton 2017). Others aim at preventing, slowing down, or even reversing the degenerative process of the IVD using biomolecule-, gene-, or cell-based strategies. In that perspective, cell-based therapies propose to inject exogenous cells to repopulate the NP and to synthesize the proteoglycan-rich matrix in order to restore the NP hydration and viscoelastic properties.

4.2 COMMITTED CELL-BASED THERAPIES

As NP cells are responsible for the unique features of the NP matrix, it seems rationale to use these cells to repair degenerated IVDs. Several studies established the beneficial effects of NP cell injection into degenerated IVD in different animal models as demonstrated by improved histological degeneration grade, disc height, and hydration (Feng et al. 2011b; Meisel et al. 2007; Okuma et al. 2000; Watanabe et al. 2003). In these studies, the autologous NP cells are isolated from the NP, which is aspirated to induce IVD degeneration, amplified, and reinjected when the IVD degeneration is established.

The pitfall of this type of therapy lies in the lack of availability of healthy committed NP cells. The low cellularity of the NP tissue, combined with the impossibility to expand these cells *in vitro* without losing their phenotype (Kluba et al. 2005), makes it unrealistic to obtain a number of NP cells high enough for a therapeutic protocol without damaging the donor IVD. The option of harvesting NP cells from the degenerative disc to treat, chosen in published clinical trials (see Section 4.5), is hampered by the fact that these cells display the trademarks of the pathology such as a premature senescence, a catabolic phenotype, and a lower proliferative capability

(Le Maitre, Freemont, and Hoyland 2007a). To overcome these hurdles, transplantation of committed cells from sources other than the IVD but able to produce a tissue matrix similar to that of the NP has also been tested in animal models of IVD degeneration. Specifically, chondrocytes derived from auricular (Gorensek et al. 2004) and juvenile articular cartilage (Acosta et al. 2011) were implanted for IVD repair in a rabbit and porcine model, respectively. Results showed that these cells survived and synthesized a cartilage-like matrix but no improvement on disc height and hydration was reported.

Adult stem cells appear, therefore, an attractive alternative cell source for IVD repair due to the low morbidity associated with their harvesting procedure, their ease of expansion *in vitro*, and their differentiation potential into chondrocyte, a cell type akin to NP cells.

4.3 ADULT STEM CELLS

Stem cells are classically defined as unspecialized cells that can self-renew and give rise to differentiated cell types during embryogenesis. In postnatal organs/tissues, adult stem cells are fundamental to tissue homeostasis and injury repair when triggered with the appropriate signals (Verfaillie 2002). These functions make them highly attractive in regenerative medicine and tissue engineering therapeutic approaches as regenerative cell source. Today, evidence is accumulating that a small subpopulation of adult stem cells are present in many differentiated tissues. Compared to pluripotent embryonic stem cells, which can be maintained continuously in culture for more than 300–400 cell doublings and can differentiate into all somatic cell types (Mikkola et al. 2006; Thomson et al. 1998), adult stem cells exhibit a lower degree of self-renewal and differentiation potential. Moreover, they are committed into a lineage-specific differentiation pathway and would theoretically be capable of producing only a limited range of specialized cells according to the embryonic origin of their tissue of origin. As they can be directly isolated from the patient, implantation of autologous adult stem cells, however, offers great promises in replacing nonfunctioning or lost cells and regenerating diseased and damaged tissues, including IVD.

First studies on skeletal tissue regeneration have focused on the use of adult stem cells derived from the bone marrow stroma. Initially identified as a subpopulation of bone marrow cells with osteogenic potential by Friedenstein and collaborators in the 1960s (Friedenstein et al. 1968), they were given several terms, including "stromal stem cells," "mesenchymal stem cells," and more recently, "multipotent mesenchymal stromal cells," also known as MSCs. Bone-marrow-derived MSCs (BM-MSCs) may differentiate into stromal cells supporting the hematopoiesis as well as into cells of the skeletal connective tissue such as osteoblasts, chondrocytes, tenocytes, bone marrow adipocytes, and vascular smooth muscle cells (Dennis et al. 1999; Friedenstein et al. 1968; Pittenger et al. 1999; Sacchetti et al. 2007). The BM-MSCs present the advantages of being easily collected, isolated, and expanded *in vitro*.

In 2000s, adipose tissue has also been introduced as a new source for adult stem cells (Zuk et al. 2001). These adipose-tissue-derived MSCs (AT-MSCs), also called adipose-tissue-derived stem cells (ADSCs), can be easily and repeatable harvested in large quantities (5×10^5 cells could be obtained from 400 to 600 mg adipose tissue; Zhu et al. 2008) using minimally invasive and low-morbidity techniques, which is one of the prime requirements for a regenerative cell source for cell-based therapies and tissue engineering. AT-MSCs also display and maintain strong proliferation ability and multidifferentiation potential after several passages (Zhu et al. 2008).

Since these pioneering studies, other MSCs have been isolated from a variety of tissues, such as umbilical cord, muscle, dental pulp, and virtually any connective tissues (Porada, Zanjani, and Almeida-Porad 2006). One theory for the varied locations of MSCs is that these cells derive from perivascular cells (i.e., pericytes) (Caplan 2015) and can therefore be found in any vascularized tissue. In 2006, the International Society for Cellular Therapy recommended that cells should fulfill the following criteria to be considered as MSCs: (i) the cells must be plastic adherent when maintained under standard culture conditions; (ii) they must express CD73, CD90, and CD105 markers and should not express CD34, CD45, Human Leukocyte Antigen–antigen D Related (HLA–DR), CD14 or CD11b, and CD79α or CD19; and (iii) they should be able to differentiate into osteoblasts, chondrocytes, and adipocytes *in vitro* (Dominici et al. 2006). However, more recently, researchers in the MSC field have shown that MSCs from various adult and neonatal tissues have similar, but not identical, functional potential (Hass et al. 2011); therefore, these commonly described markers appear not distinctive enough and may not be sufficient for defining the cellular phenotype and biological function of MSCs (Keating 2012).

Risbud et al. (2007) provided evidence that the IVD also contains skeletal progenitor cells exhibiting the immunophenotype and the multidifferentiation potential typical of MSCs. Since then, various studies have validated the presence of IVD-derived MSCs (IVD-MSCs) in several species, such as human, pig, or rabbit (Blanco et al. 2010; Liu et al. 2011, 2014; Risbud et al. 2007; Sakai et al. 2012; Tao et al. 2013; Wang et al. 2016) (for a recent review, see Li et al. 2015). By selecting angiopoietin-1 receptor (Tie2)-positive cells, Sakai and collaborators further validated the presence of single-cell, highly proliferative, and multipotent progenitors in human, murine, and bovine NPs (Sakai et al. 2012; Tekari et al. 2016). Although MSCs could be derived from all IVD compartments (i.e., NP (Blanco et al. 2010; Risbud et al. 2007; Sakai et al. 2012; Tao et al. 2013; Wang et al. 2016), AF (Liu et al. 2014; Risbud et al. 2007; Wang et al. 2016), and end plates (Liu et al. 2011; Wang et al. 2016)), the perichondrium, a region in the outer zone of the AF and adjacent to the epiphyseal plate, has been proposed as the physiological niche of the IVD-MSCs (Henriksson et al. 2009a; Shi et al. 2015).

Another promising adult stem cell population with high regenerative potential for IVD repair is the induced pluripotent stem cells (iPSCs). These cells are adult, somatic cells that are genetically reprogrammed to express genes that are typical of embryonic stem cells (Takahashi and Yamanaka 2006), leading to an embryonic stem cell-like state, with extensive proliferative capability and pluripotent differentiation potential. Two recent studies aimed to obtain notochordal-like cells, as progenitors of NP cells, from iPSCs and to differentiate these cells into NP cells using either a laminin-rich environment (Chen et al. 2013) or the natural NP (Liu, Rahaman, and Bal 2014) extracellular matrix. Both studies provided evidence of the possibility to generate cells expressing *type II collagen* and *cytokeratin-8* genes and to produce glycosaminoglycans and collagens. The iPSCs could therefore represent, in the future, a promising alternative to scarce, autologous NP cells, although major concerns relating to completeness of reprogramming, epigenetic changes, and genomic instability remain to be addressed before their eventual application to clinical care.

Adult stem cells have initially emerged as key elements of regenerative medicine therapies because of their differentiation potential properties into mature cells of different cell lineages. Since then, however, it has been established that these cells are also endowed with a highly paracrine secretory potential of a broad selection of cytokines, chemokines, and growth factors, in addition to components of the extracellular matrix, making them clinically highly attractive. Of particular interest are the antiapoptotic, proangiogenic, and immunomodulatory effects that have been demonstrated for both BM-MSCs and AT-MSCs (Baraniak and McDevitt 2010; Bertolini et al. 2012; Caplan and Correa 2011). An emerging body of data suggests that the soluble factors released by MSCs play a central role in the mechanism of action of most, if not all, of their systemic effects (Horwitz and Dominici 2008). Through their trophic activities, they establish a regenerative microenvironment to support the repair of injured tissues.

4.4 ADULT STEM CELL-BASED THERAPIES

Due to the potential of adult stem cells to differentiate into chondrocytes, a cell type that shares common features with NP cells, they represent an attractive source of regenerative cells for IVD repair. Among the different tissue sources, BM-MSCs have been the most investigated in IVD degeneration models in both animal and preclinical studies (for recent reviews, see Sakai and Andersson 2015 and Vadalà et al. 2016). Sakai and collaborators were the pioneers in this field to demonstrate a positive effect of the transplantation of autologous BM-MSCs into degenerated IVDs in a rabbit model of induced IVD degeneration (Sakai et al. 2003, 2005, 2006). They observed significant improvements in disc height, hydration, and glycosaminoglycan content. The benefit of BM-MSC injection for IVD repair was further confirmed in mice (Yang et al. 2009), rats (Allon et al. 2010; Crevensten et al. 2004), rabbits (Cai et al. 2015; Hee et al. 2010; Ho et al. 2008; Leung et al. 2014; Subhan et al. 2014; Tao et al. 2016; Vadalà et al. 2012; Yang et al. 2010), dogs (Hang et al. 2017; Hiyama et al. 2008; Serigano et al. 2010), pigs (Bendtsen et al. 2011; Henriksson et al. 2009b), goats (Zhang et al. 2011), and sheep (Ghosh et al. 2012), largely demonstrating their regenerative potential.

AT-MSCs were also used as an easily accessible and expandable alternative source of adult stem cells in mouse (Marfia et al. 2014); rat (Jeong et al. 2010); rabbit (Chun et al. 2012), and dog (Ganey et al. 2009) models. Higher disc height and hydration, as well as better histologic degradation grading, in AT-MSC-treated IVDs compared to untreated IVDs suggested that this type of adult stem cells could also be suitable for IVD repair. Due to their later discovery, IVD-MSCs were only recently evaluated in rabbit models (Chen et al. 2016; Wang et al. 2014). Similar to the results obtained with other sources of MSCs, the implantation of IVD-MSCs improved disc hydration and NP matrix gene expression. An increase in disc height, however, was observed only in one study (Wang et al. 2014). Other tissue origins of adult stem cells, such as synovial tissue of the knee (Miyamoto et al. 2010) or umbilical cord tissue (Tam et al. 2014), have also been assessed for the healing of degenerated IVD, broadening the potential sources of regenerative cells. However, further *in vivo* studies are needed to definitively establish the actual therapeutic efficacy of these alternative cell types.

It is important to note that in the vast majority of these animal models, IVD degeneration was induced by the experimenter. In fact, only a very limited number of animal species shows a spontaneous IVD degeneration as seen in humans. In the nonspontaneously degenerative models (e.g., mouse, rat, and rabbit), the IVD degeneration is induced either by genetic mutation, supra-physiologic loading of IVDs, chemical digestion of the NP matrix, or, most commonly, by AF stabbing or partial nucleotomy. All these traumatic procedures result in progressive degenerative changes in both NP and inner AF over time, with, for instance, a mild degeneration observed 4 weeks postinjury in rabbits (Okuma et al. 2000; Sakai et al. 2003). The main limitation of these induced models is that the degenerated IVD still benefits from a physiological nutrient supply from the surrounding tissues and from a healthy tissue microenvironment. In contrast, in humans, where the degenerative process is progressive, the context is much more unfavorable and involves altered nutrient supply, catabolic/anabolic unbalance, and NP cell senescence. The relevance of using these animal models to evaluate therapies for human IVD degeneration is therefore under debate among scientists (Alini et al. 2008). The animal models used for IVD degeneration are presented in detail in Chapter 2.

Despite these limitations, the preclinical studies provided evidence that, in general, MSC transplantation can decelerate or arrest the course of IVD degeneration: disc height and hydration, assessed by X-ray imaging and T2-weighted magnetic resonance imaging (MRI), respectively; were improved, while histologic degradation grading, as well as glycosaminoglycan content and NP matrix-related gene expression, was enhanced, after cell injection compared to untreated controls. In 2015, these outcomes were investigated in a meta-analysis of 22 studies using BM-MSCs and AT-MSCs for IVD regeneration (Wang et al. 2015), whose results are presented in Table 4.1. Noteworthy, among these numerous studies, only a very few reported a complete IVD regeneration, with disc height and glycosaminoglycan content similar to the healthy controls (Hiyama et al. 2008; Sakai et al. 2006). All these outcomes, however, demonstrated the proof of efficacy of adult stem cell therapy for IVD degeneration in preclinical animal models, providing a foundation for testing in human studies.

TABLE 4.1
Significant Effects of BM-MSC and AT-MSC Injection in Animal Models of IVD Degeneration

Animal Model	Study Number	SMD	95% CI	p Value
		Disc Height Index		
Dog	2	6.10	4.97; 7.23	<0.001
Pig	1	2.50	1.53; 3.46	<0.001
Rabbit	6	4.01	2.89; 5.14	<0.001
Rat	3	3.17	0.13; 6.21	0.04
Sheep	1	−2.75	−4.40; −1.09	0.001
		MRI T2 Signal		
Dog	2	2.48	0.92; 4.03	0.002
Rabbit	8	2.42	1.11; 3.73	<0.001
Rat	3	2.05	1.03; 3.06	<0.001
Sheep	1	2.53	0.95; 4.11	0.002
		Histologic Disc Degeneration Grade		
Dog	2	−3.88	−6.98; −0.77	0.01
Rabbit	4	−2.41	−3.15; −1.67	<0.001
Rat	4	−3.61	−5.79; −1.42	0.001
Sheep	1	−0.75	−1.93; 0.43	0.21
		Type II Collagen Expression		
Dog	1	2.41	0.86; 3.95	0.002
Pig	1	5.67	4.02; 7.31	<0.001
Rabbit	5	3.38	0.37; 6.39	0.03
Rat	2	4.26	−0.94; 9.45	0.11

Source: Adapted from Wang, F., Shi, R., Cai, F., Wang, Y.-T., Wu, X.-T., *Stem Cells Dev.*, 24, 2479–2495, 2015.

Note: Standardized mean difference (SMD) between the control group and the adult stem cell transplantation group and related 95% confidence interval (CI) were calculated for each study using Cohen's method. A positive SMD indicates an increase of the selected outcome in the adult stem cell transplantation group compared to the control group.

4.5 CLINICAL STUDIES

To date, only a few clinical trials targeting IVD repair or regeneration have been published using transplantation of either committed or adult stem cells. These studies are summarized in Table 4.2.

In 2002, a study (EuroDISC) was conducted to evaluate the safety and efficacy of autologous disc chondrocyte transplant (ADCT) combined with discectomy (surgical removal of herniated IVD material). An interim analysis was published

Adult Stem Cells for Intervertebral Disc Repair

TABLE 4.2
Published Clinical Studies Using Cell-Based Therapies for IVD Regeneration

References	Study Design	Mode	Cell Type	Cell Preparation	Number of Patients	Follow-Up (Years)	Main Outcomes
Committed Cell-Based Therapies
Hohaus et al. 2008; Meisel et al. 2006, 2007	Controlled, nonblinded phase 1	Autologous	NP cells from hernia discectomy	NP fragments digested and expanded for 2 weeks	28	2	Reduced low back pain; Improved hydration in IVDs adjacent to treated IVDs; No change in disc height
Coric et al. 2013	Nonrandomized phase 1	Allogenic	Juvenile articular chondrocytes	Expansion for two passages; Dose: $1–2 \times 10^7$ in 1–2 mL; Carrier: Fibrin gel	15	1	Improved pain scores; Improvement on MRI (10/13)
Mochida et al. 2015	Pilot phase 1	Autologous	Activated NP cells	Coculture with autologous MSCs for 3 days; Dose: 10^6 cells in 702 μL; In suspension in saline	9	3	No clinical adverse events; Improved pain scores; Mild improvement on MRI
Adult Stem Cell-Based Therapies
Yoshikawa et al. 2010	Case series	Autologous	BM-MSC	Dose: 10^5 cells in 1 mL; Carrier: collagen sponges	2	2	Reduced low back pain; Improved hydration
Orozco et al. 2011	Pilot phase 1	Autologous	BM-MSC	Dose: 10^7 cells in 1 mL; In suspension in Ringer-lactate sol, 0.5% Alb, 5 mM glucose	10	1	Important analgesic effect; Improved disability at 3 m; Increased hydration at 12 m; No change in disc height

(*Continued*)

TABLE 4.2 (CONTINUED)
Published Clinical Studies Using Cell-Based Therapies for IVD Regeneration

References	Study Design	Mode	Cell Type	Cell Preparation	Number of Patients	Follow-Up (Years)	Main Outcomes
Adult Stem Cell-Based Therapies							
Pettine et al. 2015	Open label pilot phase 1	Autologous	BM-concentrate	Dose: $1.2 \pm 0.1 \times 10^8$ nucleated cells in 2–3 mL (with 2713 ± 491 CFU-F/mL) In suspension	26	1	Reduced low back pain (21/26) Increased hydration (8/26) Best results with CFU-F dose >2000/mL
Elabd et al. 2016	Case series	Autologous	BM-MSC	Hypoxic (5% O_2) culture Dose: $15-52 \times 10^6$ in 0.25–1 mL In suspension in autologous platelet lysate	5	4–6	Reduced IVD protrusion (4/5) Improved strength and mobility Potential correlation with MSC dose
Pang et al. 2014	Case series	Allogenic	Umbilical cord MSC	Dose: 10^7 in 1 mL In suspension	2	2	Reduced low back pain (2/2) Increased hydration (1/2)

on the 2-year follow-up of 28 patients. The authors concluded that the ADCT/discectomy treated group showed a clinically significant decrease in the Oswestry disability index and reduced disc desiccation on MRI compared to the discectomy only group (Hohaus et al. 2008; Meisel et al. 2006, 2007). No significant difference in disc height index, however, was observed between the groups. The efficacy of allogeneic chondrocytes, harvested from the articular surface of young cadaveric donor tissue (ISTO Technologies), was also evaluated in a prospective study (Coric et al. 2013). Pain and disability improved for all recipients during the 12 months of follow-up and 10/13 of the evaluated patients had enhanced MRI hydration values. A phase 2, randomized, double-blind study evaluating this treatment, as compared to a placebo, has been recently terminated but not yet published (NCT01771471).

More recently, Mochida et al. (2015) reported a study on the efficacy of the transplantation of autologous NP cells, harvested from IVD removed during spinal fusion and activated by coculture with autologous BM-MSCs for 3 days, in the degenerated lumbar IVD. MRI results indicated that the disc degeneration did not worsen during the 3-year follow-up, with a mild improvement noted in 1 case.

There have also been several case studies to assess the efficacy of adult stem cell transplantation into the IVD, to treat chronic low back pain. Yoshikawa et al. (2010) were the first to investigate the potential of autologous BM-MSCs in two patients suffering from lumbar spinal stenosis. Both patients had significant symptomatic relief 2 years after surgery. Subsequently, Orozco et al. (2011) conducted a pilot study on 10 patients. The feasibility and safety of BM-MSC transplantation were confirmed and patients exhibited rapid improvement of pain and disability; MRI evaluations showed significant increase in water content at 1-year follow-up, but no improvement was observed in disc height index.

Because hypoxic preconditioning is known to have beneficial biological effects on MSCs, including the up-regulation of chondrocyte specific genes such as *aggrecan*, *type II collagen*, and *Sox-9* and activation of hypoxia-inducible genes such as prosurvival cues (Shang et al. 2014), Elabd et al. assessed the therapeutic efficacy of BM-MSCs expanded under 5% O_2 hypoxic conditions prior to transplantation in five patients. After a 4- to 6-year follow-up, all patients reported improvement in pain, and four out of five showed a reduction in posterior protrusion and a limited reduction of disc height (Elabd et al. 2016).

The effects of bone marrow concentrate, in treating moderate to severe discogenic low back pain were also reported in a prospective case series by Pettine et al. (2015). Significant improvement in pain scores was demonstrated in 21 out of 26 patients at 3, 6, and 12 months post-treatment, and 8 out of 20 patients showed a one-level improvement in Pfirrmann MRI grading after 1 year. Interestingly, the greater improvements were observed in patients who received a higher colony-forming unit-fibroblast (CFU-F, synonymous with MSC) dose (>2000/mL), suggesting that the effectiveness of the therapy is cell-dose dependent, although its exact mechanism remains unclear.

During the last decade, a dozen of additional unpublished clinical trials have been registered at the US National Health Institutes or at the European Union Clinical Trials Register and are listed in Table 4.3. Among these study proposals, one terminated study aimed to assess juvenile chondrocyte transplantation (#NCT01771471) and one ongoing study aims to evaluate NP cell transplantation (#NCT01640457/

TABLE 4.3
Unpublished Clinical Studies Using Cell-Based Therapies for IVD Regeneration

Sponsor	Study Dates Start	Study Dates Completion	Study Design	Mode	Cell Type	Cell Preparation	Number of Patients	Follow-Up (Years)	Status	Trial ID
Committed Cell-Based Therapies										
ISTO Technologies, Inc.	2012	2020	Randomized, double blind, placebo controlled phase 2	Allogenic	Juvenile chondrocytes	Dose: NS Carrier: Fibrin	44	2	Completed	NCT01771471[a]
Tetec AG	2012	2021	Prospective, randomized, multicenter phases 1 and 2	Autologous	Disc chondrocytes (Novocart Disc plus)	Dose: $0.5–2 \times 10^6$ Carrier: modified albumin, hyaluronic acid gel	120	5	Ongoing	NCT01640457[a] 2010-023830-22[b]
Adult Stem Cell-Based Therapies										
Biostar	2012	2014	Prospective, open, nonrandomized phases 1 and 2	Autologous	AT-MSC	Dose: 4×10^7 in 1 mL	8	0.5	Recruiting	NCT01643681[a]
Bioheart, Inc.	2014	2017	Open label, nonrandomized, multicenter	Autologous	AT-MSC	Dose: patient-specific In suspension	100	0.5	Recruiting	NCT02097862[a]

(Continued)

TABLE 4.3 (CONTINUED)
Unpublished Clinical Studies Using Cell-Based Therapies for IVD Regeneration

Sponsor	Study Dates Start	Study Dates Completion	Study Design	Mode	Cell Type	Cell Preparation	Number of Patients	Follow-Up (Years)	Status	Trial ID
Adult Stem Cell-Based Therapies										
Inbo Han	2015	2017	Open label phase 1	Autologous	AT-MSC	Dose: 2 or 4×10^7 In Tissuefill Hyaluronic acid derivatives	10	1	Recruiting	NCT02338271[a]
The Foundation for Spinal Research, Education and Humanitarian Care, Inc.	2013	2018	Prospective cohort clinical study	Autologous or allogenic	BM-MSC	Dose: NS In suspension	100	2	Recruiting	NCT02529566[a]
Århus University Hospital	2013	NS	Single blind, randomized, controlled phase 2	Autologous	BM-MSC	NS	34	2	Ongoing	2012-003160-44[b]

(Continued)

TABLE 4.3 (CONTINUED)
Unpublished Clinical Studies Using Cell-Based Therapies for IVD Regeneration

Sponsor	Study Dates Start	Study Dates Completion	Study Design	Mode	Cell Type	Cell Preparation	Number of Patients	Follow-Up (Years)	Status	Trial ID
\multicolumn{11}{l}{Adult Stem Cell-Based Therapies}
Mesoblast, Ltd.	2011	2015	Prospective, double blind, randomized, controlled phase 2	Allogenic	STRO-3 selected BM-MSC	Two doses: 6×10^6 or 18×10^6 Carrier: hyaluronic acid	100	3	Completed	NCT01290367[a] (Mesoblast Inc. 2017)
Red de Terapia Celular	2013	2017	Prospective, open, nonrandomized phases 1 and 2	Allogenic	BM-MSC	Dose: $10 \pm 5 \times 10^6$ In suspension	25	1	Completed	NCT01860417[a]
Mesoblast, Ltd.	2015	2020	Prospective, multicenter, randomized, double-blind, placebo-controlled phase 3	Allogenic	STRO-3 selected BM-MSC	Dose: 6×10^6 with or w/o Hyaluronic acid	360	2	Recruiting	NCT02412735[a]

Note: NS, not specified.
[a] US National Library of Medicine: ClinicalTrials.gov
[b] EU Clinical Trials Register: clinicaltrialregister.eu

#2010-023830-22), five studies aim to assess the efficacy of BM-MSCs (#NCT02529566, #2012-003160-44, #NCT01290367, #NCT01860417, and #NCT02412735), and three studies assess the efficacy of AT-MSCs (#NCT01643681, #NCT02097862, and #NCT02338271).

The utility of allogeneic adult stem cells remains an intense area of investigation due to their advantages over autologous stem cells with regard to time, cost, and quality assurance (Zhang et al. 2015). Two completed (#NCT01290367 and #NCT01860417) and two ongoing (#NCT02529566 and #NCT02412735) clinical trials have been conducted for evaluating the effects of allogenic BM-MSCs. Notably, a phase 2 clinical trial study on 100 patients assessed the potency of allogenic mesenchymal precursor cells (MPCs) (Mesoblast, Melbourne, Australia), commercially available cells derived from specific STRO-3 positive selected BM-MSC population, which has been demonstrated to possess a higher proliferation capacity, higher expression levels of early stem cell markers, and tri-lineage differentiation potential (See et al. 2011). The trial results showed that 41% and 35% of patients who received 6 and 18 million MPCs, respectively, reached target criteria for treatment success (i.e., a 50% reduction in the pain score and a 15-point reduction in the Oswestry disability index) after 24 months. In comparison, only 18% and 13% of the patients treated with hyaluronic acid or saline reached these criteria, respectively (Mesoblast Inc. 2017). These promising outcomes have supported an ongoing 360-patient phase 3 trial, making it the largest human IVD cell-based therapy study (#NCT02412735).

All these clinical studies highlight the efficacy of either committed- or adult stem-cell-based therapy in reducing pain and disability in a large majority of the enrolled patients. Improvements on disc hydration, however, were not systematically reported and no improvements on disc height were found. This latter clinical outcome was also observed in most of the animal studies and may be related to the experimental observation that transplanted cells do not illicit a high enough production of proteoglycan-rich matrix to reestablish a normal disc height. It is, therefore, the general belief that further research advancements on adult stem cell-based protocols would undoubtedly improve the clinical treatment of IVD degeneration. Although other issues, still unknown, may be identified in the future, currently known aspects (focused on the use of adult stem cells) that may contribute to the success of this therapeutic procedure will be discussed in the sections that follow.

4.6 CHALLENGES FOR ADULT STEM CELL THERAPIES FOR IVD REPAIR

4.6.1 Do Transplanted Adult Stem Cells Survive in the IVD?

The adult IVD is devoid of blood vessels and relies on both oxygen and nutrient supplies by diffusion from surrounding tissues, mainly through the end plates (Ogata and Whiteside 1981; Shirazi-Adl, Taheri, and Urban 2010). Thus, cells located within the core of the NP can be as far as 8 mm away from a blood vessel. Moreover, solute transport through the collagen- and proteoglycan-rich matrix is restricted, even for small molecules (Rajasekaran et al. 2004), resulting in low oxygen (Huang and Gu 2008; Holm et al. 1981; Sélard, Shirazi-Adl, and Urban 2003; Soukane, Shirazi-Adl, and Urban et al. 2007) and glucose (Holm, Selstam, and Nachemson 1982; Jackson et al. 2011) contents within the NP. As waste products are cleared by the same route, lactic acid, an end metabolite of the anaerobic glycolysis, accumulates in the NP, causing a local tissue acidification (Holm, Selstam, and Nachemson 1982; Huang and Gu 2008; Kitano et al. 1993; Soukane, Shirazi-Adl, and Urban et al. 2007). Additionally, the NP is a hyperosmotic tissue (Antoniou et al. 1996; Kitano et al. 1993), due to its high content in proteoglycans. As a result, the healthy NP represents a challenging surrounding where the native NP cells are simultaneously exposed to hypoxia, nutrient deprivation, low pH, hyperosmolarity, and high mechanical loading. This environment worsens as degeneration progresses, especially in species with spontaneous and progressive IVD degeneration, such as humans. Decreased permeability of the end plates (Benneker et al. 2005; Rajasekaran et al. 2004) and diminished blood flow at the end plate margins (Holm and Nachemson 1988; Kauppila et al. 1997) are, indeed, strongly associated with IVD degeneration. This reduced vascularization leads, in turn, to a further lowering of oxygen (Bartels et al. 1998; Huang and Gu 2008) and glucose (Jackson et al. 2011) contents as well as a decrease in pH (Bartels et al. 1998; Diamant, Karlsson, and Nachemson 1968; Huang and Gu 2008; Kitano et al. 1993). Osmolarity also lessens on account of the proteoglycan degradation during degeneration (Antoniou et al. 2004; Iatridis et al. 2007; Kitano et al. 1993). Table 4.4 presents the microenvironment characteristics of normal and pathological NPs.

The monitoring of the viability of labeled MSCs delivered into IVD in different animal models proved, surprisingly, that these cells are capable of surviving in this harsh surrounding (Allon et al. 2010; Cai et al. 2015; Chun et al. 2012; Crevensten et al. 2004; Ganey et al. 2009; Hee et al. 2010; Henriksson et al. 2009b; Hiyama et al. 2008; Ho et al. 2008; Leung et al. 2014; Marfia et al. 2014; Sakai et al. 2003, 2005, 2006; Serigano et al. 2010; Subhan et al. 2014; Tao et al. 2016; Vadalà et al. 2012; Yang et al. 2009, 2010; Zhang et al. 2011). It is, however, difficult to conclude on the capacity of the surviving MSCs thriving in the IVD environment. Their number has, indeed, been reported to both increase (Crevensten et al. 2004; Henriksson et al. 2009b; Leung et al. 2014; Marfia et al. 2014; Sakai et al. 2005, 2006) and decrease (Miyamoto et al. 2010; Sakai et al. 2003; Tao et al. 2016; Yang et al. 2010) over time. Moreover, such viability outcome, observed in animal models with induced IVD degeneration, may not translate to humans who have a progressive IVD degeneration, characterized by a further decrease in pH, oxygen, and glucose levels, as shown in Table 4.4. So far, only *in vitro* studies have assessed MSC viability and functions in microenvironments representative of a human degenerative IVD. The results showed that low glucose content (Deschepper et al. 2013; Farrell et al. 2015) and low pH (Han et al. 2014; Naqvi and Buckley 2016; Wuertz, Godburn, and Iatridis 2009; Wuertz et al. 2008) are detrimental to MSCs, while hypoxia favors their discogenic differentiation (Feng et al. 2011a; Li et al. 2014; Naqvi and Buckley 2015; Risbud et al. 2004; Stoyanov et al. 2011). Although these studies are "only" conducted *in vitro*, long-term adult stem cell engraftment, in the severe environment, found in spontaneously degenerative IVD remains to be demonstrated.

TABLE 4.4
Microenvironment Characteristics in Normal and Pathological NPs

Parameters	IVD State	Models	Values	References
Oxygen	Normal	Dog	0.3–8.0 kPa	Holm et al. 1981
		Numerical model	1–5 kPa	Sélard, Shirazi-Adl, and Urban 2003
		Numerical model	0.5 kPa	Huang and Gu 2008
		Numerical model	0.55 kPa	Soukane, Shirazi-Adl, and Urban et al. 2007
	Pathological	Human (scoliosis)	0.9–6.7 kPa	Bartels et al. 1998
		Human (IVD degeneration)	2.7–10.7 kPa	Bartels et al. 1998
		Numerical model	0 kPa	Huang and Gu 2008
Glucose	Normal	Dog	1.0–3.0 µmol/g	Holm, Selstam, and Nachemson 1982
		Numerical model	1.8 mM	Jackson et al. 2011
	Pathological	Numerical model	0.346 mM	Jackson et al. 2011
Lactic acid pH	Normal	Dog	2.0–6.0 µmol/g	Holm, Selstam, and Nachemson 1982
		Human	pH 7.14	Kitano et al. 1993
		Numerical model	pH 7.1	Huang and Gu 2008
		Numerical model	pH 7.0–7.3	Soukane, Shirazi-Adl, and Urban et al. 2007
	Pathological	Human (scoliosis)	4–6 mmol/L	Bartels et al. 1998
		Human (IVD degeneration)	2–6 mmol/L	Bartels et al. 1998
		Human (prolapses)	pH 6.7–7.6	Diamant, Karlsson, and Nachemson 1968
		Human (IVD degeneration)	pH 6.6	Kitano et al. 1993
		Numerical model	pH 6.1	Huang and Gu 2008
Osmolality	Normal	Human (2–5 years)	H_2O = 87% ww GAG = 650 µg/mg dw	Antoniou et al. 1996
		Human (10 years)	H_2O = 81.1% ww GAG = 274 µg/mg dw	Kitano et al. 1993
		Human (33 years)	H_2O = 78.5% ww GAG = 366 µg/mg dw	Kitano et al. 1993
	Pathological	Human (IVD degeneration)	H_2O = 74.5% ww GAG = 141 µg/mg dw	Kitano et al. 1993
		Human (IVD degeneration)	H_2O = 75.7% ww GAG = 250 µg/mg dw	Iatridis et al. 2007

(Continued)

TABLE 4.4 (CONTINUED)
Microenvironment Characteristics in Normal and Pathological NPs

Parameters	IVD State	Models	Values	References
		Thompson grade 2/3		
		Human (IVD degeneration)	H_2O = 75% ww GAG = 86 µg/mg dw	Antoniou et al. 2004
		Thompson grade 5		

Note: ww, wet weight; dw, dry weight; GAG, glycosaminoglycans.

4.6.2 WHAT IS THE MECHANISM OF ACTION OF TRANSPLANTED ADULT STEM CELLS?

While the animal and clinical studies highlight the potential for adult stem-cell therapy in IVD degeneration, the mechanisms of action behind this regenerative effect are still uncertain. A better understanding of these mechanisms will allow improving and optimizing stem cell-based therapies.

Transplanted adult stem cells may participate in IVD repair in several ways. First, they may promote IVD repair directly via their differentiation into NP cells and, consequently, their production of NP matrix. Several *in vitro* studies, showed that under appropriate stimuli, MSCs express *type II collagen* and *aggrecan* genes, and synthesize a glycosaminoglycan-rich matrix (see Section 4.6.3). Such differentiation potential was confirmed in *in vivo* studies based on positive type II collagen and aggrecan immunostaining of prelabeled MSCs injected into degenerated NPs (Marfia et al. 2014; Sakai et al. 2005; Yang et al. 2009). Secondly, adult stem cells may participate in IVD repair indirectly by stimulating the local NP cell proliferation and matrix production. Numerous *in vitro* studies provided evidence that NP cells displayed an improved phenotype and a higher proliferative and glycosaminoglycan synthesis capacities when cocultured with MSCs (Miyamoto et al. 2010; Naqvi and Buckley 2015; Niu et al. 2009; Richardson et al. 2006; Watanabe et al. 2010; Yamamoto et al. 2004; Yang et al. 2008). Although several of these studies demonstrated that the beneficial effects are mediated through paracrine, soluble factors, it appears that cell-to-cell contact improves NP cell stimulation by MSCs (Richardson et al. 2006; Watanabe et al. 2010; Yamamoto et al. 2004). Finally, the anti-inflammatory and immunomodulatory activity attributed to MSCs can also contribute to their beneficial effects on IVD degeneration. Indeed, numerous reports have shown that adult MSCs regulate the adaptive and innate immune systems through cell-to-cell interactions between MSCs and lymphocytes as well as through production of soluble factors (Gao et al. 2016; Zhao, Ren, and Han 2016).

4.6.3 Should the Adult Stem Cells Be Predifferentiated?

Most of the *in vivo* studies have been performed using undifferentiated, unpreconditioned MSCs, except for a few of them, which specifically sought to evaluate the benefits of hypoxic (Elabd et al. 2016) or NP cell coculture (Allon et al. 2010) preconditioning before transplantation on the MSC therapeutic efficacy. Many *in vitro* research efforts, however, have been devoted to predifferentiating MSCs into NP cells in hope of improving their aptness to survive and produce NP matrix in the harsh IVD environment. Such stem cell differentiation toward the NP cell phenotype also aims to obtain a stabilized, committed cell phenotype and therefore avoid potential ossifications that could be observed after leakage of undifferentiated BM-MSCs from injected NPs (Sobajima et al. 2008; Vadalà et al. 2012). Literature reports have demonstrated that MSC differentiation into NP-like cells can be induced by growth factors, most commonly transforming growth factor β (Bertolo et al. 2012; Clarke et al. 2014; Colombier et al. 2016; Gantenbein-Ritter et al. 2011; Steck et al. 2005) or growth differentiation factor-5 or -6 (Clarke et al. 2014; Colombier et al. 2016; Gantenbein-Ritter et al. 2011; Peroglio et al. 2013; Stoyanov et al. 2011), or by coculture with IVD cells (de Vries et al. 2015; Kim et al. 2009; Lu et al. 2007; Niu et al. 2009; Potier et al. 2014; Richardson et al. 2006; Ruan et al. 2012; Strassburg et al. 2010; Vadalà et al. 2008). Interestingly, hypoxia has also proven to facilitate discogenic MSC differentiation induced with transforming growth factor β (Feng et al. 2011a; Li et al. 2014; Naqvi and Buckley 2015; Risbud et al. 2004; Stoyanov et al. 2011).

As IVD is considered as a fibrocartilaginous tissue, most of the aforementioned studies used major markers of chondrocytic cells (i.e., gene expression of *Sox9, type II collagen*, and *aggrecan* and glycosaminoglycan synthesis) to characterize stem cell discogenic differentiation. Hyaline cartilage and NP, however, have a distinct embryological origin and display significant differences in tissue composition. For instance, the ratio of glycosaminoglycans to collagens is much higher, both at the protein and at the gene levels, in NP than in hyaline cartilage (e.g., 27:1 for young, human NP vs. 5:1 for articular cartilage at the protein level) (Clouet et al. 2009; Minogue et al. 2010a; Mwale, Roughley, and Antoniou 2004). Using chondrocyte markers to characterize NP-differentiated stem cell is somewhat an oversimplification, and it is therefore difficult to conclude on the true efficacy of the proposed differentiation protocols. Accordingly, recent efforts have been undertaken to identify more specific markers of healthy NP cells. So far, no specific marker has been found, but immunohistological analyses have shown that young, human NPs exhibit high expression of cytokeratin-8/-18/-19 (Rutges et al. 2010; Stosiek, Kasper, and Karsten 1988; Weiler et al. 2010). Microarray investigations confirmed that these markers are much highly expressed in NP than in articular cartilage or in AF, as described in Table 4.5. Interestingly, these studies also revealed a high expression of Brachyury (T) in NP cells, an unequivocal marker of a notochordal origin. Table 4.5 discloses the recommendations for the phenotyping of young, healthy NP cells established in 2015 by the Spine Research Interest Group at the 2014 Annual ORS Meeting (Risbud et al. 2015).

TABLE 4.5
Potential Phenotypic Markers for Healthy NP Cells Found in at Least Three Species and Included in the Recommendation of the Spine Research Interest Group at the 2014 Annual ORS Meeting Marker Recommended

NP Markers	Species	Comparison	Level Gene	Protein	References
Brachyury (T)	Rat	NP vs. AF	×	×	Tang et al. 2012
	Cow	NP vs. AC	×		Minogue et al. 2010b
	Human	NP vs. AC	×		Minogue et al. 2010b
Cytokeratin 8	Rat	NP vs. AC/AF	×		Lee et al. 2007
	Cow	NP vs. AC	×		Minogue et al. 2010b
	Human	<40 vs. >40 years		×	Stosiek, Kasper, and Karsten 1988
	Human	<30 vs. >30 years		×	Weiler et al. 2010
Cytokeratin 18	Dog	NP vs. AC	×	×	Sakai et al. 2009
	Cow	NP vs. AC	×		Minogue et al. 2010b
	Human	<40 vs. >40 years		×	Stosiek, Kasper, and Karsten 1988
	Human	<30 vs. >30 years		×	Weiler et al. 2010
Cytokeratin 19	Rat	NP vs. AC/AF	×	×	Lee et al. 2007
	Cow	NP vs. AC	×		Minogue et al. 2010b
	Human	<40 vs. >40 years		×	Stosiek, Kasper, and Karsten 1988
	Human	<30 vs. >30 years		×	Weiler et al. 2010
	Human	<30 vs. >30 years			Rutges et al. 2010
Stabilized HIF1 under normoxia	Rat	NP vs. AF		×	Risbud et al. 2006
	Sheep	NP vs. AF		×	Risbud et al. 2006
	Human	NP vs. AF		×	Risbud et al. 2006

Note: AC, articular cartilage.

4.6.4 What Is the Best Adult Stem Cell Source?

One issue of debate into cell therapy for IVD degeneration pertains to the selection of the best tissue origin for adult stem cells. Currently, there is an increasing consensus stating that adult stem cells originated from a particular tissue preferentially differentiate into the type of cells residing in this tissue. In accordance with this theory, IVD-MSCs displayed a higher glycosaminoglycan synthesis and *aggrecan* and *type II collagen* gene expression than their bone marrow- or adipose tissue-derived counterparts under standard chondrogenic culture conditions (Liu et al. 2011; Mizrahi et al. 2013; Shi et al. 2015). IVD-MSCs also demonstrated a higher proliferative capacity and NP matrix-related gene expression than AT-MSCs in conditions mimicking the IVD microenvironment (Han et al. 2014; Li et al. 2014). IVD could therefore be the most promising source of stem cells for IVD repair. When considering clinical applications of tissue engineering or cell-based therapies, however, the regenerative cell source should fulfill several criteria that include (i) minimal donor site morbidity, (ii) easy access to tissue source, and (iii) high frequency of regenerative cells. In that regard, IVD-MSCs display the same impediments as NP cells, which have a high donor site morbidity due to the anatomic position of the IVD-MSC niche, and a paucity of IVD-MSCs that further decreases with age (Sakai et al. 2012; Yasen et al. 2013; Zhao et al. 2017) and IVD degeneration grade (Jia et al. 2017; Mizrahi et al. 2013). When considering autologous cell-based therapies, MSCs derived from either bone marrow or adipose tissue, remain therefore, better candidates as they fulfill the aforementioned criteria for a regenerative cell source. It is worth noting, however, that IVD-MSCs still constitute an interesting cell source to repopulate the NP through recruitment with physiological morphogens and/or mitotic growth factors. This therapeutic approach is discussed in Chapter 6.

Interestingly, delivery of allogeneic (Acosta et al. 2011; Hee et al. 2010; Leung et al. 2014; Sobajima et al. 2008; Subhan et al. 2014; Vadalà et al. 2012; Yang et al. 2009, 2010; Zhang et al. 2011) as well as xenogeneic (Allon et al. 2010; Chen et al. 2016; Henriksson et al. 2009b; Jeong et al. 2010; Marfia et al. 2014; Tao et al. 2016; Wang et al. 2014; Wei et al. 2009) MSCs into NPs were used in small and large animal models, and results showed that the injected MSCs could survive up to several months and participate in IVD repair. Additionally, no inflammatory cells infiltrated the injected NP, even though the animals had not received immunosuppressive treatment, providing evidence of the immune-privileged aspect of this avascular biologic structure. The feasibility of using allogenic MSCs, as source of regenerative cells, for IVD repair is further supported by a few completed and ongoing phase 2/3 clinical studies using allogenic BM-MSCs, and showing an improvement in pain and disability scores (#NCT01290367, #NCT01860417, #NCT02529566, and #NCT02412735) (see Section 4.5). The interest in using allogenic MSCs resides in the possibility to create "off-the-shelf" regenerative products, prepared following good manufacturing practices, that contain well-characterized MSCs. Such therapeutic strategy, however, also presents several disadvantages linked to the ethical aspect of donor tissue collection and to the potential pathogen transmission.

4.6.5 What Is the Optimal Dose of Adult Stem Cells?

Another unanswered question relates to the optimal number of adult stem cells to inject in order to achieve IVD repair. A wide range of dosage of MSCs, indeed, has been used and reported to successfully lessen IVD degeneration in both preclinical and clinical studies. In a rabbit model, for example, the number of injected MSCs reported in the literature ranges from 5000 (Leung et al. 2014) to 4 million (Yang et al. 2010) cells, with both doses leading to improved disc height and MRI grading. The few studies actually investigating a potential dose effect, unfortunately, do not help to a better comprehension as they reported conflicting results. There are some indications that a low dose of MSCs leads to a more sustained recovery in rabbit (Leung et al. 2014) and sheep (Ghosh et al. 2012) models (5000 and 0.5 million cells, respectively). Similarly, a recent report on a phase 2 clinical trial; comparing injection of two different doses (6 million or 18 million cells) of immunoselected allogeneic BM-MSCs in combination with hyaluronic acid, indicated best clinical outcomes with the lowest cell dosage (Mesoblast Inc. 2017). This could be explained by the fact that the limited nutrient supply of IVD can only support a restricted number of exogenous cells. Still, another study disclaimed no clear effects of different MSC doses (0.1 million, 1 million, and 10 million cells) on disc height and hydration of canine, degenerated IVDs, although the AF structure appeared to be better maintained with the higher doses (Serigano et al. 2010). To add to the confusion, two other clinical trials described a potential correlation between injected cell number (15 million to 52 million BM-MSCs or BM concentrate containing < or > 2000 CFU-F/mL) and clinical scores (Elabd et al. 2016; Pettine et al. 2015). All together, these data indicate that supplementary studies are needed to determine the optimal MSC number for IVD regeneration.

4.6.6 How Should Adult Stem Cells Be Delivered?

The choice of a suitable cell carrier remains an important question for cell-based therapies as it may play several roles in the therapy's success. The carrier may support the MSC survival and function in the IVD microenvironment and act as a framework to guide tissue ingrowth. By locally retaining the injected MSCs within the NP, the carrier may also limit the MSC leakage and the resulting osteophyte formation, a drawback observed with MSCs injected with saline solution only (Sobajima et al. 2008; Vadalà et al. 2012). Last but not the least, the carrier may provide an immediate mechanical support to the IVD, provided that the biomaterial has sufficient load-bearing properties. So far, adult stem cells have often been delivered *in vivo* within a hydrogel, such as fibrin (Acosta et al. 2011; Allon et al. 2010; Yang et al. 2010), hyaluronan (Chun et al. 2012; Crevensten et al. 2004; Ganey et al. 2009; Ghosh et al. 2012; Omlor et al. 2014; Subhan et al. 2014), or attelo-collagen (Hang et al. 2017; Hee et al. 2010; Sakai et al. 2003, 2005, 2006), but a number of other synthetic and natural biomaterials are currently proposed as cell carriers for IVD repair and are discussed in Chapter 5.

One of the major stumbling blocks in cell-based therapies for IVD repair, however, is the cell delivery protocol into the NP structure, which commonly relies on a trans-annular puncture. Since AF has a limited healing potential, most of the proposed therapies may actually initiate or worsen IVD degeneration. AF puncture is, after all, one of the validated methods used to induce IVD degeneration in animal models. Consistency of degeneration initiation, however, appears to be related to needle size, with larger ones resulting in a greater degeneration (Elliott et al. 2008; Hsieh et al. 2009; Keorochana et al. 2010). Similar observations have been noticed in asymptomatic patients submitted to discograms: over a 10-year period, greater IVD degeneration was observed in patients who had discograms with a 22-G needle compared with those who had discograms with a 25-G needle (Carragee et al. 2009). Although small needle size can be used to minimize the incidence of IVD degeneration after AF puncture, the risk to potentially harm patients persists and complementary methods should be sought. One option is to capitalize on the homing property of intravenously injected MSCs toward injured tissues (Cornelissen et al. 2015) to bypass a direct cell injection and keep the AF intact. This approach is discussed in Chapter 6. Another option is to repair the AF puncture after cell injection. Attempts to close punctured AF by sutures or annuloplasty devices have shown a tendency to improve, although not significantly, IVD healing after hernia discectomy (Ahlgren et al. 2000; Bailey et al. 2013; Bron et al. 2010). More recently, tissue engineering approaches have also been proposed for repairing damaged AF (for recent reviews, see Guterl et al. 2013 and Sharifi et al. 2015). Interestingly, MSCs, in association with pentosan polysulfate and a gelatin/fibrin scaffold, were used to close damaged AF in a sheep model and to successfully reduce the resulting IVD degeneration (Oehme et al. 2014).

4.6.7 WHAT IS THE TARGET PATIENT PROFILE?

A last concern of IVD regenerative therapies refers to the selection of patients who may benefit from adult stem cell-based therapies.

Ideally, regenerative therapies should be applied to patients with early to moderate stages of IVD degeneration, when the adjacent structures (e.g., facet joints) are still intact and are not a source of pain by themselves and when the resident IVD cells are still able to respond to the provided exogenous stimuli (e.g., biomolecules, cells). Today, however, it is still difficult to diagnose such mild conditions, especially in asymptomatic patients, due to the lack of appropriate diagnostic tools. As this active field of research has been progressing, we could anticipate having, in the near future, screening techniques for genetic predisposition of IVD degeneration or imaging techniques sensitive enough to detect the early signs of IVD degeneration that will allow treating patients in the earliest stages of the pathology.

Realistically, adult stem cell-based therapies could be useful to treat young patients diagnosed with discogenic pain and a moderate degeneration of IVD whose IVD and adjacent structure integrity remain preserved. Segments adjacent to IVD submitted to spinal fusion or arthroplasty could also be cell-treated to prevent their further degeneration, both the cell injection and the fusion/IVD replacement steps being carried out during the same surgical procedure. A final potential application of cell-based therapies is AF, and potentially NP, repair after hernia discectomy.

4.7 CONCLUDING REMARKS

As demonstrated by the animal and clinical studies presented in this chapter, adult stem cell-based therapies for IVD repair have made significant progress over the last decade and appear as a promising therapeutic for the complex disease that is IVD degeneration. This field of research is still in its infancy, but several clinical trials have already disclosed promising outcomes on cell therapy efficacy for reducing back pain and disability in treated patients. The clinical outcomes, however, also indicated only a marginal improvement in disc hydration and no significant change in disc height. From the animal studies, it appears that adult stem cell transplantation decelerated or arrested the course of IVD degeneration but did not reverse the degenerative process. Such limitation prompts the scientific community to seek further research improvements on adult stem cell-based protocols so that they can be clinically useful for the treatment of back pain.

Among the numerous issues that the scientists must address, the fate of the injected cells should be clarified. Whether stem cell preconditioning or predifferentiation is needed for their prolonged survival and proper function in the harsh IVD microenvironment should also be elucidated. Likewise, innovative cell carriers that enhance stem cell survival and function, but also contribute to the biomechanical function of the IVD, should be developed. Furthermore, new AF sealing techniques should also be established and tested to minimize leakage of injected cells and improve IVD repair.

The rate of advancement in the field of IVD repair, however, suggests that all these hurdles will be overcome in time. In the meantime, the IVD will remain a challenging and intricate tissue to repair that will keep puzzling and fascinating biologists and engineers for many years.

REFERENCES

Acosta, F.L., L. Metz, H.D. Adkisson et al. 2011. "Porcine intervertebral disc repair using allogeneic juvenile articular chondrocytes or mesenchymal stem cells." *Tissue Eng Part A* 17 (23–24):3045–55.

Ahlgren, B.D., W. Lui, H.N. Herkowitz, M.M. Panjabi, and J.P. Guiboux. 2000. "Effect of anular repair on the healing strength of the intervertebral disc: A sheep model." *Spine (Phila Pa 1976)* 25 (17):2165–70.

Alini, M., S.M. Eisenstein, K. Ito et al. 2008. "Are animal models useful for studying human disc disorders/degeneration?" *Eur Spine J* 17 (1):2–19.

Allon, A.A., N. Aurouer, B.B. Yoo et al. 2010. "Structured coculture of stem cells and disc cells prevent disc degeneration in a rat model." *Spine J* 10 (12):1089–97.

Antoniou, J., C.N. Demers, G. Beaudoin, T. Goswami, and F. Mwale. 2004. "Apparent diffusion coefficient of intervertebral discs related to matrix composition and integrity." *Magn Reson Imaging* 22:963–72.

Antoniou, J., T. Steffen, F. Nelson et al. 1996. "The human lumbar intervertebral disc: Evidence for changes in the biosynthesis and denaturation of the extracellular matrix with growth, maturation, aging, and degeneration." *J Clin Invest* 98 (4):996–1003.

Bailey, A., A. Araghi, S. Blumenthal, and G. Huffmon. 2013. "Prospective, multicenter, randomized, controlled study of annular repair in lumbar discectomy: Two-year follow-up." *Spine (Phila Pa 1976)* 38 (14):1161–9.

Baraniak, PR., and T.C. McDevitt. 2010. "Paracrine actions in stem cells and tissue regeneration." *Regen Med* 5 (1):121–43.

Bartels, E.M., J.C.T. Fairbank, C.P. Winlove, and J.P.G. Urban. 1998. "Oxygen and lactate concentrations measured *in vivo* in the intervertebral discs of patients with scoliosis and back pain." *Spine (Phila Pa 1976)* 23 (1):1–8.

Battié, M.C., and T. Videman. 2006. "Lumbar disc degeneration: Epidemiology and genetics." *J Bone Joint Surg Am.* 88-A (Suppl 2):3–9.

Bendtsen, M., C.E. Bunger, X. Zou, C. Foldager, and H.S. Jorgensen. 2011. "Autologous stem cell therapy maintains vertebral blood flow and contrast diffusion through the endplate in experimental intervertebral disc degeneration." *Spine (Phila Pa 1976)* 36 (6):E373–9.

Benneker, L.M., P.F. Heini, M. Alini, S.E. Anderson, and K. Ito. 2005. "2004 Young Investigator Award Winner: Vertebral endplate marrow contact channel occlusions and intervertebral disc degeneration." *Spine (Phila Pa 1976)* 30 (2):167–73.

Bertolini, F., V. Lohsiriwat, J.-Y. Petit, and M.G. Kolonin. 2012. "Adipose tissue cells, lipotransfer and cancer: A challenge for scientists, oncologists and surgeons." *Biochim Biophys Acta Rev Cancer* 1826 (1):209–14.

Bertolo, A., M. Mehr, N. Aebli et al. 2012. "Influence of different commercial scaffolds on the *in vitro* differentiation of human mesenchymal stem cells to nucleus pulposus-like cells." *Eur Spine J* 21 (Suppl 6):S826–38.

Blanco, J.F., I.F. Graciani, F.M. Sanchez-Guijo et al. 2010. "Isolation and characterization of mesenchymal stromal cells from human degenerated nucleus pulposus: Comparison with bone marrow mesenchymal stromal cells from the same subjects." *Spine (Phila Pa 1976)* 35 (26):2259–65.

Boos, N., S. Weissbach, H. Rohrbach et al. 2002. "Classification of age-related changes in lumbar intervertebral discs." *Spine (Phila Pa 1976)* 27 (23):2631–44.

Bowles, R.D., and L.A. Setton. 2017. "Biomaterials for intervertebral disc regeneration and repair." *Biomaterials* 129:54–67.

Bron, J.L., A.J. Van Der Veen et al. 2010. "Biomechanical and *in vivo* evaluation of experimental closure devices of the annulus fibrosus designed for a goat nucleus replacement model." *Eur Spine J* 19 (8):1347–55.

Cai, F., X.-T. Wu, X.-H. Xie et al. 2015. "Evaluation of intervertebral disc regeneration with implantation of bone marrow mesenchymal stem cells (BMSCs) using quantitative T2 mapping: A study in rabbits." *Int Orthop* 39 (1):149–59.

Caplan, A.I. 2015. "Adult mesenchymal stem cells: When, where, and how." *Stem Cells Int* 2015:628767.

Caplan, A.I., and D. Correa. 2011. "The MSC: An injury drugstore." *Cell Stem Cell* 9 (1):11–5.

Carragee, E., A. Don, E. Hurwitz et al. 2009. "Does discography cause accelerated progression of degeneration changes in the lumbar disc: A ten-year matched cohort study." *Spine (Phila Pa 1976)* 34 (21):2338–45.

Chen, J., E.J. Lee, L. Jing et al. 2013. "Differentiation of mouse induced pluripotent stem cells (iPSCs) into nucleus pulposus-like cells *in vitro*." *PLoS One* 8 (9):1–14.

Chen, X., L. Zhu, G. Wu et al. 2016. "A comparison between nucleus pulposus-derived stem cell transplantation and nucleus pulposus cell transplantation for the treatment of intervertebral disc degeneration in a rabbit model." *Int J Surg* 28 (253):77–82.

Chun, H.-J., Y.S. Kim, B.K. Kim et al. 2012. "Transplantation of human adipose–derived stem cells in a rabbit model of traumatic degeneration of lumbar discs." *World Neurosurg* 78 (3-4):364–71.

Clarke, L.E., J.C. McConnell, M.J. Sherratt et al. 2014. "Growth differentiation factor 6 and transforming growth factor-beta differentially mediate mesenchymal stem cell differentiation, composition, and micromechanical properties of nucleus pulposus constructs." *Arthritis Res Ther* 16 (2):R67.

Clouet, J., G. Grimandi, M. Pot-Vaucel et al. 2009. "Identification of phenotypic discriminating markers for intervertebral disc cells and articular chondrocytes." *Rheumatology (Oxford)* 48 (11):1447–50.

Colombier, P., J. Clouet, C. Boyer et al. 2016. "TGF-β1 and GDF5 act synergistically to drive the differentiation of human adipose stromal cells towards nucleus pulposus-like cells." *Stem Cells* 34 (3):653–67.

Coric, D., K. Pettine, A. Sumich, and M.O. Boltes. 2013. "Prospective study of disc repair with allogeneic chondrocytes presented at the 2012 Joint Spine Section Meeting." *J Neurosurg Spine* 18 (1):85–95.

Cornelissen, A.S., M.W. Maijenburg, M.A. Nolte, and C. Voermans. 2015. "Organ-specific migration of mesenchymal stromal cells: Who, when, where and why?" *Immunol Lett* 168 (2):159–69.

Crevensten, G., A.J.L. Walsh, D. Ananthakrishnan et al. 2004. "Intervertebral disc cell therapy for regeneration: Mesenchymal stem cell implantation in rat intervertebral discs." *Ann Biomed Eng* 32 (3):430–4.

de Vries, S.A., E. Potier, M. van Doeselaar et al. 2015. "Conditioned medium derived from notochordal cell-rich nucleus pulposus tissue stimulates matrix production by canine nucleus pulposus cells and bone marrow derived stromal cells." *Tissue Eng Part A* 21 (5–6):1077–84.

Dennis, J.E., A. Merriam, A. Awadallah et al. 1999. "A quadripotential mesenchymal progenitor cell isolated from the marrow of an adult mouse." *J Bone Min Res.* 14 (5):700–9.

Deschepper, M., M. Manassero, K. Oudina et al. 2013. "Proangiogenic and prosurvival functions of glucose in human mesenchymal stem cells upon transplantation." *Stem Cells.* 31 (3):526–35.

Diamant, B., J. Karlsson, and A. Nachemson. 1968. "Correlation between lactate levels and pH in discs in patients with lumbar rhizopaties." *Experientia* 24 (12):1195–6.

Dominici, M., K. Le Blanc, I. Mueller et al. 2006. "Minimal criteria for defining multipotent mesenchymal stromal cells. The International Society for Cellular Therapy position statement." *Cytotherapy* 8 (4):315–7.

Elabd, C., C.J. Centeno, J.R. Schultz et al. 2016. "Intra-discal injection of autologous, hypoxic cultured bone marrow-derived mesenchymal stem cells in five patients with chronic lower back pain: A long-term safety and feasibility study." *J Transl Med* 14:253.

Elliott, D., C. Yerramalli, J. Beckstein et al. 2008. "The effect of relative needle diameter in puncture and sham injection animal models of degeneration." *Spine (Phila Pa 1976)* 33 (6):588–96.

Farrell, M.J., J.I. Shin, L.J. Smith, and R.L. Mauck. 2015. "Functional consequences of glucose and oxygen deprivation on engineered mesenchymal stem cell-based cartilage constructs." *Osteoarthritis Cartilage* 23 (1):134–42.

Feng, G., X. Jin, J. Hu et al. 2011a. "Effects of hypoxias and scaffold architecture on rabbit mesenchymal stem cell differentiation towards a nucleus pulposus-like phenotype." *Biomaterials* 32 (32):8182–9.

Feng, G., X. Zhao, H. Liu et al. 2011b. "Transplantation of mesenchymal stem cells and nucleus pulposus cells in a degenerative disc model in rabbits: A comparison of 2 cell types as potential candidates for disc regeneration." *J Neurosurg Spine* 14 (3):322–9.

Freeman, B.J.C., and J. Davenport. 2006. "Total disc replacement in the lumbar spine: A systematic review of the literature." *Eur Spine J* 15 (suppl 3):S439–47.

Freemont, A.J. 2009. "The cellular pathobiology of the degenerate intervertebral disc and discogenic back pain." *Rheumatology (Oxford)* 48 (1):5–10.

Friedenstein, A.J., K.V. Petrakova, A.I. Kurolesova, and G.P. Frolova. 1968. "Heterotopic of bone marrow. Analysis of precursor cells for osteogenic and hematopoietic tissues." *Transplantation* 6 (2):230–47.

Ganey, T., W.C. Hutton, T. Moseley, M.H. Hedrick, and H.J. Meisel. 2009. "Intervertebral disc repair using adipose tissue-derived stem and regenerative cells. Experiments in a canine model." *Spine (Phila Pa 1976)* 34 (21):2297–304.

Gantenbein-Ritter, B., L.M. Benneker, M. Alini, and S. Grad. 2011. "Differential response of human bone marrow stromal cells to either TGF-β (1) or rhGDF-5." *Eur Spine J* 20 (6):962–71.

Gao, F., S.M. Chiu, D.A.L. Motan et al. 2016. "Mesenchymal stem cells and immunomodulation: Current status and future prospects." *Cell Death Dis* 7 (1):e2062.

Ghiselli, G., J.C. Wang, N.N. Bhatia, W.K. Hsu, and E.G. Dawson. 2004. "Adjacent segment degeneration in the lumbar spine." *J Bone Joint Surg Am* 86 (7):1497–503.

Ghosh, P., R. Moore, B. Vernon-Roberts et al. 2012. "Immunoselected STRO-3+ mesenchymal precursor cells and restoration of the extracellular matrix of degenerate intervertebral discs." *J Neurosurg Spine* 16 (5):479–88.

Gorensek, M., C. Jaksimovic, N. Kregar-Velikonja et al. 2004. "Nucleus pulposus repair with cultured autologous elastic cartilage derived chondrocytes." *Cell Mol Biol Lett* 9 (2):363–73.

Gruber, H.E., B. Gordon, C. Williams et al. 2009. "A new small animal model for the study of spine fusion in the sand rat: Pilot studies." *Lab Anim* 43 (3):272–7.

Guterl, C.C., E.Y. See, S.B. Blanquer et al. 2013. "Challenges and strategies in the repair of ruptured annulus fibrosus." *Eur Cells Mater* 25:1–21.

Han, B., H.C. Wang, H. Li et al. 2014. "Nucleus pulposus mesenchymal stem cells in acidic conditions mimicking degenerative intervertebral discs give better performance than adipose tissue-derived mesenchymal stem cells." *Cells Tissues Organs* 199 (5–6): 342–52.

Hang, D., F. Li, W. Che et al. 2017. "One-stage positron emission tomography and magnetic resonance imaging to assess mesenchymal stem cell survival in a canine model of intervertebral disc degeneration." *Stem Cells Dev* 18 (26):1334–43.

Hass, R., C. Kasper, S. Bohm, and R. Jacobs. 2011. "Different populations and sources of human mesenchymal stem cells (MSC): A comparison of adult and neonatal tissue-derived MSC." *Cell Commun Signal* 9:12.

Hee, H.T., H.D. Ismail, C.T. Lim, J.C.H. Goh, and H.K. Wong. 2010. "Effects of implantation of bone marrow mesenchymal stem cells, disc distraction and combined therapy on reversing degeneration of the intervertebral disc." *J Bone Joint Surg Br* 92 (5):726–36.

Henriksson, H., M. Thornemo, C. Karlsson et al. 2009a. "Identification of cell proliferation zones, progenitor cells and a potential stem cell niche in the intervertebral disc region: A study in four species." *Spine (Phila Pa 1976)* 34 (21):2278–87.

Henriksson, H.B., T. Svanvik, M. Jonsson et al. 2009b. "Transplantation of human mesenchymal stems cells into intervertebral discs in a xenogeneic porcine model." *Spine (Phila Pa 1976)* 34 (2):141–8.

Hiyama, A., J. Mochida, T. Iwashina et al. 2008. "Transplantation of mesenchymal stem cells in a canine disc degeneration model." *J Orthop Res* 26 (5):589–600.

Ho, G., V.Y.L. Leung, K.M.C. Cheung, and D. Chan. 2008. "Effect of severity of intervertebral disc injury on mesenchymal stem cell-based regeneration." *Connect Tissue Res* 49 (1):15–21.

Hohaus, C., T.M. Ganey, Y. Minkus, and H.J. Meisel. 2008. "Cell transplantation in lumbar spine disc degeneration disease." *Eur Spine J* 17 (Suppl. 4):492–503.

Holm, S., A. Maroudas, J.P.G. Urban, G. Selstam, and A. Nachemson. 1981. "Nutrition of the intervertebral disc: Solute transport and metabolism." *Connect Tissue Res* 8 (2):101–9.

Holm, S., and A. Nachemson. 1988. "Nutrition of the intervertebral disc: Acute effects of cigarette smoking. An experimental animal study." *Ups J Med Sci* 83 (1):91–9.

Holm, S., G. Selstam, and A. Nachemson. 1982. "Carbohydrate metabolism and concentration profiles of solutes in the canine lumbar intervertebral disc." *Acta Physiol Scand* 115:147–56.

Horwitz, E.M., and M. Dominici. 2008. "How do mesenchymal stromal cells exert their therapeutic benefit?" *Cytotherapy* 10 (8):771–4.

Hsieh, A., D. Hwang, D. Ryan, A. Freeman, and H. Kim. 2009. "Degenerative annular changes induced by puncture are associated with insufficiency of disc biomechanical function." *Spine (Phila Pa 1976)* 34 (10):998–1005.

Huang, C.-Y.C., and W.Y. Gu. 2008. "Effects of mechanical compression on metabolism and distribution of oxygen and lactate in intervertebral disc." *J Biomech* 41 (6):1184–96.

Hunter, C.J., J.R. Matyas, and N.A. Duncan. 2003. "The notochordal cell in the nucleus pulposus: A review in the context of tissue engineering." *Tissue Eng* 9 (4):667–77.

Iatridis, J.C., J.J. MacLean, M. O'Brien, and I. Stokes. 2007. "Measurements of proteoglycan and water content distribution in human lumbar intervertebral discs." *Spine (Phila Pa 1976)* 32 (14):1493–7.

Jackson, A.R., C.-Y.C. Huang, M.D. Brown, and W.Y. Gu. 2011. "3D finite element analysis of nutrient distributions and cell viability in the intervertebral disc: Effects of deformation and degeneration." *J Biomech Eng* 133 (9):091006.

Jeong, J.H., J.H. Lee, E.S. Jin et al. 2010. "Regeneration of intervertebral discs in a rat disc degeneration model by implanted adipose-tissue-derived stromal cells." *Acta Neurochir (Wien)* 152 (10):1771–7.

Jia, Z., P. Yang, Y. Wu et al. 2017. "Comparison of biological characteristics of nucleus pulposus mesenchymal stem cells derived from non-degenerative and degenerative human nucleus pulposus." *Exp Ther Med* 13:3574–80.

Kauppila, L., T. McAlindon, S. Evans et al. 1997. "Disc degeneration/back pain and calcification of the abdominal aorta. A 25-year follow-up study in Framingham." *Spine (Phila Pa 1976)* 22 (14):10642–7.

Keating, A. 2012. "Mesenchymal stromal cells: New directions." *Cell Stem Cell*. 10 (6):709–16.

Keorochana, G., J.S. Johnson, C.E. Taghavi et al. 2010. "The effect of needle size inducing degeneration in the rat caudal disc: Evaluation using radiograph, magnetic resonance imaging, histology, and immunohistochemistry." *Spine J* 10 (11):1014–23.

Kepler, C.K., R.K. Ponnappan, C.A. Tannoury, M.V. Risbud, and D.G. Anderson. 2013. "The molecular basis of intervertebral disc degeneration." *Spine J* 13 (3):318–30.

Kim, D.H., S.-H. Kim, S.-J. Heo et al. 2009. "Enhanced differentiation of mesenchymal stem cells into NP-like cells via 3D co-culturing with mechanical stimulation." *J Biosci Bioeng* 108 (1):63–7.

Kitano, T., J. Zerwekh, Y. Usui et al. 1993. "Biochemical changes associated with the symptomatic human intervertebral disk." *Clin Orthop Relat Res* 293:372–7.

Kluba, T., T. Niemeyer, C. Gaissmaier, and T. Gründer. 2005. "Human annulus fibrosis and nucleus pulposus cells of the intervertebral disc: Effect of degeneration and culture system on cell phenotype." *Spine (Phila Pa 1976)* 30 (24):2743–48.

Le Maitre, C.L., A.J. Freemont, and J.A. Hoyland. 2004. "Localization of degradative enzymes and their inhibitors in the degenerate human intervertebral disc." *J Pathol* 204 (1):47–54.

Le Maitre, C.L., A. Freemont, and J. Hoyland. 2007a. "Accelerated cellular senescence in degenerate intervertebral discs: A possible role in the pathogenesis of intervertebral disc degeneration." *Arthritis Res Ther* 9 (3):R45.

Le Maitre, C.L., A. Pockert, D.J. Buttle, A.J. Freemont, and J.A. Hoyland. 2007b. "Matrix synthesis and degradation in human intervertebral disc degeneration." *Biochem Soc Trans* 35 (Pt 4):652–5.

Lee, C.R., D. Sakai, T. Nakai et al. 2007. "A phenotypic comparison of intervertebral disc and articular cartilage cells in the rat." *Eur Spine J* 16 (12):2174–85.

Leung, V.Y.L., D.M.K. Aladin, F. Lv et al. 2014. "Mesenchymal stem cells reduce intervertebral disc fibrosis and facilitate repair." *Stem Cells* 32:2164–77.

Li, H., Y. Tao, C. Liang et al. 2014. "Influence of hypoxia in the intervertebral disc on the biological behaviors of rat adipose-and nucleus pulposus-derived mesenchymal stem cells." *Cells Tissues Organs* 198 (4):266–77.

Li, Z., M. Peroglio, M. Alini, and S. Grad. 2015. "Potential and limitations of intervertebral disc endogenous repair." *Curr Stem Cell Res Ther* 10 (4):329–38.

Liu, C., Q. Guo, J. Li et al. 2014. "Identification of rabbit annulus fibrosus-derived stem cells." *PLoS One* 9 (9):e108239.

Liu, L.T., B. Huang, C.Q. Li et al. 2011. "Characteristics of stem cells derived from the degenerated human intervertebral disc cartilage endplate." *PLoS One* 6 (10): e26285.

Liu, Y., M.N. Rahaman, and B.S. Bal. 2014. "Modulating notochordal differentiation of human induced pluripotent stem cells using natural nucleus pulposus tissue matrix." *PLoS One* 9 (7):e100885.

Lu, Z.F., B. Zandieh Doulabi, P.I. Wuisman, R.A. Bank, and M.N. Helder. 2007. "Differentiation of adipose stem cells by nucleus pulposus cells: Configuration effect." *Biochem Biophys Res Commun* 359 (4):991–6.

Marfia, G., R. Campanella, S.E. Navone et al. 2014. "Potential use of human adipose mesenchymal stromal cells for intervertebral disc regeneration: A preliminary study on biglycan-deficient murine model of chronic disc degeneration." *Arthritis Res Ther* 16 (5):457.

Maroudas, A., R.A. Stockwell, A. Nachemson, and J. Urban. 1975. "Factors involved in the nutrition of the human lumbar intervertebral disc: Cellularity and diffusion of glucose *in vitro*." *J Anat* 120 (Pt 1):113–30.

Meisel, H.J., T. Ganey, W.C. Hutton et al. 2006. "Clinical experience in cell-based therapeutics: Intervention and outcome." *Eur Spine J* 15 (Suppl. 3):397–405.

Meisel, H.J., V. Siodla, T. Ganey et al. 2007. "Clinical experience in cell-based therapeutics: Disc chondrocyte transplantation. A treatment for degenerated or damaged intervertebral disc. *Biomol Eng* 24 (1 Spec. Issue):5–21.

Mesoblast Inc. 2017. *Durable Three-Year Outcomes in Degenerative Disc Disease After a Single Injection of Mesoblast's Cell Therapy*. March 15, 5–7. https://globenewswire.com/news-release/2017/03/15/937833/0/en/Durable-Three-Year-Outcomes-In-Degenerative-Disc-Disease-After-a-Single-Injection-of-Mesoblast-s-Cell-Therapy.html

Mikkola, M., C. Olsson, J. Palgi et al. 2006. "Distinct differentiation characteristics of individual human embryonic stem cell lines." *BMC Dev Biol* 6:40.

Minogue, B.M., S.M. Richardson, L.A.H. Zeef, A.J. Freemont, and J.A. Hoyland. 2010a. "Characterization of the human nucleus pulposus cell phenotype and evaluation of novel marker gene expression to define adult stem cell differentiation." *Arthritis Rheum* 62 (12):3695–705.

Minogue, B.M., S.M. Richardson, L.A.H. Zeef, A.J. Freemont, and J.A. Hoyland. 2010b. "Transcriptional profiling of bovine intervertebral disc cells: Implications for identification of normal and degenerate human intervertebral disc cell phenotypes." *Arthritis Res Ther* 12 (1):R22.

Miyamoto, T., T. Muneta, T. Tabuchi et al. 2010a. "Intradiscal transplantation of synovial mesenchymal stem cells prevents intervertebral disc degeneration through suppression of matrix metalloproteinase-related genes in nucleus pulposus cells in rabbits." *Arthritis Res Ther* 12 (6):R206.

Mizrahi, O., D. Sheyn, W. Tawackoli et al. 2013. "Nucleus pulposus degeneration alters properties of resident progenitor cells." *Spine J* 13 (7):803–14.

Mochida, J., D. Sakai, Y. Nakamura et al. 2015. "Intervetrval disc repair with activated nucleus pulposus cell transplantation: A three-year, propesctive clinical study of its safety." *Eur Cells Mater* 29:202–12.

Mwale, F., P. Roughley, and J. Antoniou. 2004. "Distinction between the extracellular matrix of the nucleus pulposus and hyaline cartilage: A requisite for tissue engineering of intervertebral disc." *Eur Cells Mater* 8:58–63.

Naqvi, S.M., and C.T. Buckley. 2015. "Extracellular matrix production by nucleus pulposus and bone marrow stem cells in response to altered oxygen and glucose microenvironments." *J Anat* 227 (6):757–66.

Naqvi, S.M., and C.T. Buckley. 2016. "Bone marrow stem cells in response to intervertebral disc-like matrix acidity and oxygen concentration: Implications for cell-based regenerative therapy." *Spine (Phila Pa 1976)* 41 (9):743–50.

Niu, C.-C., L.-J. Yuan, S.-S. Lin, L.-H. Chen, and W.-J. Chen 2009. "Mesenchymal stem cell and nucleus pulposus cell coculture modulates cell profile." *Clin Orthop Relat Res.* 467 (12):3263–72.

Oehme, D., P. Ghosh, S. Shimmon et al. 2014. "Mesenchymal progenitor cells combined with pentosan polysulfate mediating disc regeneration at the time of microdiscectomy: A preliminary study in an ovine model." *J Neurosurg Spine* 20 (June):657–69.

Ogata, K., and L.A. Whiteside. 1981. "1980 Volvo award winner in basic science. Nutritional pathways of the intervertebral disc. An experimental study using hydrogen washout technique." *Spine (Phila Pa 1976)* 6 (3):211–6.

Okuma, M., J. Mochida, K. Nishimura, K. Sakabe, and K. Seiki. 2000. "Reinsertion of stimulated nucleus pulposus cells retards intervertebral disc degeneration: An *in vitro* and *in vivo* experimental study." *J Orthop Res.* 18 (6):988–97.

Omlor, G.W., J. Fischer, K. Kleinschmitt et al. 2014. "Short-term follow-up of disc cell therapy in a porcine nucleotomy model with an albumin–hyaluronan hydrogel: *In vivo* and *in vitro* results of metabolic disc cell activity and implant distribution." *Eur Spine J* 23:1837–47.

Orozco, L., R. Soler, C. Morera et al. 2011. "Intervertebral disc repair by autologous mesenchymal bone marrow cells: A pilot study." *Transplantation* 92 (7):822–8.

Pang, X., H. Yang, and B. Peng. 2014. "Human umbilical cord mesenchymal stem cell transplantation for the treatment of chronic discogenic low back pain." *Pain Physician* 17 (3):525–30.

Peroglio, M., D. Eglin, L.M. Benneker, M. Alini, and S. Grad. 2013. "Thermoreversible hyaluronan-based hydrogel supports *in vitro* and *ex vivo* disc-like differentiation of human mesenchymal stem cells." *Spine J* 13 (11):1627–39.

Pettine, K.A., M.B. Murphy, R.K. Suzuki, and T.T. Sand. 2015. "Percutaneous injection of autologous bone marrow concentrate cells significantly reduces lumbar discogenic pain through 12 months." *Stem Cells* 33 (1):146–56.

Pittenger, M.F., A.M. Mackay, S.C. Beck et al. 1999. "Multilineage potential of adult human mesenchymal stem cells." *Science* 284 (5411):143–7.

Porada, C.D., E.D. Zanjani, and G. Almeida-Porad. 2006. "Adult mesenchymal stem cells: A pluripotent population with multiple applications." *Curr Stem Cell Res Ther* 1 (3): 365–69.

Potier, E., S. de Vries, M. van Doeselaar, and K. Ito. 2014. "Potential application of notochordal cells for intervertebral disc regeneration: An *in vitro* assessment." *Eur Cells Mater.* 28:68–80.

Raj, P.P. 2008. "Intervertebral disc: Pathophysiology-treatment." *Pain Pract* 8 (1):18–44.

Rajasekaran, S., J.N. Babu, R. Arun et al. 2004. "ISSLS prize winner: A study of diffusion in human lumbar discs: A serial magnetic resonance imaging study documenting the influence of the endplate on diffusion in normal and degenerate discs." *Spine (Phila Pa 1976)* 29 (23):2654–67.

Richardson, S.M., R.V. Walker, S. Parker et al. 2006. "Intervertebral disc cell-mediated mesenchymal stem cell differentiation." *Stem Cells* 24 (3):707–16.

Risbud, M.V., T.J. Albert, A. Guttapalli et al. 2004. "Differentiation of mesenchymal stem cells towards a nucleus pulposus-like phenotype *in vitro*: Implications for cell-based transplantation therapy." *Spine (Phila Pa 1976)* 29 (23):2627–32.

Risbud, M.V., A. Guttapalli, D.G. Stokes et al. 2006. "Nucleus pulposus cells express HIF-1 alpha under normoxic culture conditions: A metabolic adaptation to the intervertebral disc microenvironment." *J Cell Biochem* 98 (1):152–9.

Risbud, M.V., Z.R. Schoepflin, F. Mwale et al. 2015. "Defining the phenotype of young healthy nucleus pulposus cells: Recommendations of the spine research interest group at the 2014 annual ORS meeting." *J Orthop Res* 33 (3):283–93.

Risbud, M., A. Guttapalli, T. Tsai et al. 2007. "Evidence for skeletal progenitor cells in the degenerate human intervertebral disc." *Spine (Phila Pa 1976)* 32 (23):2537–44.

Ruan, D., Y. Zhang, D. Wang et al. 2012. "Differentiation of human Wharton's jelly cells toward nucleus pulposus-like cells after coculture with nucleus pulposus cells *in vitro*." *Tissue Eng Part A* 18 (1–2):167–75.

Rutges, J., L.B. Creemers, W. Dhert et al. 2010. "Variations in gene and protein expression in human nucleus pulposus in comparison with annulus fibrosus and cartilage cells: Potential associations with aging and degeneration." *Osteoarthritis Cartilage* 18 (3):416–23.

Sacchetti, B., A. Funari, S. Michienzi et al. 2007. "Self-renewing osteoprogenitors in bone marrow sinusoids can organize a hematopoietic microenvironment." *Cell* 131 (2):324–36.

Sakai, D., and G.B.J. Andersson. 2015. "Stem cell therapy for intervertebral disc regeneration: Obstacles and solutions." *Nat Rev Rheumatol* 11 (4):243–56.

Sakai, D., J. Mochida, T. Iwashina et al. 2006. "Regenerative effects of transplanting mesenchymal stem cells embedded in atelocollagen to the degenerated intervertebral disc." *Biomaterials* 27 (3):335–45.

Sakai, D., J. Mochida, T. Iwashina et al. 2005. "Differentiation of mesenchymal stem cells transplanted to a rabbit degenerative disc model: Potential and limitations for stem-cell therapy in disc regeneration." *Spine (Phila Pa 1976)* 30 (21):2379–87.

Sakai, D., J. Mochida, Y. Yamamoto et al. 2003. "Transplantation of mesenchymal stem cells embedded in Atelocollagen gel to the intervertebral disc: A potential therapeutic model for disc degeneration." *Biomaterials* 24 (20):3531–41.

Sakai, D., T. Nakai, J. Mochida, M. Alini, and S. Grad. 2009. "Differential phenotype of intervertebral disc cells: Microarray and immunohistochemical analysis of canine nucleus pulposus and annulus fibrosus. *Spine (Phila Pa 1976)* 34 (14):1448–56.

Sakai, D., Y. Nakamura, T. Nakai et al. 2012. "Exhaustion of nucleus pulposus progenitor cells with aging and degeneration of the intervertebral disc." *Nat Commun* 3:1264.

See, F., T. Seki, P.J. Psaltis et al. 2011. "Therapeutic effects of human STRO-3-selected mesenchymal precursor cells and their soluble factors in experimental myocardial ischemia." *J Cell Mol Med* 15 (10):2117–29.

Sélard, E., A. Shirazi-Adl, and J.P.G. Urban. 2003. "Finite element study of nutrient diffusion in the human intervertebral disc." *Spine (Phila Pa 1976)* 28 (17):1945–53.

Serigano, K., D. Sakai, A. Hiyama et al. 2010. "Effect of cell number on mesenchymal stem cell transplantation in a canine disc degeneration model." *J Orthop Res* 28 (10):1267–75.

Shang, J., H. Liu, J. Li, and Y. Zhou. 2014. "Roles of hypoxia during the chondrogenic differentiation of mesenchymal stem cells." *Curr Stem Cell Res Ther* 9 (2):141–7.

Sharifi, S., S. Bulstra, D.W. Grijpma, and R. Kuijer. 2015. "Treatment of the degenerated intervertebral disc; closure, repair and regeneration of the annulus fibrosus." *J Tissue Eng Regen Med* 9:1120–32.

Shi, R., F. Wang, X. Hong et al. 2015. "The presence of stem cells in potential stem cell niches of the intervertebral disc region: An *in vitro* study on rats." *Eur Spine J* 24 (11):2411–24.

Shirazi-Adl, A., M. Taheri, and J.P.G. Urban. 2010. "Analysis of cell viability in intervertebral disc: Effect of endplate permeability on cell population." *J Biomech* 43 (7):1330–36.

Sobajima, S., G. Vadala, A. Shimer et al. 2008. "Feasibility of a stem cell therapy for intervertebral disc degeneration." *Spine J* 8 (6):888–96.

Soukane, D., A. Shirazi-Adl, and J.P.G. Urban. 2007. "Computation of coupled diffusion of oxygen, glucose and lactic acid in an intervertebral disc." *J Biomech* 40:2645–54.

Steck, E., H. Bertram, R. Abel et al. 2005. "Induction of intervertebral disc-like cells from adult mesenchymal stem cells." *Stem Cells* 23 (3):403–11.

Stosiek, P., M. Kasper, and U. Karsten. 1988. "Expression of cytokeratin and vimentin in nucleus pulposus cells." *Differentiation* 39 (1):78–81.

Stoyanov, J.V., B. Gantenbein-Ritter, A. Bertolo et al. 2011. "Role of hypoxia and growth and differentiation factor-5 on differentiation of human mesenchymal stem cells towards intervertebral nucleus pulposus-like cells." *Eur Cells Mater* 21:533–47.

Strassburg, S., S.M. Richardson, A.J. Freemont, and J.A. Hoyland. 2010. "Co-culture induces mesenchymal stem cell differentiation and modulation of the degenerate human nucleus pulposus cell phenotype." *Regen Med* 5 (5):701–11.

Subhan, R.A., K. Puvanan, M.R. Murali et al. 2014. "Fluoroscopy assisted minimally invasive transplantation of allogenic mesenchymal stromal cells embedded in hystem reduces the progression of nucleus pulposus degeneration in the damaged interverbal disc: A preliminary study in rabbits." *Sci World J* 2014.

Takahashi, K., and S. Yamanaka. 2006. "Induction of pluripotent stem cells from mouse embryonic and adult fibroblast cultures by defined factors." *Cell* 126 (4):663–76.

Tam, V., I. Rogers, D. Chan, V.Y.L. Leung, and K.M.C. Cheung. 2014. "A comparison of intravenous and intradiscal delivery of multipotential stem cells on the healing of injured intervertebral disk." *J Orthop Res* 32 (6):819–25.

Tang, X., L. Jing, and J. Chen. 2012. "Changes in the molecular phenotype of nucleus pulposus cells with intervertebral disc aging." *PLoS One* 7 (12):e52020.

Tao, H., Y. Lin, G. Zhang, R. Gu, and B. Chen. 2016. "Experimental observation of human bone marrow mesenchymal stem-cell transplantation into rabbit intervertebral discs." *Biomed Rep* 5 (3):357–60.

Tao, Y.Q., C.Z. Liang, H. Li et al. 2013. "Potential of co-culture of nucleus pulposus mesenchymal stem cells and nucleus pulposus cells in hyperosmotic microenvironment for intervertebral disc regeneration." *Cell Biol Int* 37 (8):826–34.

Tekari, A., S.C.W. Chan, D. Sakai, S. Grad, and B. Gantenbein. 2016. "Angiopoietin-1 receptor Tie2 distinguishes multipotent differentiation capability in bovine coccygeal nucleus pulposus cells." *Stem Cell Res Ther* 7 (1):75.

Thomson, J.A., J. Itskovitz-Eldor, S.S. Shapiro et al. 1998. "Embryonic stem cell lines derived from human blastocysts." *Science*. 282 (5391):1145–7.

Trout, J.J., J.A. Buckwalter, and K.C. Moore. 1982. "Ultrastructure of the human intervertebral disc: II. Cells of the nucleus pulposus." *Anat Rec* 204 (4):307–14.

Vadalà, G., F. Russo, L. Ambrosio, M. Loppini, and V. Denaro. 2016. "Stem cells sources for intervertebral disc regeneration." *World J Stem Cells* 8 (5):185–201.

Vadalà, G., G. Sowa, M. Hubert et al. 2012. "Mesenchymal stem cells injection in degenerated intervertebral disc: Cell leakage may induce osteophyte formation." *J Tissue Eng Regen Med* 6 (5):348–55.

Vadalà, G., R.K. Studer, G. Sowa et al. 2008. "Coculture of bone marrow mesenchymal stem cells and nucleus pulposus cells modulate gene expression profile without cell fusion." *Spine (Phila Pa 1976)* 33 (8):870–6.

Verfaillie, C.M. 2002. "Adult stem cells: Assessing the case for pluripotency." *Trends Cell Biol* 12 (11):502–8.

Wang, F., R. Shi, F. Cai, Y.-T. Wang, and X.-T. Wu. 2015. "Stem cell approaches to intervertebral disc regeneration: Obstacles from the disc microenvironment." *Stem Cells Dev* 24 (21):2479–95.

Wang, H., Y. Zhou, T.W. Chu et al. 2016. "Distinguishing characteristics of stem cells derived from different anatomical regions of human degenerated intervertebral discs." *Eur Spine J* 25 (9):2691–704.

Wang, H., Y. Zhou, B. Huang et al. 2014. "Utilization of stem cells in alginate for nucleus pulposus tissue engineering." *Tissue Eng Part A* 20 (5–6):908–20.

Watanabe, K., J. Mochida, T. Nomura et al. 2003. "Effect of reinsertion of activated nucleus pulposus on disc degeneration: An experimental study on various types of collagen in degenerative discs." *Connect Tissue Res* 44 (2):104–8.

Watanabe, T., D. Sakai, Y. Yamamoto et al. 2010. "Human nucleus pulposus cells significantly enhanced biological properties in a coculture system with direct cell-to-cell contact with autologous mesenchymal stem cells." *J Orthop Res* 28 (5):623–30.

Wei, A., H. Tao, S.A. Chung et al. 2009. "The fate of transplanted xenogeneic bone marrow-derived stem cells in rat intervertebral discs." *J Orthop Res* 27 (3):374–9.

Weiler, C., A.G. Nerlich, R. Schaaf et al. Boos. 2010. "Immunohistochemical identification of notochordal markers in cells in the aging human lumbar intervertebral disc." *Eur Spine J* 19 (10):1761–70.

Wuertz, K., K. Godburn, and J.C. Iatridis. 2009. "MSC response to pH levels found in degenerating intervertebral discs." *Biochem Biophys Res Commun* 379 (4):824–9.

Wuertz, K., K.E. Godburn, C. Neidlinger-Wilke, J. Urban, and J.C. Iatridis. 2008. "Behavior of mesenchymal stem cells in the chemical microenvironment of the intervertebral disc." *Spine (Phila Pa 1976)* 33 (17):1843–9.

Yamamoto, Y., J. Mochida, D. Sakai et al. 2004. "Upregulation of the viability of nucleus pulposus cells by bone marrow-derived stromal cells: Significance of direct cell-to-cell contact in coculture system." *Spine (Phila Pa 1976)* 29 (14):1508–14.

Yang, F., L.V.Y. Leung, K.D.K. Luk, D. Chan, and K.M.C. Cheung. 2009. "Mesenchymal stem cells arrest intervertebral disc degeneration through chondrocytic differentiation and stimulation of endogenous cells." *Mol Ther* 17 (11):1959–66.

Yang, H., J. Wu, J. Liu et al. 2010. "Transplanted mesenchymal stem cells with pure fibrinous gelatin-transforming growth factor-b1 decrease rabbit intervertebral disc degeneration." *Spine J* 10 (9):802–10.

Yang, S.-H., C.-C. Wu, T.T.-F. Shih, Y.-H. Sun, and F.-H. Lin. 2008. "*In vitro* study on interaction between human nucleus pulposus cells and mesenchymal stem cells through paracrine stimulation." *Spine (Phila Pa 1976)* 33 (18):1951–7.

Yasen, M., Q. Fei, W.C. Hutton et al. 2013. "Changes of number of cells expressing proliferation and progenitor cell markers with age in rabbit intervertebral discs." *Acta Biochim Biophys Sin (Shanghai)* 45 (5):368–76.

Yoshikawa, T., Y. Ueda, K. Miyazaki, M. Koizumi, and Y. Takakura. 2010. "Disc regeneration therapy using marrow mesenchymal cell transplantation: A report of two case studies." *Spine (Phila Pa 1976)* 35 (11):E475–80.

Zhang, J., X. Huang, H. Wang et al. 2015. "The challenges and promises of allogeneic mesenchymal stem cells for use as a cell-based therapy." *Stem Cell Res Ther* 6 (1):234.

Zhang, Y., S. Drapeau, S.A. Howard, E.J.M.A. Thonar, and D.G. Anderson. 2011. "Transplantation of goat bone marrow stromal cells to the degenerating intervertebral disc in a goat disc injury model." *Spine (Phila Pa 1976)* 36 (5):372–7.

Zhang, Y., T. Guo, X. Guo, and S. Wu. 2009. "Clinical diagnosis for discogenic low back pain." *Int J Biol Sci* 5 (7):647–58.

Zhao, Q., H. Ren, and Z. Han. 2016. "Mesenchymal stem cells: Immunomodulatory capability and clinical potential in immune diseases." *J Cell Immunother* 2 (1):3–20.

Zhao, Y., Z. Jia, S. Huang et al. 2017. "Age-related changes in nucleus pulposus mesenchymal stem cells: An *in vitro* study in rats." *Stem Cells Int* 2017:6761572. doi: 10.1155/2017/6761572.

Zhu, Y., T. Liu, K. Song et al. 2008. "Adipose-derived stem cell: A better stem cell than BMSC." *Cell Biochem Funct* 26 (6):664–75.

Zuk, P.A., M. Zhu, H. Mizuno et al. 2001. "Multilineage cells from human adipose tissue: Implications for cell-based therapies." *Tissue Eng* 7 (2):211–28.

5 Materials for Cell Delivery in Degenerated Intervertebral Disc

Joana Silva-Correia, Joaquim Miguel Oliveira, and Rui Luís Reis

CONTENTS

5.1 Introduction 137
5.2 Materials for Nucleus Pulposus 138
    5.2.1 Natural Materials 140
    5.2.2 Synthetic Materials 142
5.3 Materials for AF 145
    5.3.1 Natural Materials 146
    5.3.2 Synthetic Materials 147
5.4 Concluding Remarks 148
Acknowledgments 149
References 149

5.1 INTRODUCTION

Several polymers have been investigated as potential cell delivery carriers for intervertebral disc (IVD) regeneration, which can be divided into natural and synthetic biopolymers and can be used in the form of hydrogels or solid scaffolds (Silva-Correia et al. 2013a; Schutgens et al. 2015). Biomaterials provide a stable three-dimensional (3D) environment for cell proliferation and extracellular matrix (ECM) synthesis, by serving as a load bearing structure to support IVD biomechanical forces (Whatley and Wen 2012). Ideally, it should degrade in parallel with neo-tissue formation, without forming toxic subproducts. Also, biomaterials enable better reproduction of the native tissue by providing clearly defined microstructures and macrostructures. Although some approaches to IVD regeneration consider the use of cell-free materials and growth factors, cell delivery to the degenerated disc by using a biomaterial carrier seems to be among the most promising research directions. As the native disc has a low cellular content, and in the process of IVD degeneration the existing cells become senescent, cell-based tissue engineering approaches focus on the transplantation of different cell types, including IVD cells, mesenchymal stem cells (MSCs), and other chondrocytic cells (Grad et al. 2010).

The advantages of using polymer biomaterials include: inexpensive production, facility and flexibility of the manufacturing process, and accessibility to an extensive diversity of properties. Scaffolds composed by natural and synthetic polymers have been investigated for *in vitro* cell–biomaterial interaction and *in vivo* implantation in animal models of IVD degeneration. Some of the most studied and promising natural and synthetic polymer-based biomaterials for cell delivery in the degenerated IVD will be discussed herein and are overviewed in Table 5.1.

5.2 MATERIALS FOR NUCLEUS PULPOSUS

Cell-based strategies aimed to repair the degenerated IVD (in early stages of degeneration) focus mainly on the development of suitable biomaterials that can substitute and aid in the regeneration of the central nucleus (nucleus pulposus [NP]) (Priyadarshani, Li, and Yao 2016). Degeneration of the IVD starts with NP loss of hydration and structure, resulting in malfunctioning and ultimately leading to tears in the outer structure of the IVD, i.e., the annulus fibrosus (AF) (Bertagnoli et al. 2005). So, when selecting a suitable biomaterial to act as a cell carrier to the NP, the specific properties and functionality of the damaged tissue and the therapeutic strategy to follow must be considered (Silva et al. 2017). Hydrogels are the most investigated candidates to serve as cell carriers for NP regeneration (Nguyen and West 2002; Pereira et al. 2013). As 3D networks of hydrophilic polymers, they can retain a large quantity of water within their structure without dissolving, thus mimicking the highly hydrated NP tissue (80% of water) and providing an adequate environment for cell proliferation and matrix production (Périé, Korda, and Iatridis 2005). Moreover, the cell-loaded material should improve disc biomechanics and height in order to be as close as possible to those of the native NP tissue (Roughley et al. 2006). One of the properties that need to be determined before using hydrogels in NP regeneration is their mechanical performance (Silva-Correia et al. 2013a). The mechanical properties of the hydrogels can be tuned by means of altering the crosslinking method, which can either be physical or chemical, or the crosslinking density, but mechanical properties must be in balance with cell biocompatibility (Reitmaier et al. 2012; Silva-Correia et al. 2013b). In the clinical setting, the only current way to approach the NP with biomaterials is by injection. By using an injection needle with a small diameter, it is possible to administer the hydrogel by using minimally invasive techniques to simplify and accelerate the surgical practice (Boyd and Carter 2006; Gantenbein-Ritter and Sakai 2011). By using this approach, only a small hole resulting from needle insertion is created in the AF. In addition, hydrogels can be polymerized in situ only after injection, thus presenting less risk of extrusion or migration, and sustaining the cells at the site they were delivered (Silva et al. 2017). Hydrogels can withstand

TABLE 5.1
Overview of Natural and Synthetic Biomaterials Currently Investigated as Cell Carriers for the Regeneration of NP and AF

	Biomaterial Examples	Targeted Tissue	Cells	References
Natural origin	Collagen	NP, AF	Primary NP cells, MSC	Bowles et al. 2010; Halloran et al. 2008; Sakai et al. 2006; Sato et al. 2003; Wilke et al. 2006; Zhou et al. 2016
	Alginate	NP, AF	Primary NP cells, AC	Bron et al. 2011; Chou and Nicoll 2009; Chou et al. 2009; Leone et al. 2008; Mizuno et al. 2004; Shao and Hunter 2007
	Carboxymethyl-cellulose	NP	Primary NP cells, MSC	Gupta, Cooper, and Nicoll 2011; Reza and Nicoll 2010
	Gellan gum	NP	Primary NP cells, NC, MSC	Khang et al. 2015; Silva-Correia et al. 2011, 2013b, 2013c; Tsaryk et al. 2017
	Chitosan	NP	MSC; primary NP cells	Bertolo et al. 2012; Dang et al. 2006; Richardson et al. 2008; Roughley et al. 2006
	Silk	NP, AF, total IVD	Primary NP cells, primary AF cells, chondrocytes, MSC	Chang et al. 2007; 2010; Du et al. 2014; Hu et al. 2012; Park et al. 2011, 2012; Zeng et al. 2014
	Hyaluronan	NP	Primary NP cells, MPC	Frith et al. 2013; Peroglio et al. 2012; Su, Chen, and Lin 2010
Synthetic	PLA/PGA	NP, AF	Primary NP cells, primary AF cells, MSC	Helen and Gough 2008; Mizuno et al. 2004; Richardson et al. 2006; Ruan et al. 2010
	PCL	AF	Primary AF cells, MSC	Driscoll et al. 2013; Koepsell et al. 2011; Nerurkar, Elliott, and Mauck 2007; Wismer et al. 2014; van Uden et al. 2015
	PEG	NP	MPC, primary NP cells	Bowles et al. 2010; Frith et al. 2013; Francisco et al. 2013, 2014
	PU	NP, AF	Primary AF cells	Li et al. 2016; Wismer et al. 2014

Note: AC, articular chondrocytes; NC, nasal chondrocytes; MPC, mesenchymal precursor cell.

complex biomechanical loads, and by being crosslinked in situ, irregular defects on the site of injury can be completely filled (Silva-Correia et al. 2013a). Polymerization in situ can occur under mild conditions by physical crosslinking methods (e.g., ionic-crosslinked) or by chemical crosslinking (e.g., photo- and enzymatically-crosslinked), which confers hydrogels' improved stability and biomechanics (Van Tomme, Storm, and Hennink 2008). Several types of hydrogels have been studied for cell-based NP regeneration, which can be divided into natural and synthetic biomaterials.

5.2.1 NATURAL MATERIALS

Currently, there is a growing interest in the study of natural-origin hydrogels for NP regeneration due to their reduced manufacturing cost and high biocompatibility (Malafaya, Silva, and Reis 2007; Puppi et al. 2010; Shogren and Bagley 1999). Most of the natural hydrogels are extracted and need to be purified before being used in tissue engineering purposes. Although they require purification steps that sometimes involve toxic solvents and harsh reagents, they are often less expensive as compared to synthetic hydrogels. Natural-origin hydrogels present numerous biological advantages as previously described, but they generally need to be tuned for enhanced physical properties, such as improved solubility and suitable rate of degradation to allow an adequate tissue regeneration (Temenoff and Mikos 2000). The most promising natural-origin hydrogels being investigated for cell-based regeneration of the NP are: collagen (Halloran et al. 2008; Sakai et al. 2006; Wilke et al. 2006; Zhou et al. 2016), alginate (Bron et al. 2011; Chou and Nicoll 2009; Chou et al. 2009; Leone et al. 2008; Mizuno et al. 2004), carboxymethylcellulose (Gupta, Cooper, and Nicoll 2011; Reza and Nicoll 2010), gellan gum (Khang et al. 2015; Silva-Correia et al. 2011, 2013b, 2013c; Tsaryk et al. 2017), chitosan (Bertolo et al. 2012; Dang et al. 2006; Richardson et al. 2008; Roughley et al. 2006), silk (Du et al. 2014; Hu et al. 2012; Park et al. 2011; Zeng et al. 2014), and hyaluronan (Chen et al. 2013; Frith et al. 2013; Peroglio et al. 2012; Su, Chen, and Lin 2010). Each one of these natural hydrogels presents advantages and disadvantages for NP regeneration (Silva et al. 2017). For example, Peroglio et al. (2012) demonstrated that hyaluronan-based hydrogels, with thermoreversible properties, can be easily injected and obtained by using a mild gelling mechanism. Hyaluronan grafted poly (*N*-isopropylacrylamide) (HA-pNIPAM) hydrogels were studied for their potential to be used as injectable carriers for NP cells. In this work, it was shown that *in vitro* culturing of a selected HA-pNIPAM hydrogel formulation with bovine NP cells, promoted maintenance of cell phenotype and ECM production after 1 week. In addition, the cell-loaded hydrogel could be injected through a 22-gauge needle in the NP and maintained cells viable after 1 week of *ex vivo* culturing. However, at high concentration or high molecular weight, this polymer loses its ability to withstand shear forces, and thus, it is often mixed with other natural-origin hydrogels, such as gelatin, or with synthetic polymers, like polyethylene glycol (PEG), to provide viscoelastic properties (Schutgens et al. 2015). Also, native hyaluronan served as the basis for several chemical modifications in order to obtain biomaterials (HYADD3 and HYAFF120) with enhanced mechanical and rheological performances similar to the native NP tissue (Gloria et al. 2010). These hydrogels were considered suitable for use in minimally invasive techniques, since their biomechanical behaviour was not affected by injection (Gloria et al. 2010).

Chitosan is a polymer composed of glucosamine and N-acetyl glucosamine and can be obtained from chitin isolated from crustaceans' shells (Suh and Matthew 2000). Roughley et al. (2006) showed in an *in vitro* study that chitosan hydrogels allow an efficient entrapment of IVD cells and/or ECM proteins, inhibiting the release of the produced proteoglycans into the culture medium. One of the major drawbacks in the use of chitosan for tissue engineering applications is that it needs to be neutralized before application, and neutralization causes immediate crosslinking (Schutgens et al. 2015). Several functional modifications have been attempted to enhance the biological, physical, and mechanical properties of chitosan polymer (Suh and Matthew 2000). For instance, Dang et al. (2006) described the modification of chitosan via conjugation with hydroxybutyl groups, which enabled obtaining a water-soluble and thermoresponsive polymer. At low temperature, the thermosensitive hydroxybutyl chitosan can be dissolved in water, and as the temperature increases up to 37°C, it rapidly forms a gel. Dang and coworkers have demonstrated the attractiveness of this thermoresponsive polymer for IVD tissue engineering by showing that MSCs and human NP and AF cells obtained from degenerated discs could proliferate with no deleterious effect on metabolic activity or production of ECM.

Alginate has been investigated for several tissue engineering applications and it is one of the most widely studied natural hydrogels for cell-based NP tissue engineering (Bron et al. 2011; Baer et al. 2001; Chou and Nicoll 2009; Chou et al. 2009). Its 3D structure is formed by crosslinking, induced by the presence of divalent cations such as calcium (Chou and Nicoll 2009). This type of crosslinking appears to be unstable over time, which is caused by a depletion of ions either by diffusion or cell consumption (Baer et al. 2001; Bron et al. 2011). Thus, one of the disadvantages of using alginate hydrogels obtained by ionic crosslinking is the limited control over the mechanical properties, swelling, and degradation. Attempts have been made to overcome this disadvantage by chemical modification, in order to improve structural stability and adjust the mechanics and biodegradability of alginate hydrogels (Jeon et al. 2009). Another drawback is that a complete elimination of the immunogenic contaminants in the raw material is still not optimized and standardized (Dusseault et al. 2006; Tam et al. 2006). Instead, different procedures for alginate purification have been developed, which, besides modifying a polymer's hydrophilicity and viscosity, results in high variability between purified alginates from different studies. In fact, alginate hydrogels have been shown to elicit immunological reaction in mice even after additional purification procedures have been applied (Dusseault et al. 2006). Bron et al. (2011) demonstrated that by changing the percentage of alginate to 2% (w/v), the resulting hydrogels displayed similar stiffness as compared to the native NP. Although alginate hydrogels presented low adhesiveness to cells, the phenotype of NP cells was maintained when cultured for 4 weeks. In another work, photo-crosslinked alginate was compared to non-photo-crosslinked material in terms of mechanical properties and ECM synthesis *in vivo* (Chou and Nicoll 2009; Chou et al. 2009). Chou and coauthors have demonstrated that photo-crosslinking led to enhanced mechanics and matrix production by the preloaded bovine NP cells.

Despite advances having been made in the improvement of alginate, the biomaterial still presents poor biocompatibility and has some limitations in terms of long-term structural integrity.

Gellan gum-based hydrogels for NP regeneration has been proposed by the first time by Silva-Correia et al. (2011). Gellan gum is a polysaccharide originated by bacterial fermentation and the structural repeating units of glucose, rhamnose, and glucuronic acid. In that study, functionalization of gellan gum has been achieved by incorporation of methacrylate units into the backbone of the polymer (Silva-Correia et al. 2011). This allowed improving the water solubility and processability at physiological temperature and to obtain a photo-crosslinkable gellan gum-based hydrogel with a more stable structure (Figure 5.1), as compared to the original form, and without the disadvantages of ionic polymerization (long-term loss of stability and structural integrity *in vivo*) (Oliveira et al. 2010). Actually, the photo-crosslinked gellan gum-based hydrogel presented improved mechanical properties as compared to the nonmodified gellan gum, with values of storage modulus (1 Hz) being similar to the native human NP, as demonstrated by dynamic mechanical analysis (Silva-Correia et al. 2011, 2013b). Moreover, its mechanical performance was maintained when loaded with cells and during *in vitro* culturing (Figure 5.1). In *in vitro* and *in vivo* studies, methacrylated gellan gum (GG-MA) hydrogels were shown to be biocompatible and nonangiogenic and supported human NP cellular encapsulation and viability (Silva-Correia et al. 2011, 2012, 2013b; Tsaryk et al. 2017). One advantage of this hydrogel is that it can be easily applied by injection using minimally invasive techniques and then photo-crosslinked in situ, allowing production of stable and reproducible gellan gum-based hydrogels (Silva-Correia et al. 2013b).

5.2.2 SYNTHETIC MATERIALS

Synthetic materials are mainly based on polymer networks that can absorb large amounts of water (Schutgens et al. 2015). Contrary to natural hydrogels, synthetic hydrogels can be easily tuned for a specific tissue engineering application, allowing for more predictable and reproducible chemical and physical properties (Place et al. 2009). For example, the degradation rate can be easily modified to be adequate to the tissue regeneration rate. In addition, synthetic polymers can simply be mixed with other polymers in order to increase the range of possibilities regarding its properties (Place et al. 2009). Synthetic materials with a high degree of purity have low immunogenicity and toxicity and present a low risk of infection. However, they present the disadvantages of having a high cost of production and lacking the bioactivity typical of the natural biomaterials. Some examples of synthetic polymers described in the literature as potential cell carriers for NP tissue engineering are polylactides/glycolides (PLA/PGA) (Richardson et al. 2006; Ruan et al. 2010), PEG (Bowles et al. 2010; Frith et al. 2013; Francisco et al. 2013, 2014), and polyurethane (PU) (Li et al. 2016). There are fewer works describing the use of only synthetic-based materials for NP tissue engineering, when compared to the natural-based materials. However, they still arouse interest due to the possibility of being combined with a natural-based polymer, gathering in an advanced material, the advantages of both synthetic- and natural-origin materials (Schutgens et al. 2015).

Materials for Cell Delivery in Degenerated Intervertebral Disc 143

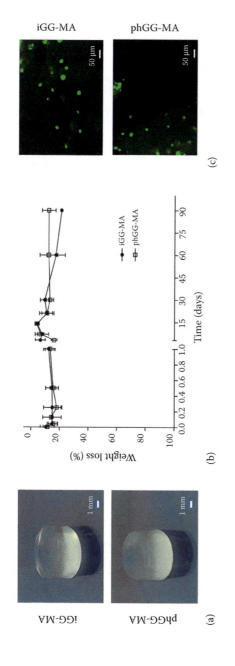

FIGURE 5.1 Methacrylated gellan gum hydrogels obtained by ionic- (iGG-MA) or photo-crosslinking (phGG-MA) for NP cell delivery. (a) Photographs of iGG-MA and phGG-MA hydrogels. (b) Long-term degradation of iGG-MA and phGG-MA hydrogels in PBS (pH 7.4; 37°C and 60 rpm; $n = 9$). (c) Cellular viability of human NP cells cultured for 21 days in iGG-MA and phGG-MA hydrogels. *(Continued)*

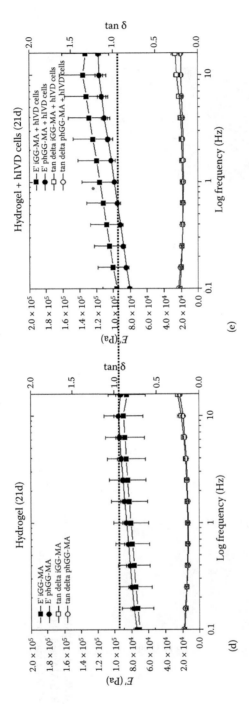

FIGURE 5.1 (CONTINUED) Methacrylated gellan gum hydrogels obtained by ionic-(iGG-MA) or photo-crosslinking (phGG-MA) for NP cell delivery. (d) Dynamic mechanical analysis of acellular and cell-loaded iGG-MA and phGG-MA hydrogels after immersion in culture medium for 21 days, showing an increase in storage modulus (e) in the cell-loaded materials. (Adapted from Silva-Correia, J., Oliveira, J.M., Caridade, S.G. et al., *J. Tissue Eng. Regen. Med.*, 5, E97–E107, 2011; Silva-Correia, J., Gloria, A., Oliveira, M.B. et al., *J. Biomed. Mater. Res. Part A*, 101, 3438–3446, 2013; Silva-Correia, J., Zavan, B., Vindigni, V. et al., *Adv. Healthc. Mater.*, 2, 568–575, 2013. With permission.)

One example of the use of synthetic materials for NP regeneration is poly (lactic-co-glycolic acid) scaffolds, composed by PLA and PGA (70:30), which were investigated for their regenerative capacity in a canine animal model (Ruan et al. 2010). It was demonstrated that after 4 weeks, the scaffolds completely degraded but the disc height was maintained since the autologous NP cells previously seeded onto the scaffolds were able to produce cartilaginous-like tissue (Ruan et al. 2010).

PEG-based hydrogels display biomechanical properties similar to articular cartilage, regarding the determination of compression modulus, tensile strength, and hydrostatic swelling (Nguyen et al. 2012). However, they are hardly used alone in cellular approaches due to their bio-inert nature, which does not provide an optimal environment for cell adhesion and matrix formation (Zhu 2010). To circumvent this disadvantage, PEG hydrogels, as well as other synthetic hydrogels, can be functionalized with molecules (e.g., cell adhesive peptides) (Benoit and Anseth 2005) or blended with natural hydrogels that do allow for adhesion (e.g., hyaluronan, silk) (Collin et al. 2011; Neo et al. 2014; Zhu 2010). It has been shown that porcine NP cells encapsulated in softer PEG hydrogels functionalized with laminin had enhanced production of collagen and glycosaminoglycans as compared to PEG hydrogels (Francisco et al. 2013, 2014). Biodegradation can also be achieved in PEG-based hydrogels, by the incorporation of enzyme-sensitive peptides (Zhu 2010).

PU scaffolds are currently being investigated for cartilage repair and other biomedical applications due to their high biocompatibility and easiness of processing into different structures (e.g., hydrogels and electrospun scaffolds) (Hung et al. 2014; Schutgens et al. 2015). They are usually biodegradable *in vivo*, giving origin to water and carbon dioxide as by-products (Santerre et al. 2005). The biomechanical performance of PU hydrogels seems promising, since the evaluation performed in human cadavers by spine injection in the NP demonstrated that the hydrogels could withstand compressive loads by transmitting the forces to the AF in a similar way as the native tissue (Dahl et al. 2010). Some disadvantages on the use of PU as cell carrier biomaterial for NP regeneration include absence of data regarding NP cell–material interactions and their intrinsic noninjectability (Schutgens et al. 2015). When considering an approach to the NP, the AF must be left intact and healthy as much as possible. As these materials are naturally noninjectable, different approaches other than injection must be pursued (e.g., incision through the endplate) (Li et al. 2016).

5.3 MATERIALS FOR AF

Although major efforts have been undertaken to develop cell-based biomaterials for NP regeneration, since disc degeneration generally starts in this tissue, other approaches are directed to regeneration of the AF compartment (Bron et al. 2009; Silva-Correia et al. 2013a) or even combining materials for both NP and AF tissue engineering in a biphasic composite. Contrary to the biomaterials used for cell delivery to the NP, which are mainly based on hydrogels, the materials being investigated for AF tissue engineering are solid scaffolds, with much harder and stronger properties (Sharifi et al. 2015).

As the AF is a highly organized tissue, the development of biomaterials that mimic this structure is a more complex task. The perfect material for AF regeneration must support the growth of cells and allow producing new tissue. Moreover, it should present mechanical properties similar to native AF tissue and must have some degree of adhesion to the surrounding tissue in order to withstand physiological strain (Sharifi et al. 2015). Most of the scaffolds being developed for AF regeneration are processed in the form of porous scaffolds or electrospun nanofibers (Nerurkar, Elliott, and Mauck 2007). Electrospinning technique allows forming sets of aligned polymeric nanofibers, where cells can attach and produce an oriented and more organized matrix, thus creating a multilamellar structure that reproduces the hierarchical organization of AF structure (Silva-Correia et al. 2013a). Other strategies that have been followed to mimic the aligned fibers orientation observed in AF, rely on collagen contraction (Bowles et al. 2010), silk fiber winding (Park et al. 2012), and the insertion of an oriented honeycomb structure into scaffolds (Sato et al. 2003). Some of the materials under study have been shown to accomplish the requirements of an ideal scaffold for AF regeneration, namely, be biocompatible and immunogenic, to close the AF gap and retain the NP, and permit resident cells, or other delivered cells, to proliferate and maintain/differentiate into the AF phenotype (Bron et al. 2009). However, the main problem is still the inexistence of an adequate surgical approach for delivery of the AF substitute and its fixation *in vivo*.

5.3.1 Natural Materials

Several natural-origin materials have been investigated as cell carriers for AF tissue engineering, among which are collagen (Bowles et al. 2010; Sato et al. 2003), alginate (Shao and Hunter 2007), and silk (Chang et al. 2007, 2010; Du et al. 2014; Park et al. 2012).

Collagen has been one of the materials used for developing AF substitutes, since it is one of the main constituents of the annulus. It has been exploited for AF tissue engineering in its pure form or chemically modified, alone or combined with other biomaterials or molecules (Bowles et al. 2010; Sato et al. 2003). In particular, collagen type I has the ability to organize into fibrillar hydrogels, which has been shown to support the attachment and growth of sheep primary AF cells, as well as enhance proteoglycan production. Bowles et al. (2010) observed that the AF cells presented spindle-shaped morphology and aligned in a similar way to the observed in IVD. Although collagen fiber alignment was produced in the circumferential direction, it did not allow reproducing the alternating 60° angles of orientation from lamellae to lamellae.

As the strongest natural fiber found in nature, the use of silk for AF tissue engineering may present advantages over other natural biopolymers. Chang et al. (2007, 2010) studied the effect of different porosities of silk scaffolds on the adhesion and proliferation of seeded bovine AF cells. Spreading and proliferation of AF cells on the porous scaffolds and accumulation of ECM were observed, which were not significantly improved by modification with arginine–glycine–aspartate peptides, although it affected cell morphology.

5.3.2 Synthetic Materials

Some examples of synthetic materials used in cell-based AF tissue engineering are PLA/PGA (Helen and Gough 2008; Mizuno et al. 2004), polycaprolactone (PCL) (Driscoll et al. 2013; Koepsell et al. 2011; Nerurkar, Elliott, and Mauck 2007; van Uden et al. 2015; Wismer et al. 2014), and PU (Wismer et al. 2014).

PLA and PGA have often been used in AF tissue engineering applications, as these polymers can undergo hydrolytic degradation through the ester bonds (Sharifi et al. 2015). Mizuno et al. (2004, 2006) seeded PLA/PGA scaffolds with ovine AF cells in a composite IVD structure and implanted it subcutaneously in nude mice for 12 weeks. It was observed that the gross morphology and histological observation of the composite scaffolds were very similar to those of native IVD. However, the degradation of the material caused accumulation of acidic products, which are not easily removed, due to AF avascular nature. This acidic environment contributes to ECM degradation and cell damage.

Another biodegradable polymer that has also shown success in emulating the alternating orientation of collagen fibers of the native AF is PCL. PCL presents good biomechanical properties, effortless thermoplastic processing, and a very slow degradation (months to years) that also occurs by hydrolysis of ester linkages (Dash and Konkimalla 2012). In the development of AF tissue engineered scaffolds, PCL has been often processed through electrospinning in an attempt to reproduce the morphological and mechanical properties of the native AF (Figure 5.2) (Nerurkar et al. 2009; Nerurkar, Elliott, and Mauck 2007). Nerurkar et al. (2009) have shown that MSCs seeded in electrospun PCL scaffolds arranged in a multilamellar architecture

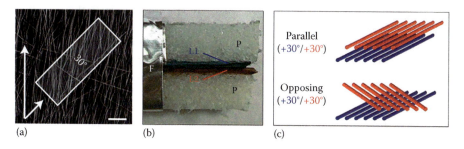

FIGURE 5.2 Fabrication of PCL-based tissue constructs with alternating orientation lamellae. (a) Scaffolds were excised 30° from the prevailing fiber direction of electrospun PCL nanofibers mats to replicate the oblique collagen orientation within a single lamella of the AF (scale bar: 25 μm). (b) At zero weeks, MSC-seeded scaffolds were formed into bilayers between pieces of porous polypropylene and wrapped with a foil sleeve. (P: porous polypropylene; F: foil; L1/2: lamella 1/2); (c) Bilayers were oriented with either parallel (+30°/+30°) or opposing (+30°/−30°) fiber alignment relative to the long axis of the scaffold. (Adapted from Nerurkar, N.L., Baker, B.M., Sen, S. et al. *Nat. Mater.*, 8, 986–992, 2009. With permission.)

similar to that of native AF, produced an organized, collagen-rich ECM after 10 weeks of *in vitro* culturing. More recently, PCL has been used to fabricate custom-tailored rabbit AF scaffolds (van Uden et al. 2015) and patient-specific scaffolds that mimic the AF of the human IVD (Oner et al. 2017), with different internal architectures by 3D printing. Differently designed structures were produced showing submacro- to macro-porosities (nine porosity patterns) (van Uden et al. 2015). It has been shown that the bioprinted noncytotoxic PCL scaffolds presented values of compressive stiffness higher than those of human IVD, and these properties were maintained after hydration.

5.4 CONCLUDING REMARKS

Cell-based approaches to address IVD regeneration may be dependent on the use of biomaterials since they provide immediate mechanical support and instigate cells in the IVD to differentiate into a desired phenotype and synthesize appropriate ECM. Natural materials present several advantages, such as biocompatibility and safety, as well as well-known degradation routes. They generally provide a more adequate environment for cell proliferation and tissue regeneration while being degraded in parallel. The drawbacks of natural-origin polymers include the need for tuning materials' properties, which are usually difficult to control. On the other hand, synthetic materials enable adjusting their properties in an easy way and their manufacturing process is simple and reproducible. They are often nonadhesive, hindering cells' interaction, and their degradation products are sometimes toxic to cells. Generally, synthetic biomaterials have been studied in blends in order to gather properties of different materials, such as biocompatibility, biomechanics, and biodegradation.

Cell-carrier materials for NP tissue engineering are generally injectable, to allow use in minimally invasive approaches using needles with small diameters and limiting the used volume. The use of solid scaffolds in the NP is limited and an alternative approach to access the NP through the endplate must be considered. In contrast, AF regeneration is more reasonable to be performed with solid scaffolds that could resist to the loads in the spine and provide more adhesion to the surrounding structures. Nonetheless, cell delivery to the IVD seems to be more feasible by combining a biomaterial for NP tissue engineering with other solid scaffold relevant for AF regeneration, i.e., by using composite scaffolds, especially in severe stages of degeneration. Although a viable functional implant for a total IVD regeneration has not been developed yet through the current strategies, this task may not be unachievable in the near future. With the emerging field of 3D printing technologies applied to tissue engineering and regenerative medicine evolving rapidly in the last decades, we might be closer to print cell-loaded implants with high structural complexity and reproducible shape that allow compliance with patient specificity, for a successful, complete IVD regeneration.

ACKNOWLEDGMENTS

The authors would like to acknowledge the financial support provided by the Portuguese Foundation for Science and Technology (FCT) through the project B-FABULUS (PTDC/BBB-ECT/2690/2014). The FCT distinctions attributed to J. Silva-Correia (IF/00115/2015) and J. Miguel Oliveira (IF/01285/2015) under the Investigator FCT program are also greatly acknowledged.

REFERENCES

Baer, A.E., J.Y. Wang, V.B. Kraus, and L.A. Setton. 2001. "Collagen gene expression and mechanical properties of intervertebral disc cell-alginate cultures." *J Orthop Res* 19:2–10.

Benoit, D.S., and K.S. Anseth. 2005. "The effect on osteoblast function of colocalized RGD and PHSRN epitopes on PEG surfaces." *Biomaterials* 26:5209–20.

Bertagnoli, R., C.T. Sabatino, J.T. Edwards et al. 2005. "Mechanical testing of a novel hydrogel nucleus replacement implant." *Spine J* 5:672–81.

Bertolo, A., M. Mehr, N. Aebli et al. 2012. "Influence of different commercial scaffolds on the *in vitro* differentiation of human mesenchymal stem cells to nucleus pulposus-like cells." *Eur Spine J* 21 (Suppl 6):S826–38.

Bowles, R.D., R.M. Williams, W.R. Zipfel, and L.J. Bonassar. 2010. "Self-assembly of aligned tissue-engineered annulus fibrosus and intervertebral disc composite via collagen gel contraction." *Tissue Eng Part A* 16:1339–48.

Boyd, L.M., and A.J. Carter. 2006. "Injectable biomaterials and vertebral endplate treatment for repair and regeneration of the intervertebral disc." *Eur Spine J* 15:414–21.

Bron, J.L., M.N. Helder, H.-J. Meisel, B.J. Van Royen, and T.H. Smit. 2009. "Repair, regenerative and supportive therapies of the annulus fibrosus: Achievements and challenges." *Eur Spine J* 18:301–13.

Bron, J.L., L.A. Vonk, T.H. Smit, and G.H. Koenderink. 2011. "Engineering alginate for intervertebral disc repair." *J Mech Behav Biomed Mater* 4:1196–205.

Chang, G., H.J. Kim, D. Kaplan, G. Vunjak-Novakovic, and R.A. Kandel. 2007. "Porous silk scaffolds can be used for tissue engineering annulus fibrosus." *Eur Spine J* 16:1848–57.

Chang, G., H.J. Kim, G. Vunjak-Novakovic, D.L. Kaplan, and R. Kandel. 2010. "Enhancing annulus fibrosus tissue formation in porous silk scaffolds." *J Biomed Mater Res Part A* 92:43–51.

Chen, Y.C., W.Y. Su, S.H. Yang, A. Gefen, and F.H. Lin. 2013. "In situ forming hydrogels composed of oxidized high molecular weight hyaluronic acid and gelatin for nucleus pulposus regeneration." *Acta Biomater* 9:5181–93.

Chou, A.I., S.O. Akintoye, and S.B. Nicoll. 2009. "Photo-crosslinked alginate hydrogels support enhanced matrix accumulation by nucleus pulposus cells *in vivo*." *Osteoarthritis Cartilage* 17:1377–84.

Chou, A.I., and S.B. Nicoll. 2009. "Characterization of photo-crosslinked alginate hydrogels for nucleus pulposus cell encapsulation." *J Biomed Mater Res A* 91:187–94.

Collin, E.C., S. Grad, D.I. Zeugolis et al. 2011. "An injectable vehicle for nucleus pulposus cell-based therapy." *Biomaterials* 32:2862–70.

Dahl, M.C., M. Ahrens, J.E. Sherman, and E.O. Martz. 2010. "The restoration of lumbar intervertebral disc load distribution: A comparison of three nucleus replacement technologies." *Spine (Phila Pa 1976)* 35:1445–53.

Dang, J.M., D.D. Sun, Y. Shin-Ya et al. 2006. "Temperature-responsive hydroxybutyl chitosan for the culture of mesenchymal stem cells and intervertebral disk cells." *Biomaterials* 27:406–18.

Dash, T.K., and V.B. Konkimalla. 2012. "Poly-small je, Ukrainian-caprolactone based formulations for drug delivery and tissue engineering: A review." *J Control Release* 158:15–33.

Driscoll, T.P., R.H. Nakasone, S.E. Szczesny, D.M. Elliott, and R.L. Mauck. 2013. "Biaxial mechanics and inter-lamellar shearing of stem-cell seeded electrospun angle-ply laminates for annulus fibrosus tissue engineering." *J Orthop Res* 31:864–70.

Du, L., M. Zhu, Q. Yang et al. 2014. "A novel integrated biphasic silk fibroin scaffold for intervertebral disc tissue engineering." *Mater Lett* 117:237–240.

Dusseault, J., S.K. Tam, M. Menard et al. 2006. "Evaluation of alginate purification methods: Effect on polyphenol, endotoxin, and protein contamination." *J Biomed Mater Res Part A* 76:243–51.

Francisco, A.T., P.Y. Hwang, C.G. Jeong et al. 2014. "Photo-crosslinkable laminin-functionalized polyethylene glycol hydrogel for intervertebral disc regeneration." *Acta Biomater* 10:1102–11.

Francisco, A.T., R.J. Mancino, R.D. Bowles et al. 2013. "Injectable laminin-functionalized hydrogel for nucleus pulposus regeneration." *Biomaterials* 34:7381–8.

Frith, J.E., A.R. Cameron, D.J. Menzies et al. 2013. "An injectable hydrogel incorporating mesenchymal precursor cells and pentosan polysulphate for intervertebral disc regeneration." *Biomaterials* 34:9430–40.

Gantenbein-Ritter, B., and D. Sakai. 2011. "Biomaterials for intervertebral disc regeneration." In *Comprehensive Biomaterials, Volume 6*, ed. P. Ducheyne, K.E. Healy, D.W. Hutmacher, D.W. Grainger, and C.J. Kirkpatrick, 161–9. United Kingdom: Elsevier.

Gloria, A., A. Borzacchiello, F. Causa, and L. Ambrosio. 2010. "Rheological characterization of hyaluronic acid derivatives as injectable materials toward nucleus pulposus regeneration." *J Biomater Appl* 26:745–59.

Grad, S., M. Alini, D. Eglin et al. 2010. "Cells and biomaterials for intervertebral disc regeneration." *Synth Lect Tissue Eng* 2:1–104.

Gupta, M.S., E.S. Cooper, and S.B. Nicoll. 2011. "Transforming growth factor-beta 3 stimulates cartilage matrix elaboration by human marrow-derived stromal cells encapsulated in photo-crosslinked carboxymethylcellulose hydrogels: Potential for nucleus pulposus replacement." *Tissue Eng Part A* 17:2903–10.

Halloran, D.O., S. Grad, M. Stoddart et al. 2008. "An injectable cross-linked scaffold for nucleus pulposus regeneration." *Biomaterials* 29:438–447.

Helen, W., and J.E. Gough. 2008. "Cell viability, proliferation and extracellular matrix production of human annulus fibrosus cells cultured within PDLLA/Bioglass® composite foam scaffolds *in vitro*." *Acta Biomater* 4:230–43.

Hu, J., B. Chen, F. Guo et al. 2012. "Injectable silk fibroin/polyurethane composite hydrogel for nucleus pulposus replacement." *J Mater Sci Mater Med* 23:711–22.

Hung, K.C., C.S. Tseng, and S.H. Hsu. 2014. "Synthesis and 3D printing of biodegradable polyurethane elastomer by a water-based process for cartilage tissue engineering applications." *Adv Healthc Mater* 3:1578–87.

Jeon, O., K.H. Bouhadir, J.M. Mansour, and E. Alsberg. 2009. "Photo-crosslinked alginate hydrogels with tunable biodegradation rates and mechanical properties." *Biomaterials* 30:2724–34.

Khang, G., S.K. Lee, H.N. Kim et al. 2015. "Biological evaluation of intervertebral disc cells in different formulations of gellan gum-based hydrogels." *J Tissue Eng Regen Med* 9:265–75.

Koepsell, L., L. Zhang, D. Neufeld, H. Fong, and Y. Deng. 2011. "Electrospun nanofibrous polycaprolactone scaffolds for tissue engineering of annulus fibrosus." *Macromol Biosci* 11:391–9.

Leone, G., P. Torricelli, A. Chiumiento, A. Facchini, and R. Barbucci. 2008. "Amidic alginate hydrogel for nucleus pulposus replacement." *J Biomed Mater Res Part A* 84:391–401.

Li, Z., G. Lang, X. Chen et al. 2016. "Polyurethane scaffold with in situ swelling capacity for nucleus pulposus replacement." *Biomaterials* 84:196–209.

Malafaya, P.B., G.A. Silva, and R.L. Reis. 2007. "Natural-origin polymers as carriers and scaffolds for biomolecules and cell delivery in tissue engineering applications." *Adv Drug Deliv Rev* 59:207–233.

Mizuno, H., A.K. Roy, C.A. Vacanti et al. 2004. "Tissue-engineered composites of annulus fibrosus and nucleus pulposus for intervertebral disc replacement." *Spine (Phila Pa 1976)* 29:1290–7; discussion 1297–8.

Mizuno, H., A.K. Roy, V. Zaporojan et al. 2006. "Biomechanical and biochemical characterization of composite tissue-engineered intervertebral discs." *Biomaterials* 27:362–70.

Neo, P.Y., P. Shi, J.C. Goh, and S.L. Toh. 2014. "Characterization and mechanical performance study of silk/PVA cryogels: Towards nucleus pulposus tissue engineering." *Biomed Mater* 9:065002.

Nerurkar, N.L., B.M. Baker, S. Sen et al. 2009. "Nanofibrous biologic laminates replicate the form and function of the annulus fibrosus." *Nat Mater* 8:986–992.

Nerurkar, N.L., D.M. Elliott, and R.L. Mauck. 2007. "Mechanics of oriented electrospun nanofibrous scaffolds for annulus fibrosus tissue engineering." *J Orthop Res* 25:1018–28.

Nguyen, K.T., and J.L. West. 2002. "Photopolymerizable hydrogels for tissue engineering applications." *Biomaterials* 23:4307–14.

Nguyen, Q.T., Y. Hwang, A.C. Chen, S. Varghese, and R.L. Sah. 2012. "Cartilage-like mechanical properties of poly (ethylene glycol)-diacrylate hydrogels." *Biomaterials* 33:6682–90.

Oliveira, J.T., T.C. Santos, L. Martins et al. 2010. "Gellan gum injectable hydrogels for cartilage tissue engineering applications: *In vitro* studies and preliminary *in vivo* evaluation." *Tissue Eng Part A* 16:343–53.

Oner, T., I.F. Cengiz, M. Pitikakis et al. 2017. "3D segmentation of intervertebral discs: From concept to the fabrication of patient-specific scaffolds." *J 3D Print Med* 1:91–101.

Park, S.-H., H. Cho, E.S. Gil et al. 2011. "Silk-fibrin/hyaluronic acid composite gels for nucleus pulposus tissue regeneration." *Tissue Eng Part A* 17:2999–3009.

Park, S.H., E.S. Gil, B.B. Mandal et al. 2012. "Annulus fibrosus tissue engineering using lamellar silk scaffolds." *J Tissue Eng Regen Med* 6 (Suppl 3):s24–33.

Pereira, D.R., J. Silva-Correia, J.M. Oliveira, and R.L. Reis. 2013. "Hydrogels in acellular and cellular strategies for intervertebral disc regeneration." *J Tissue Eng Regen Med* 7:85–98.

Périé, D., D. Korda, and J.C. Iatridis. 2005. "Confined compression experiments on bovine nucleus pulposus and annulus fibrosus: Sensitivity of the experiment in the determination of compressive modulus and hydraulic permeability." *J Biomech* 38:2164–71.

Peroglio, M., S. Grad, D. Mortisen et al. 2012. "Injectable thermoreversible hyaluronan-based hydrogels for nucleus pulposus cell encapsulation." *Eur Spine J* 21 (Suppl 6):S839–49.

Place, E.S., J.H. George, C.K. Williams, and M.M. Stevens. 2009. "Synthetic polymer scaffolds for tissue engineering." *Chem Soc Rev* 38:1139–51.

Priyadarshani, P., Y. Li, and L. Yao. 2016. "Advances in biological therapy for nucleus pulposus regeneration." *Osteoarthritis Cartilage* 24:206–12.

Puppi, D., F. Chiellini, A.M. Piras, and E. Chiellini. 2010. "Polymeric materials for bone and cartilage repair." *Prog Polym Sci* 35:403–40.

Reitmaier, S., U. Wolfram, A. Ignatius et al. 2012. "Hydrogels for nucleus replacement—Facing the biomechanical challenge." *J Mech Behav Biomed Mater* 14:67–77.

Reza, A.T., and S.B. Nicoll. 2010. "Characterization of novel photo-crosslinked carboxymethylcellulose hydrogels for encapsulation of nucleus pulposus cells." *Acta Biomater* 6:179–86.

Richardson, S.M., J.M. Curran, R. Chen et al. 2006. "The differentiation of bone marrow mesenchymal stem cells into chondrocyte-like cells on poly-L-lactic acid (PLLA) scaffolds." *Biomaterials* 27:4069–78.

Richardson, S.M., N. Hughes, J.A. Hunt, A.J. Freemont, and J.A. Hoyland. 2008. "Human mesenchymal stem cell differentiation to NP-like cells in chitosan-glycerophosphate hydrogels." *Biomaterials* 29:85–93.

Roughley, P., C. Hoemann, E. DesRosiers et al. 2006. "The potential of chitosan-based gels containing intervertebral disc cells for nucleus pulposus supplementation." *Biomaterials* 27:388–96.

Ruan, D.-K., H. Xin, C. Zhang et al. 2010. "Experimental intervertebral disc regeneration with tissue-engineered composite in a canine model." *Tissue Eng Part A* 16:2381–9.

Sakai, D., J. Mochida, T. Iwashina et al. 2006. "Regenerative effects of transplanting mesenchymal stem cells embedded in atelocollagen to the degenerated intervertebral disc." *Biomaterials* 27:335–45.

Santerre, J.P., K. Woodhouse, G. Laroche, and R.S. Labow. 2005. "Understanding the biodegradation of polyurethanes: From classical implants to tissue engineering materials." *Biomaterials* 26:7457–70.

Sato, M., M. Kikuchi, M. Ishihara et al. 2003. "Tissue engineering of the intervertebral disc with cultured annulus fibrosus cells using atelocollagen honeycomb-shaped scaffold with a membrane seal (ACHMS scaffold)." *Med Biol Eng Comput* 41:365–71.

Schutgens, E.M., M.A. Tryfonidou, T.H. Smit et al. 2015. "Biomaterials for intervertebral disc regeneration: Past performance and possible future strategies." *Eur Cells Mater* 30:210–31.

Shao, X., and C.J. Hunter. 2007. "Developing an alginate/chitosan hybrid fiber scaffold for annulus fibrosus cells." *J Biomed Mater Res Part A* 82:701–10.

Sharifi, S., S.K. Bulstra, D.W. Grijpma, and R. Kuijer. 2015. "Treatment of the degenerated intervertebral disc; closure, repair and regeneration of the annulus fibrosus." *J Tissue Eng Regen Med* 9:1120–32.

Shogren, R.L., and E.B. Bagley. 1999. "Natural polymers as advanced materials: Some research needs and directions." In *Biopolymers*, ed. S.H Iman, R.V. Greene, and B.R. Zaidi, 2–11. Washington, DC: American Chemical Society.

Silva-Correia, J., S.I. Correia, J.M. Oliveira, and R.L. Reis. 2013a. "Tissue engineering strategies applied in the regeneration of the human intervertebral disk." *Biotechnol Adv* 31:1514–31.

Silva-Correia, J., A. Gloria, M.B. Oliveira et al. 2013b. "Rheological and mechanical properties of acellular and cell-laden methacrylated gellan gum hydrogels." *J Biomed Mater Res Part A* 101:3438–46.

Silva-Correia, J., V. Miranda-Gonçalves, A.J. Salgado et al. 2012. "Angiogenic potential of gellan-gum-based hydrogels for application in nucleus pulposus regeneration: *In vivo* study." *Tissue Eng Part A* 18:1203–12.

Silva-Correia, J., J.M. Oliveira, S.G. Caridade et al. 2011. "Gellan gum-based hydrogels for intervertebral disc tissue-engineering applications." *J Tissue Eng Regen Med* 5:E97–107.

Silva-Correia, J., B. Zavan, V. Vindigni et al. 2013c. "Biocompatibility evaluation of ionic- and photo-crosslinked methacrylated gellan gum hydrogels: *in vitro* and *in vivo* study." *Adv Healthc Mater* 2:568–75.

Silva, S.S., E.M. Fernandes, S. Pina et al. 2017. "Natural-origin materials for tissue engineering and regenerative medicine." In *Comprehensive Biomaterials II, Volume 2*, ed. P. Ducheyne, K.E. Healy, D.W. Hutmacher, D.W. Grainger, and C.J. Kirkpatrick, 228–52. United Kingdom: Elsevier.

Su, W.Y., Y.C. Chen, and F.H. Lin. 2010. "Injectable oxidized hyaluronic acid/adipic acid dihydrazide hydrogel for nucleus pulposus regeneration." *Acta Biomater* 6:3044–55.

Suh, J.K., and H.W. Matthew. 2000. "Application of chitosan-based polysaccharide biomaterials in cartilage tissue engineering: A review." *Biomaterials* 21:2589–98.

Tam, S.K., J. Dusseault, S. Polizu et al. 2006. "Impact of residual contamination on the biofunctional properties of purified alginates used for cell encapsulation." *Biomaterials* 27:1296–305.

Temenoff, J.S., and A.G. Mikos. 2000. "Review: Tissue engineering for regeneration of articular cartilage." *Biomaterials* 21:431–40.

Tsaryk, R., J. Silva-Correia, J.M. Oliveira et al. 2017. "Biological performance of cell-encapsulated methacrylated gellan gum-based hydrogels for nucleus pulposus regeneration." *J Tissue Eng Regen Med* 11:637–48.

Van Tomme, S.R., G. Storm, and W.E. Hennink. 2008. "In situ gelling hydrogels for pharmaceutical and biomedical applications." *Int J Pharm* 355:1–18.

van Uden, S., J. Silva-Correia, V.M. Correlo, J.M. Oliveira, and R.L. Reis. 2015. "Custom-tailored tissue engineered polycaprolactone scaffolds for total disc replacement." *Biofabrication* 7:015008.

Whatley, B.R., and X. Wen. 2012. "Intervertebral disc (IVD): Structure, degeneration, repair and regeneration." *Mater Sci Eng C* 32:61–77.

Wilke, H.-J., F. Heuer, C. Neidlinger-Wilke, and L. Claes. 2006. "Is a collagen scaffold for a tissue engineered nucleus replacement capable of restoring disc height and stability in an animal model?" *Eur Spine J* 15:433–8.

Wismer, N., S. Grad, G. Fortunato et al. 2014. "Biodegradable electrospun scaffolds for annulus fibrosus tissue engineering: Effect of scaffold structure and composition on annulus fibrosus cells *in vitro*." *Tissue Eng Part A* 20:672–82.

Zeng, C., Q. Yang, M. Zhu et al. 2014. "Silk fibroin porous scaffolds for nucleus pulposus tissue engineering." *Mater Sci Eng C* 37:232–40.

Zhou, X., Y. Tao, J. Wang et al. 2016. "Three-dimensional scaffold of type II collagen promote the differentiation of adipose-derived stem cells into a nucleus pulposus-like phenotype." *J Biomed Mater Res Part A* 104:1687–93.

Zhu, J. 2010. "Bioactive modification of poly(ethylene glycol) hydrogels for tissue engineering." *Biomaterials* 31:4639–56.

6 Cell Recruitment for Intervertebral Disc

*Catarina Leite Pereira, Sibylle Grad,
Mário Adolfo Barbosa, and Raquel M. Gonçalves*

CONTENTS

6.1 Introduction and Motivation 155
6.2 IVD Anatomy 156
6.3 IVD Degeneration 158
6.4 Cell Recruitment for Endogenous Repair/Regeneration 159
    6.4.1 Stem Cell Homing and Migration 160
        6.4.1.1 Stem Cell Migration: Chemokines, Receptors,
            and Growth Factors 162
    6.4.2 Immune Cell Migration: Chemokines, Receptors,
        and Growth Factors 164
6.5 Strategies to Enhance Cell Migration 165
    6.5.1 Stem Cell Recruitment: Chemokine Delivery Systems 165
        6.5.1.1 Chemokine Receptor Overexpression in Stem Cells 168
    6.5.2 Immune Cell Migration: Chemokine Delivery Systems 168
6.6 Cell Migration in the IVD—A New Paradigm? 169
    6.6.1 IVD Degeneration: Involved Cytokines and Chemokines 169
    6.6.2 Cell Recruitment in the IVD 171
6.7 Conclusions 173
Acknowledgments 174
References 174

6.1 INTRODUCTION AND MOTIVATION

The intervertebral disc (IVD) is a fascinating organ that combines biology and physics in a perfect manner. Due to its avascular nature, the IVD undergoes an early degenerative process when compared to other tissues in the human body. This degenerative process is often recognized as the main contributor for low back pain and radicular leg pain (Kuslich, Ulstrom, and Michael 1991), despite most of the degenerative discs being described as asymptomatic (Boos et al. 1995; Jensen et al. 1994; Videman et al. 2003).

Back pain affects approximately 632 million people globally (Vos et al. 2012), and it is estimated that 84% of the population will suffer from low back pain at some point in their lives (Walker 2000). This represents a substantial socioeconomic burden, as a consequence of the high costs of health care, diminished productivity, and absenteeism from work (Martin et al. 2008). Multiple conditions can result in back pain; however, disc degeneration is accounted for 40% of the cases (Cheung et al. 2009), which represents a large percentage and therefore consolidates the relevance of studying its nature and the underlying causes of IVD degeneration.

Like other cartilaginous tissues, the IVD has been considered to have very poor self-repair capacity, demanding a need for development of novel strategies for its regeneration. The treatment and regeneration of degenerated IVD still represent a significant challenge in medical science, due to its nature and all the biomechanics involved in its functions. Biology and engineering, two complex sciences, have been pursuing, in the past years, the unraveling of the nature of the IVD, in an attempt to discover in cells and/or materials a perfect combination that closely resembles such a particular organ and might contribute to its regeneration.

One of the hallmarks of IVD degeneration is the alteration of cell metabolism, cell death, and senescence that results in the loss of tissue integrity and function failure. Cell-based therapies to repopulate the degenerated IVD have been widely explored in the past years, while cell recruitment strategies still remain in the shadow, with poor investment, due to IVD nature and its well-known poor endogenous repair capacity. In this chapter, we will review some of the concepts behind cell migration and recruitment, the molecules involved, as well as strategies to improve this natural mechanism, and finally, what's known within the IVD research field context.

6.2 IVD ANATOMY

The IVD is a critical organ in the spine, providing the flexibility and the capacity to absorb biomechanical forces. The IVD structure corresponds to one-third of the total human spine length (Pattappa et al. 2012). Besides the 23 discs that compose the spine, the human *columna vertebralis* consists of 33 bony vertebras positioned in five different regions: cervical (7), thoracic (12), lumbar (5), sacrum, and coccyx (9) (Figure 6.1). The different structures of the spine represent distinct functions. The vertebra's main function is to provide rigidity and support the trunk and extremities, protect the spinal cord and *cauda equina*, and anchor the *erector spinae* and other muscles, whereas the IVD provides flexibility and movement (Devereaux 2007). Throughout evolution, spine adaptation represented a major advantage for humans and other primates by providing the adoption of the upright bipedal stance, which conferred several gains, namely, in vision and in the development of other bones and structures in the human body. The discs are articulating structures between the vertebral body, which allow movement (flexion, extension and rotation) and act as a shock-absorbing organ in the spine; lack of this capacity would result in a rigid vertebral column (Roberts et al. 2006). Macroscopically, the healthy adult IVD is composed of different and interrelated tissues, the central nucleus pulposus (NP), the surrounding annulus fibrosus (AF), and the cartilaginous endplates (CEPs), which

Cell Recruitment for Intervertebral Disc

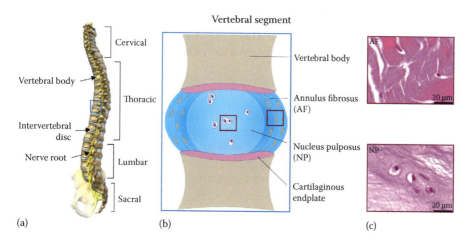

FIGURE 6.1 Vertebral spine and IVD anatomy. (a) Vertebral spine and its different regions: cervical, thoracic, lumbar, and sacral. (b) Vertebral segment: the IVD is located between two adjacent vertebras being interfaced by the CEP. The IVD is composed of the AF and the NP, which present cells with different morphologies. (c) Representative images of the AF and NP cells within the tissue (scale bar: 20 μm).

provide the connection to vertebral bodies (Figure 6.1). The normal IVD comprises a large amount of extracellular matrix (ECM) interspersed by a small number of cells that account for approximately 1% of the total IVD volume. The IVD cells are believed to be constituted at least by two phenotypically distinct populations with different cell densities, depending on the IVD area, AF ($\sim 9 \times 10^6$ cells/cm^3) or NP ($\sim 4 \times 10^6$ cells/cm^3) (Pattappa et al. 2012). The NP is a complex and heterogeneous structure composed of randomly organized collagen fibers embedded in a highly hydrated gel-like matrix rich in proteoglycans (Pattappa et al. 2012).

The NP is predominantly composed of water (70%–90%, depending on age), proteoglycans (~50% of dry weight), and collagen type II (col type II) (~20% of dry weight) (Buckwalter 1995). The AF tissue derives, as the spinal column, the vertebrae and the CEPs from the mesoderm (Roberts et al. 2006). This tissue is mainly composed of water, about 60%–80% depending on age and the region of the spine; collagen (50%–70% of dry weight); proteoglycans (10%–20% of dry weight); and noncollagenous proteins, as for instance, elastin (~25% of dry weight) (Pattappa et al. 2014). The AF consists of concentric layers of collagen type I (col type I), arranged in specific angles of 28° in the peripheral region and 44° in the central zone with respect to the transverse plane of the disc. The space between the AF layers, the interlamellar septae, contains proteoglycan aggregates and other elements, which provides interlamellar cohesion. This highly organized structure provides the AF with an anisotropic behavior, with the tensile, compressive, and shear properties differing in the axial, circumferential, and radial directions, therefore playing a key role in the mechanical function of the IVD (Pattappa et al. 2012).

The avascular nature of the IVD organ obliges that nutrition is provided by the blood supply at the disc's margins, namely, by the CEP by diffusion of nutrients and metabolites toward the center of the disc, resulting in a very low oxygen tension in the IVD. Therefore, the metabolism of IVD cells is partly anaerobic, leading to the accumulation of lactic acid and a low pH within the tissue.

6.3 IVD DEGENERATION

In the early 1970s, Schmorl and Junghanns (1971) defined the disc as the unique organ in the human body that undergoes a profound degeneration early in life (usually in the second decade). Disc degeneration is an age-related multifactorial process encompassing both genetic and environmental contributions. Although several environmental factors have been demonstrated to influence disc degeneration, such as smoking, diabetes, infection, trauma, heavy lifting, and vibration, the genetic component is recognized to play a predominant role in early disc degeneration and external factor susceptibility (Walker and Anderson 2004).

Disc degeneration is described as a progressive process, entailing significant histomorphological and biomechanical changes as a consequence of an altered cell metabolism and an unbalanced matrix synthesis (Walker and Anderson 2004). These alterations result from a cascade of disrupting events that are believed to be triggered by impaired disc nutrition, followed by an accumulation of waste products and matrix degradation (Molinos et al. 2015). Following this, the cellular microenvironment of the disc becomes progressively more hostile, characterized by an acidic pH, an up-regulation of several matrix degrading enzymes, and the release of proinflammatory cytokines (Smith et al. 2011).

Cell viability in the IVD is dependent on successful nutrition of the disc, but it is also linked to the normal aging process. Apoptosis was described to play an important role in disc degeneration, namely, in older individuals (Gruber and Hanley 1998). Adding to this, several changes in terms of cell metabolism occur and have a profound impact on the synthesis of the main matrix components, namely, aggrecan and collagen type II; this loss of matrix was observed in discs from older individuals, as reported in the study of Antoniou et al. (1996). The matrix loss observed in the course of the degenerative process is also related to a shift toward a catabolic metabolism characterized by the production of matrix degrading enzymes, such as the matrix metalloproteinases MMP-1, -3, and -9 (Crean et al. 1997; Kanemoto et al. 1996) and aggrecanases (a disintegrin and metalloproteinase with thrombospondin motifs [ADAMTS-4/5]) (Roberts et al. 2000), as well as other catabolic molecules such as the cytokines tumor necrosis factor (TNF)-α and interleukin (IL)-1 α/β (Le Maitre, Freemont, and Hoyland 2005). These molecules are chemotactic to neutrophils, induce the expression of adhesion molecules on endothelial cells, can stimulate phagocytosis, and promote the production of several other molecules that increase the rate of matrix breakdown (Risbud and Shapiro 2014; Shinmei et al. 1988). IL-6 can also be secreted by disc cells, and its expression was shown to be elevated in herniated discs. This cytokine is believed to potentiate the catabolic action of both IL-1 and TNF-α (Studer et al. 2011). Other cytokines such as IL-17 and IFN-γ are involved in disc inflammatory response, by recruiting immune cells to the injury (Risbud and Shapiro 2014). Overall, these events result in an imbalance of catabolic and anabolic activities that contribute to tissue weakness. A decrease in the hydrostatic pressure

Cell Recruitment for Intervertebral Disc

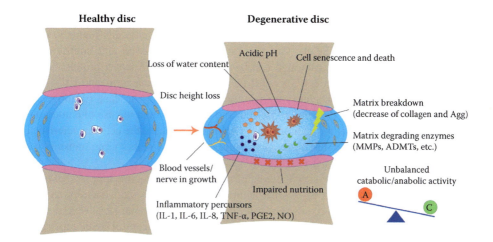

FIGURE 6.2 Disc degenerative process: changes and involved molecules. Healthy disc versus degenerative disc. Disc degeneration is characterized by an unbalanced anabolic/catabolic activity, which can be triggered by impaired nutrition (CEP blockage). The degenerative process is characterized by a decrease in the pH, by an increase in cell senescence and death, and by a catabolic environment (matrix breakdown by MMPs, ADAMTs, and inflammatory precursors) that lead to a great loss of water and, consequently, disc height. AF structure disruption due to the high hydrostatic pressure can result in blood vessel/nerve ingrowth and pain.

due to the loss of proteoglycans occurs and the disc space begins to collapse, culminating in loss of the biological function (Figure 6.2).

An important aspect that, together with other factors, might contribute to IVD degeneration is the progressive loss of notochordal cells. The NP tissue derives from the endoderm, being a remnant of the notochord (Roberts et al. 2006). Depending on the species, the notochord cells may persist to adulthood. Many animal species retain the presence of these cells (Hunter, Matyas, and Duncan 2004); however, in humans, the number of notochordal cells decreases drastically following birth, being nonidentifiable after 4 years (Roberts et al. 2006). Still, the presence or loss of notochordal cells in humans remains debatable, as the presence of these cells was demonstrated in human adult discs (Weiler et al. 2010). The loss or decrease of notochordal cells in IVD tissue might have implications on proteoglycan synthesis rate and distribution, therefore representing a key factor in the IVD degeneration process (Cappello et al. 2006).

6.4 CELL RECRUITMENT FOR ENDOGENOUS REPAIR/REGENERATION

In the past years, different regenerative strategies have been developed attempting to revert the degenerative process. Many have been focused on the application of therapeutic molecules, such as growth factors or genes, despite being limited by the low number of remaining cells found in the degenerative IVD, which might be insufficient to trigger repair, while others have been focused on the repopulation of the tissue with new cells, either endogenous NP cells or stem cells (namely, mesenchymal stem cells [MSCs]), that can integrate in the tissue and contribute to replenish the lost matrix. Although cell-based therapies represent a promise in the treatment of several disorders, they still face some ethical concerns and risks. Cell transplantation is still associated with immune rejection in the case of allogeneic cell sources (frequently the most accessible and adequate), and the use of autologous cells may lead to further tissue morbidity. Other risks concern the cell characteristics, such as differentiation status, tumorigenic potential, and diseases/virus transmission (Herberts, Kwa, and Hermsen 2011).

Alternatively, a new trend in the field of regenerative medicine appears, based on the mimicry of healing processes that naturally occur in our body. Stem/progenitor cells can be guided by intrinsic cues from their niche to a lesion site. Nevertheless, endogenous regeneration can be insufficient to achieve a successful tissue repair, but if enhanced, by maximizing the body's own regenerative capacity, it could potentially constitute an alternative strategy to achieve full tissue repair/regeneration.

6.4.1 Stem Cell Homing and Migration

Stem cell homing and migration represent similar concepts on stem cells' capacity to travel under guided navigation in our body. The homing process, designated following the original word "home," which commonly stands for residence, is used to designate the natural process by which hematopoietic cells in circulation migrate through the blood vessel and cross the vascular endothelium to the bone marrow (BM) (Lapidot, Dar, and Kollet 2005). Depending on the authors and the research field, the term homing might also be applied to the capacity of other stem cells to travel to organs in response to stress or injury, as it often happens in the regenerative medicine research field (Naderi-Meshkin et al. 2015). Still, stem cell researchers consider this as the migration/mobilization of these cells toward other tissues and sites of lesion following a stimulus, and not a homing process. This leads to the definition of home: in what concerns the progenitor cells' home/residence, it has been defined in the literature as their niche. The niche is characterized by a specific microenvironment, with appropriate anatomy and dimensions, that allows stem cells to reproduce and self-renew (Scadden 2006). The BM is the most well-known niche of stem cells, harboring both hematopoietic stem cells (HSCs), that give origin to all types of blood cells, and non-HSCs, such as MSCs, representing less than 0.01% of the BM mononuclear cells, or endothelial progenitor cells (EPCs) (Pittenger et al. 1999).

The BM is described to be the primary site of adult hematopoiesis (Kiel and Morrison 2008). MSCs, residing in the stromal component, have a key role in the support and maintenance of the HSCs' stemness, and although they can be found mostly in the BM, other niches have been described (da Silva Meirelles, Chagastelles, and Nardi 2006). MSCs can reside in multiple organs, such the adipose, skeletal, and muscle tissue, and participate in the endogenous repair process through their capacity to migrate to injury sites (Liu, Zhuge, and Velazquez 2009). MSCs are recognized as key players in this process since they are able to proliferate, differentiate, and functionally contribute to the regenerative process (Granero-Molto et al. 2008). Their mobilization to different tissues occurs following a chemotactic stimulus initiated by damage, such as trauma, fracture, inflammation, necrosis, or tumor presence (Fong, Chan, and Goodman 2011) (Figure 6.1).

The cell migration process is orchestrated by particular molecules, such as cytokines and chemokines, and involves the participation of several adhesion molecules such as selectins, chemokine receptors, and integrins (Butcher 1991; Springer 1994). Generally, it is assumed that MSCs' mobilization into the peripheral blood follows the same steps that were described for leukocyte and HSC recruitment to inflammation sites (Fox et al. 2007), although differences between the recruitment of MSCs and leucocytes have been described. These differences concern L- and E-selectin (Muller et al. 1993), which are low or absent in MSCs, and the absence of platelet/endothelial cell adhesion molecule 1/CD31 (Ruster et al. 2006). MSCs bind to endothelial cells in a P-selectin-dependent manner (rolling) (Figure 6.3), and upon activation, cell adhesion is mediated by the very-late antigen-4 (VLA-4)/vascular cell adhesion molecule (VCAM)-1 axis, while the extravasation to the matrix is mediated by matrix degrading enzymes, such as MMP-2 and membrane type 1 MMP (MT1-MMP), and tissue inhibitor of metalloproteinase 1 (Karp and Leng Teo 2009; Ries et al. 2007).

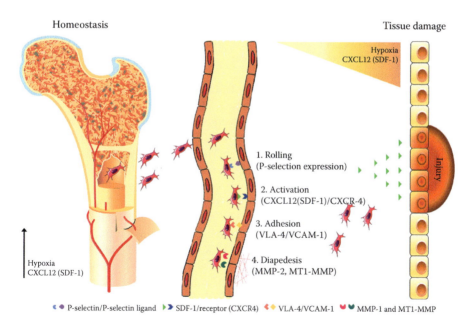

FIGURE 6.3 MSC mobilization to sites of damage. Mechanisms of MSC transendothelial migration toward injured tissue. MSCs can cross through a mechanism encompassing several phases of (1) rolling (involvement of *p*-selectin); (2) activation of chemokine receptors; (3) adhesion through adhesion molecules and their ligands (VLA-4; endothelial cells express the corresponding ligand, VCAM-1). Moreover, MSCs express the ECM-degrading enzymes, MMP-2, and membrane type (MT)1–MMP that play a role in their extravasation and, afterward, migration toward the lesion following the chemokine gradient.

6.4.1.1 Stem Cell Migration: Chemokines, Receptors, and Growth Factors

Cell recruitment is triggered by specific molecules, such as chemokines, cytokines, and growth factors, present in a chemotactic gradient (Rankin 2012). The chemokines are small peptides (8–14 kDa), grouped based on the arrangement of the terminal cysteine residues (C) that form the first disulphide bond (Zlotnik and Yoshie 2012). When adjacent, the residues are referred to as CC; when separated by an amino acid (X), they are referred to CXC. Chemokines can be further categorized based on their function: while some chemokines are involved in the leucocyte migration during a pathological condition, others are constitutively expressed in homeostatic conditions to coordinate the location and mobilization of cells (Anders, Romagnani, and Mantovani 2014). Until now, more than 40 chemokines have been described (Nomiyama, Osada, and Yoshie 2013). Chemokine signalling occurs through binding to chemokine receptors (CCR) or to atypical chemokine receptors, which are receptors that lack a specific conserved motif and do not mediate conventional signalling and directional migration (Bachelerie et al. 2014).

CXCL12, first identified as stromal cell derived factor-1α (SDF-1α) in the supernatant of BM stromal cells, is one of the most studied and important chemokines, due to its role in hematopoietic stem and progenitor cell homing and localization in the BM (Rankin 2012). Previous studies reported that SDF-1α activation occurs through two main receptors: CXCR4 and CXCR7. While SDF-1α/CXCR4 plays an important role in stem/progenitor cell migration, CXCR7 functions as a specific scavenger for SDF-1α and was described to modulate the activity of the ubiquitously expressed CXCR4, suggesting a key role in the fine-tuning of stem cell mobility (Naumann et al. 2010).

SDF-1α is a potent chemoattractant for HSCs (Aiuti et al. 1997). The importance of the SDF-1α/CXCR4 axis in HSC homing, maintenance of quiescence, and cell mobilization to the blood has been widely documented (Peled et al. 1999). Moreover, HSC mobilization is intimately associated with SDF-1α levels in the BM (Figure 6.3). Mice with CXCR4-/- chimeric BM were shown to have high levels of HSCs in the blood (Christopher et al. 2009), and the peaks of SDF-1α in the BM, which fluctuate according to the circadian variation, correlate with higher levels of HSCs in the BM and lower levels in the blood (Mendez-Ferrer et al. 2008). Other studies suggested SDF-1α as a key regulator of HSCs' quiescent state under homeostasis. Sugiyama et al. (2006) reported a reduction of HSC numbers and increased sensitivity to myelotoxic injury occurring following CXCR4 deletion in the adult mice. Tzeng et al. (2011) showed that the loss of stromal secreted SDF-1α in a murine with a conditional deletion of SDF-1α in adult stages led to a reduction of quiescent HSCs. Both studies demonstrated the importance of SDF-1α/CXCR4 signalling pathways in maintaining the quiescent HSC pool. Granulocyte colony-stimulating factor has also been described to be involved in HSC mobilization from the BM through the disruption of the SDF-1α/CXCR4 retention axis (Pitchford et al. 2009).

The SDF-1α receptor has also been described to be expressed in EPCs, and it is known that its antagonist, AMD3100, mobilizes these cells from the BM to the blood (Capoccia, Shepherd, and Link 2006; Shepherd et al. 2006). The mobilization of these cells may also occur via the vascular endothelial growth factor/vascular endothelial growth factor receptor-1 (VEGF/VEGFR2) axis, by triggering cell migration (Pitchford et al. 2009) or even through CXCR2, as demonstrated in the work of Jones et al. (2009) where this receptor was demonstrated to be critical for EPC recruitment and angiogenic response.

Mechanisms of MSC mobilization and the chemokines/receptors involved in the process have widely been studied due to the interest in these cells for tissue regeneration and repair. Like other cells, MSCs express many receptors and cell adhesion molecules that support cell homing and migration to target tissues in response to injury. Through the years, it has become generally accepted that the mobilization process of MSCs was highly dependent on the SDF-1α/CXCR4 axis (Figure 6.3). BM-derived MSCs were shown to express CXCR4 and migrate toward SDF-1α gradients *in vitro* (Ponte et al. 2007). Wynn et al. (2004) have shown that SDF-1α regulates the migration of MSCs in a dose-dependent manner, with maximal migration occurring at 30 ng/mL of SDF-1α. Cell migration was also pronounced in cells highly expressing the surface receptor CXCR4 (Wynn et al. 2004). The authors believe that mobilizing the internalized receptor and increasing its functional expression may play a key role in the mechanism of MSC engraftment to BM. Moreover, when inhibiting the receptor CXCR4, MSC migration was shown to be reduced by 46% (Wynn et al. 2004). Similar findings were found *in vivo*, where the SDF-1α/CXCR4 axis played a crucial role in BM-MSC migration to a fracture site. SDF-1α expression was increased during the healing process of live bone, suggesting it to be a key regulator in bone repair. Adding to this, the treatment of a fracture model of mice with SDF-1α improved cell migration toward the lesion, which was inhibited in the presence of a CXCR4 antagonist (Kitaori et al. 2009). Other studies further reported an increase in SDF-1α expression following facture both in rats and in plasma of human patients 2–4 days after the lesion (Kidd et al. 2010; Lee et al. 2010b), adding further evidence that the SDF-1α/CXCR4 pathway may have an important role in fracture healing (Yellowley 2013). Indirectly, MSC migration might also be triggered by hypoxia and expression of hypoxia-induced factor-1α in damage sites, which, in turn, may drive an up-regulation of SDF-1α in the tissue (Ceradini et al. 2004) (Figure 6.3).

In spite of being, by far, the most well-studied receptor of SDF-1α, the expression of CXCR4 in progenitor cells from BM is still controversial. Several studies demonstrated that CXCR4 expression is predominantly intracellular, while others defend the opposite (Honczarenko et al. 2006; Pelekanos et al. 2014). Pelekanos et al. compared the CXCR4 expression of both fetal BM-derived MSCs and adult cells. A very low expression ($3.8\% \pm 0.3\%$) of the receptor was detected by immune staining at the cell surface of fetal MSCs; however, when permeabilized, 50% to 90% of the cells stained positive for intracellular localization of CXCR4. This pattern was also found in adult MSCs, suggesting that this phenomenon is independent of cell maturity (Pelekanos et al. 2014). On the other hand, the work of Honczarenko and colleagues reported a surface expression of around 43% of the CXCR4 receptor in adult MSCs at early passages (passage 2), as assessed by flow cytometry and further confirmed by reverse transcription polymerase chain reaction (RT-PCR). However, at later passages (12-16 passages) there was a decrease in the cell surface receptors that resulted in the lack of responsiveness to chemokines. This loss was further accompanied by the decrease of several other molecules known to be involved in the migration process, such as adhesion molecules intracellular adhesion molecule (ICAM) and vascular cell adhesion molecule (VCAM) (Honczarenko et al. 2006). CXCR4 expression is dynamically regulated by external cues like hypoxia (Figure 6.3) (Schioppa et al. 2003) and can be up-regulated in adult MSCs following *in vitro* priming with a mixture of cytokines (Shi et al. 2007) or via viral transduction. Other cell surface molecules, such as integrin-β1, integrin-α4, and integrin ligands VCAM and ICAM, were found to be expressed by MSCs and also have a key role in the migration process and in the interaction with endothelial cells (Ip et al. 2007; Ruster et al. 2006). Although most of the current literature concerns the importance of SDF-1α and its receptor CXCR4, MSCs have been shown to express a wide number of chemokine receptors on the cell surface, which also play a role in the migration and guidance of the cells (Honczarenko et al. 2006), such as CXCR1, CXCR2, CXCR3, CXCR5, CXCR6, CX3CR1, CCR1, CCR3, CCR5, CCR7, CCR9, and CCR10 (Fox et al. 2007), although the heterogeneity of MSC populations does not allow the establishment of a specific repertoire of receptors and expression levels in culture, since they may vary with cell isolation, culture conditions, and passages (Honczarenko et al. 2006). Moreover, MSC expansion might lead to changes in MSC phenotype and to different chemokine receptors repertoire (Bara et al. 2014).

Besides the aforementioned cytokines, other molecules are also involved in MSC recruitment to injured tissues. Growth factors, such as platelet-derived growth factor (PDGF)-BB, PDGF-AB, epidermal growth factor (EGF), heparin-binding EGF-like growth factor, TGF-β, insulin EGF-like growth factor-1 (IGF-1), hepatocyte growth factor, basic fibroblast growth factor (bFGF), and thrombin have been reported to enhance MSC migration in appropriate concentrations (Li et al. 2007; Ozaki et al. 2007; Ponte et al. 2007). Monocyte chemotactic protein-1 (MCP-1) was also described to have a role in stimulating MSC migration in response to a primary breast tumor in mice (Dwyer et al. 2007), while MCP-3 was suggested as a potential recruitment factor for MSCs toward the heart (Schenk et al. 2007).

6.4.2 IMMUNE CELL MIGRATION: CHEMOKINES, RECEPTORS, AND GROWTH FACTORS

Similar to what has been described in the mobilization process of stem cells, chemokines and their receptors play an important role in immune cell mobilization and localized response (Surmi and Hasty 2010). These factors do not only guide the cells to target sites of infection and inflammation, but also they coordinate the interaction between these cells, thereby providing the appropriate and optimal adaptive immune response against pathogens, tumor cells, or dead cells (Sokol and Luster 2015). For example, in immune cell development, SDF-1α/CXCR4 interactions remain essential for BM retention and normal development of several immune cells, such as B cells, monocytes, macrophages, neutrophils, natural killer cells, and plasmocytoid dendritic cells (DCs) (Mercier, Ragu, and Scadden 2011). Monocyte exit from the BM seems to be dependent on CXCR4 and CCR2 in homeostatic conditions. Monocytes may further differentiate into proinflammatory (CCR2+) or anti-inflammatory (CX3CR1+) subsets. CX3CR1 expressing monocytes are thought to migrate to the periphery under CX3CL1 gradients and might develop into tissue macrophages following additional chemokine signals (Sokol and Luster 2015). Eosinophils released from the BM to the peripheral tissues are largely dependent on CCL11/CCR3 interactions, while basophil release is mainly mediated by CXCR4, although they constitutively express CXCR1, CCR1, CCR2, and CCR3 (Iikura et al. 2001; Palframan et al. 1998).

6.5 STRATEGIES TO ENHANCE CELL MIGRATION

Extending the body's inherent repair capacity by enhancing the mobilization of endogenous cells to participate in the regenerative process has been an attractive therapeutic strategy explored in diverse tissues. This goal can be achieved either by increasing the local concentration of a certain chemoattractant, using delivery systems or direct protein injection; or by overexpression of chemokine receptors in transplanted cells, to better respond to the recruitment stimuli of the lesion. Both strategies will be further discussed in the next sections.

Increasing the local concentration of chemokines in a certain tissue can be challenging due to the chemokines short half-life and rapid diffusion that can impair the formation of a gradient capable of promoting cell chemotaxis. Following these limitations, the development of delivery systems capable of protecting peptides from being cleaved by proteases and providing a sustained release became indispensable. In the past years, several chemokine delivery systems were developed for a variety of applications, such as guided immunotherapy and tissue regeneration.

6.5.1 STEM CELL RECRUITMENT: CHEMOKINE DELIVERY SYSTEMS

The great majority of the studies using chemokine delivery systems were developed for guided tissue regeneration and were focused on the SDF-1α/CXCR4 axis. These studies include *in vitro* and *in vivo* delivery of SDF-1α using different delivery systems/biomaterials (poly(lactide-co-glycolide) [PLGA], alginate, collagen, etc.) for the recruitment of transplanted or endogenous cells toward damaged tissues, such as bone, cartilage, tendons, heart, and wounds. *In vitro*, the ability of SDF-1α to be released from delivery systems and recruit MSCs has been widely demonstrated (Cross and Wang 2011; Goncalves, Antunes, and Barbosa 2012; He, Ma, and Jabbari 2010; Thieme et al. 2009). *In vivo*, numerous studies have applied the delivery of SDF-1α for tissue regeneration; some of those studies are summarized in Table 6.1.

Besides promoting cell recruitment toward the damaged tissue, in most cases, a positive effect on the tissue was reported (Table 6.1), reinforcing the role and participation of this chemokine in tissue repair and introducing new perspectives on the stimuli of endogenous repair. To date, there are no published or ongoing clinical trials using SDF-1α delivery platforms; still, there are two clinical trials, one completed and one recruiting, that use JVS-100, a nonviral gene that expresses SDF-1α for the treatment of peripheral arterial disease and critical limb ischemia (ClinicalTrials.gov identifier: NCT02544204 and NCT01410331, respectively).

SDF-1 is, by far, the most well-studied and applied chemokine to enhance stem cell migration toward a damage site; however, the delivery of other molecules has also been explored. Lee et al. (2010a) used a poly-epsilon-caprolactone scaffold for the delivery of TGF-β3 in the treatment of articular synovial joint in a rabbit

TABLE 6.1
In Vivo Studies Using SDF-1 Delivery Systems for Tissue Regeneration

Material	Chemokine	Model	Cells	Outcomes	Reference
PCL/gelatine membrane	SDF-1α	Rat cranial defect	rBM-MSCs GFP-rBM-MSCs	↑Bone formation	Ji et al. 2013
Collagen membrane scaffold	SDF-1α	Periodontal bone defect in the rat mandibular bone	Host cells	↑MSC/HSC cell engraftment ↓Inflammatory response ↑Osteoclastogenesis, angiogenesis ↑Bone regeneration	Liu et al. 2015
Collagen type I sponge scaffold	SDF-1α	Partial- and full-thickness cartilage defect in rabbit	C-hMSCs SM-hMSCs	↑Migration and adhesion of C/SM-MSCs	Zhang et al. 2013a
Ultrapurified alginate gel (UPAL gel)	SDF-1α	Full-thickness osteochondral defect in rabbit	Host cells	↑Migration of host BM MSCs ↑Histological scores ↑Compressive modulus of reparative tissues	Sukegawa et al. 2012
HA-PNIPAM hydrogel	SDF-1α	IVD nucleotomy *ex vivo*	hMSCs	↑Migration	Pereira et al. 2014
Silk-collagen sponge scaffold	SDF-1α	Achilles tendon injury model	Host cells	↑Neotendon tissue repair ↑CXCR4-expressing cells ↓Inflammatory cells	Shen et al. 2010

(Continued)

TABLE 6.1 (CONTINUED)
In Vivo Studies Using SDF-1 Delivery Systems for Tissue Regeneration

Material	Chemokine	Model	Cells	Outcomes	Reference
Alginate scaffolds	SDF-1α	Porcine wound healing model	Host cells	↑Wound healing ↓Scar formation	Rabbany et al. 2010
Gelatine hydrogel	SDF-1α + SEW2871	Mouse skin defects (wound healing)	BM-MSCs	↑MSC and macrophage recruitment ↑Wound healing	Kim and Tabata 2016
Hyaluronic acid-based hydrogel	SDF-1α (+angiogenic peptide)	Myocardial infarction model in rat	Host cells	↑Heart function ↑Angiogenesis ↓Infarct size	Song et al. 2014
HA-based hydrogel	SDF-1 analog	Myocardial infarction in rat	Host cells	↑Angiogenesis ↓Infarct size	MacArthur et al. 2013
PEGylated fibrinogen	SDF-1α	Acute myocardial infarction in mice	Host cells	↑C-kit (+) cells ↑Left ventricular function	Zhang et al. 2007
Hyaluronic acid hydrogel	SDF-1α	Myocardial infarction in mice	BM cells	↑BM cell homing	Purcell et al. 2012

Note: C-hMSCs, cartilage derived-hMSCs; GFP-rBM-MSCs, green fluorescent protein-rBM-MSCs; HA-PNIPAM, hyaluronan-poly(N-isopropylacrylamide); rBM-MSCs, rat bone marrow-MSCs; SM-hMSCs, synovial membrane-derived-hMSCs.

model. This system was proven to be efficient in recruiting host cells toward the lesion and to contribute to tissue regeneration (Lee et al. 2010a). bFGF delivery from collagen hydrogels, alone or in combination with other growth factors such as VEGF, PDGF, and nerve growth factor, has also been applied to dental pulp regeneration, resulting in the recellularization and revascularization of the damaged tissue (Kim et al. 2010).

6.5.1.1 Chemokine Receptor Overexpression in Stem Cells

CXCR4, the well-known receptor for SDF-1α/CXCL12, is highly expressed in MSCs residing in the BM, while its expression decreases with *in vitro* expansion (after passage 2) (Wynn et al. 2004). Thus, the normal laboratory routine in culturing MSCs impairs cell migration toward an SDF-1α gradient. One of the strategies to overcome this is to genetically modify MSCs to increase CXCR4 expression, thus enhancing the migration of systemically delivered MSCs toward the damaged tissue. This strategy has been applied in different studies, with promising results.

MSCs overexpressing CXCR4 via retroviral (Cheng et al. 2008) or adenoviral (Zhang et al. 2008) transduction were shown to enhance *in vivo* mobilization and engraftment of MSCs into an ischemic area, where these cells promoted neo-myoangiogenesis and alleviated early signs of left ventricular remodeling (Cheng et al. 2008; Zhang et al. 2008). Thieme et al. (2009) demonstrated both *in vivo* and *in vitro* that transient overexpression of CXCR4 in human MSCs induced by transfection enhanced SDF-1-directed chemotactic capacity to invade internal compartments of porous 3D col type I scaffolds. In the study of Ma et al. (2014), genetically modified MSCs expressing CXCR4 through lentiviral transduction migrated to the liver in larger numbers and conferred better functional recovery in damaged liver.

Although the results are encouraging, gene overexpression engages several risks: gene transfer based on viral vectors is still questionable for clinical applications (Sadelain 2004). An alternative would be transient exogenous gene transfection, which has the advantage of being rapidly degraded, limiting the duration of overexpression and thus reducing the risk of mutagenic events (Ryser et al. 2008).

6.5.2 IMMUNE CELL MIGRATION: CHEMOKINE DELIVERY SYSTEMS

Although to a small extent, the delivery of chemokines aiming to recruit immune cells has also been the focus of some studies, which reported different materials for the delivery of chemokines, such as CCL20 and CCL19, aiming to enhance cell migration. Zhao et al. combined CCL20 and PLGA microspheres to induce chemotaxis of DCs. DCs could be attracted toward the microspheres, suggesting that they could be effective for enriching DCs at an immunization site *in vivo* (Zhao et al. 2005). A similar strategy was reported by Singh, Suri, and Roy (2009) using crosslinked hydrogels comprising dextran vinyl sulfone and reactive polyethylene glycol (PEG) combined with CCL20, for DC recruitment. Sustained chemokine release from the gels attracted fourfold and sixfold more DCs than a single dose (Singh, Suri, and Roy 2009). Alginate microspheres have also been used to load and release several chemokines important in immunity, including CCL19, CCL21, CXCL12, and CXCL10 (Wang and Irvine 2011).

6.6 CELL MIGRATION IN THE IVD—A NEW PARADIGM?

6.6.1 IVD Degeneration: Involved Cytokines and Chemokines

The process of IVD degeneration is characterized by the release of several inflammatory cytokines, chemokines, and growth factors. Among the secreted proinflammatory mediators are TNF-α, IL-1 α/β, IL-6, IL-17 (Risbud and Shapiro 2014), and chemokines such as CCL5/RANTES (regulated on activation, normal T cell expressed and secreted) (Pattappa et al. 2014), which were reported to play a key role during disc degeneration and may trigger the migration of cells able to modulate inflammation and participate in tissue repair. Both IL-1 and TNF-α are known to induce significant cellular and matrix changes by increasing the expression of matrix degrading enzymes such as MMP-3, MMP-13, and ADAMTS-1, while decreasing the expression of matrix-related genes (col type I and II and agg) (Le Maitre, Freemont, and Hoyland 2005; Seguin et al. 2005) in the IVD. On the other hand, the inflammatory cytokines can also stimulate discs to produce chemotactic factors, able to recruit immune cells, such as MCP-1, CCL3, CCL4, CCL5, MCP-3, MCP-4, and CXCL10, among others.

A summary of cytokines and chemokines identified to date that are expressed during disc degeneration is presented in Table 6.2. Among several factors, CCL5/RANTES has been identified by Pattappa et al. (2014) together with CXCL6, in the conditioned media of degenerated IVD organ cultures using proteomic analysis. In migration assays, the CCL5 and CCL5/CXCL6 immunoprecipicitation resulted in a reduced MSC migration toward conditioned medium, while the CXCL6 immunoprecipicitation did not affect MSC chemotaxis. In addition, mRNA expression analysis of MSCs cultured in IVD degenerative media revealed a significant increase in the expression of the CCL5 receptors, CCR1 and CCR4. CCL5/RANTES expression in the tissue of both bovine and human discs was further confirmed by immunohistochemistry (Pattappa et al. 2014).

In the study of Kepler et al. (2013), the expression levels of CCL5/RANTES, IL-1β, IL-6, and IL-8 in painful discs removed from human patients were analyzed by quantitative RT-PCR. The authors described that CCL5/RANTES was significantly elevated in painful discs when compared to controls and this increase was correlated with an increased expression of IL-1β (Kepler et al. 2013). These results were further corroborated by a recent study by Grad et al. (2016) that investigated the levels of CCL5 and CXCL6 in the Hong Kong Disc Degeneration Population-Based Cohort of Southern Chinese. Levels of both chemokines were elevated in blood plasma samples of subjects with disc degeneration when compared to nondegenerated individuals, associating for the first time CCL5/RANTES and CXCL6 with moderate/severe lumbar disc degeneration and suggesting that these chemokines could be used as systemic biomarkers for the diagnosis and monitoring of disc degeneration (Grad et al. 2016).

CCL3, also known as macrophage inflammatory protein (MIP)-1α, was shown to be upregulated in NP cells isolated from rat and human IVD tissue following treatment with IL-1β or TNF-α. In the same study, CCL3 expression in human samples was shown to be correlated with the grade of tissue degeneration (Wang et al. 2013). This molecule induced macrophage migration after treatment with NP cells

TABLE 6.2
Cytokines and Chemokines Identified in the IVD

Molecule	Species	Model	Role	Reference
Cytokines				
IL-1	H	Degenerated and herniated discs	Chemotactic for neutrophils	Kepler et al. 2013; Risbud and Shapiro 2014
IL-6	H	Human discs	Chemotactic for T cells	Kepler et al. 2013
IL-8	H	Herniated discs	Chemotactic for neutrophils	Ahn et al. 2002; MGC Project Team et al. 2009
IL-17	H	Human discs	Chemotactic for monocytes and neutrophils	Shamji et al. 2010
TNF-α	H	Degenerated and herniated discs	Chemotactic for neutrophils	Risbud and Shapiro 2014; MGC Project Team et al. 2009
Chemokines				
CCL2/MCP-1	H	Human discs	Chemotactic for lymphocytes and mononuclear cell	Phillips et al. 2013
CCL3	H, R	NP cells	Chemotactic for immune cells	Wang et al. 2013
CCL5/RANTES	B	Organ culture IVD with CEPs	Chemotactic for MSCs	Pattappa et al. 2014
	H	Human discs	–	Kepler et al. 2013
	H	Blood plasma samples	–	Grad et al. 2016
CCL7/MCP-3	H	Herniated lumbar disc specimens	Chemotactic for immune and stem cells	Kawaguchi et al. 2002; Phillips et al. 2013
CCL20	H	Pathological discs	Chemotactic for Th17 cells	Zhang et al. 2013b
CXCL9	H	Human discs	–	Phillips et al. 2013
CXCL6	B	Organ culture IVD with CEPs	Chemotactic for immune cells and MSCs	Pattappa et al. 2014
	H	Blood plasma samples	–	Grad et al. 2016
CXCL10	H	AF cells	Chemotactic for AF cells	Hegewald et al. 2012
CXCL12/SDF-1	B	Organ culture IVD with CEPs	Chemotactic for MSCs	Pereira et al. 2014

Note: B, bovine; H, human; R, rat.

conditioned media (previously treated with IL-1β or TNF-α); migration was suggested to occur via CCR1, the primary receptor of CCL3, since its inhibition resulted in cell migration blockage (Wang et al. 2013). CCL2 and CCL3 gene expression were shown to be significantly up-regulated in both human AF and NP cells after treatment with IL-1β (Liu et al. 2017b; Wang et al. 2013). Additionally, some of those chemokines, such as CCL2, CCL7, and CXCL18, have been correlated with histological degenerative tissue changes (Phillips et al. 2013) and are known to be involved in the recruitment of immune cells to inflammatory sites (Luster 1998).

CCL20 expression in NP cells has also been reported. Both CCL20 and its receptor CCR6 have been associated with IVD degenerative conditions (Zhang et al. 2013b). CCR6, the only receptor of CCL20, is specifically expressed on the Th17 cell surface and is related to the recruitment of these cells in several inflammatory diseases; higher levels of IL-17 (T-helper 17 and Th17 associated cytokine) have been observed in patients with IVD degeneration (Shamji et al. 2010). This interaction was reinforced in a study using a rat model; Zhang et al. (2016) could show a positive correlation between the expression levels of CCL20-, CCR6-, and IL-17-producing cells, suggesting that the recruitment of these cells to degenerated IVD tissues occurred via the CCL20/CCR6 system *in vivo*. An association between AF cell migration and CXCL10 had already been established, suggesting a role of this chemokine in AF homeostasis and repair (Hegewald et al. 2012). Taken together, these results demonstrate the involvement of several chemotactic molecules in the disc degeneration pathogenesis. Moreover, most of the studies could identify cytokines and chemokines that are directly or indirectly related to the recruitment of immune cells, thereby intensifying the inflammatory response and the release of neurotrophines, promoting pain. The study of these molecules is relevant not only for the design of new therapies, by targeting symptomatic discs and the associated inflammation, but also for understanding the players in stem cell recruitment for repair.

6.6.2 Cell Recruitment in the IVD

The ability to recruit cells to several tissues by the delivery of chemokines has widely been explored in different tissues, whereas in the IVD, mainly due to its avascular nature, the migration/recruitment of cells is still at the beginning of being explored.

The results published in 2012 by Illien-Junger et al. (2012) were a game-changer in the IVD regenerative therapy, by demonstrating for the first time that MSCs could be recruited and migrate toward discs cultured under degenerative conditions, suggesting chemokine-mediated recruitment. Later, Pattappa et al. (2014) identified CCL5 and CXCL6 as the main chemokines responsible for human MSC (hMSC) migration to the IVD cultured under degenerative conditions. The relevance of this chemokine was further confirmed by Grad et al. (2016) and Kepler et al. (2013) in humans, as previously described. Together, these results have revealed tissue attempts to repair, suggesting that some inherent endogenous repair capacity could be explored. Adding to this, Henriksson et al. (2009, 2012) have been exploring stem cell niches around the IVD tissue that can constitute targets for these chemoattractant molecules or for exogenous chemoattractant therapies. A migration route from stem cell niches around the IVD toward AF and NP was proposed, by providing evidence of gradual migration of proliferating cells (BrdU+), which expressed both cell migration molecules such as SLUG, SNAI1, β1-INTEGRIN, and prechondrogenic markers such as GDF5 and SOX9 (Henriksson et al. 2012). These proliferating cells were first reported as stem cells based on the expression of certain progenitor markers, such as Notch1, Delta4, C-KIT, and Stro-1, along with the proliferation marker Ki67. The presence of these cells was confirmed in several mammals (rat, porcine, and human tissues), with similar pattern and frequency, suggesting interspecies similarity and demonstrating the presence of progenitor cells within the disc and stem cell niches in the IVD region (Henriksson et al. 2009). More recently, the same group demonstrated the presence of stem cells (CD90$^+$, OCT3/4$^+$), prechondrocytic cells (GDF5+), cells expressing migration markers (SNAI1+ and SNAI2$^+$), catabolic markers (MMP9$^+$ and MMP13$^+$), inflammatory marker (IL1R$^+$), and adhesion markers (DDR2$^+$ and β1-INTEGRIN$^+$) in the previously proposed migration route around the IVD region (Henriksson et al. 2015). Other molecules, such as MMP-9, MMP-12, DDR2, and β1-INTEGRIN suggest that ECM turnover and cellular motility occurs, demonstrating the role of ECM composition and architecture in stem cell migration (Henriksson et al. 2015). Additionally, isolation of MSC-like cells from the different areas of the IVD (AF, NP, and CEP) was recently demonstrated. Cells isolated from the three distinct tissues of the IVD were compared regarding their proliferation, passage, colony formation, migration, and invasion capacity and presented similar features, despite cells isolated from the CEP revealing higher migration capacity together with higher expression of CXCR4 (Liu et al. 2017a), suggesting that CEP-derived MSC migration might be a target to enhance treatment efficiency. Furthermore, the expression of CCR5 at the RNA level in both NP and AF cells was also demonstrated (Liu et al. 2017b). In the presence of CCL5/RANTES, increased levels of extracellular signal-regulated kinase phosphorylation were detected along with AF cell migration, suggesting that CCR5 receptor in AF cells is functional and, therefore, AF cells may have the ability to migrate in response to disc damage or inflammation (Liu et al. 2017b).

Overall, the presence of progenitor cells in different locations of the healthy and degenerated IVD opens new perspectives regarding the possibility for the mobilization/activation of these populations toward IVD repopulation and regeneration (Henriksson et al. 2009; Risbud et al. 2007; Sakai et al. 2012). These studies, by demonstrating the existence of closer stem cell niches and by supporting the idea that chemokines and chemokine-receptors might be involved in the migration process to damaged discs, have given new insights in the endogenous repair capacity of this tissue.

Still, endogenous repair has demonstrated to be insufficient in restoring IVD upon damage/degeneration; therefore, there was growing interest in strategies to enhance endogenous IVD repair capacity in the past years.

Pereira et al. (2016) investigated the potential of hMSCs seeded on the CEP to migrate toward a damaged IVD. These cells were able to migrate and contribute to ECM remodeling by improving the expression of col type II and agg in the disc. The same group has also explored the injection of a chemoattractant delivery system to enhance MSC migration toward the IVD, as previously used in different contexts aforementioned (Table 6.1). Using an *ex vivo* organ culture of nucleotomized IVDs, the sustained delivery of SDF-1 from an HA-based hydrogel was demonstrated *in vitro* and *ex vivo*. SDF-1 release significantly increased MSC migration from the CEP, toward both the NP and AF of the disc. This was not the case when SDF-1 was directly administrated to the IVD tissue, suggesting that the delivery system had a key role in protecting the chemokine against degradation and therefore guaranteeing its biological function. Moreover, the number of cells migrating toward the IVD was affected by the MSC donor's age: cells provided by older donors migrated in fewer numbers when compared to cells from younger donors (Pereira et al. 2014). Another approach was proposed by Wei et al. in 2016, aiming to improve IVD repair based on the SDF-1/CXCR4 axis, by using CXCR4-overexpressing MSCs (Wei et al. 2016). Wei et al. transplanted CXCR4-MSCs into a rabbit punctured disc and was able to track them until 16 weeks posttransplantation. Overexpression of CXCR4 promoted MSC retention in the IVD and enhanced IVD regeneration (Wei et al. 2016).

Although not addressing directly a strategy to enhance repair, Tam et al. (2014) focused on the migration of multipotent stem cells upon delivery using two different methods, in a model of mouse caudal puncture-induced degeneration, and described the advantages of one method over the other to stimulate IVD repair. Stem cells were injected intravenously or intradiscally, and both methods resulted in an increased expression of agg, although only the direct injection contributed to improve IVD morphology and disc height. This is partially explained by the limited migration to the site of lesion of cells injected intravenously, as only few cells could migrate toward the IVD. Still, a positive effect was observed using both methods, suggesting a paracrine effect of the stem cells, and although few cells reached the disc, this indicates that *in vivo*, the lesion itself might trigger cell recruitment (Tam et al. 2014). In 2015, Sakai et al. published a pilot study that aimed to assess *in vivo* the mobilization of MSCs to the disc (Sakai et al. 2015). In this study, a BM-chimeric mouse expressing enhanced green fluorescent protein (GFP) was used and disc degeneration was induced by tail-looping, thus establishing a new model of IVD degeneration. The results provided first evidence of MSCs' dynamic response following disc degeneration but also reinforced the importance of vasculature in this process, by showing that BM-MSC distribution decreased with increased distance from the vascularized areas (growth plate and CEP). In the nucleus aspiration group, the needle punch seemed to favor cell migration into the disc via the AF; still, when no AF rupture was present, the migration was more likely to occur through the CEP, which can itself also represent a barrier in limiting cellular penetration into the IVD (Sakai et al. 2015). The avascular nature of the IVD was, again, shown to be one of the biggest limitations of this tissue's self-repair and regenerative potential.

Although few studies have addressed strategies to enhance IVD repair, so far, the results obtained are encouraging and will certainly contribute to the development of new approaches for IVD repair stimulation using chemokines or growth factors that may enhance or accelerate endogenous cell migration toward the disc.

6.7 CONCLUSIONS

Stem cell recruitment for IVD degeneration is a very recent field of research, and therefore, there are still several questions that remain to be explored. Although the delivery of bioactive molecules in the IVD to stimulate stem cell migration and repopulation of the degenerated tissue represents an alternative strategy to the common cell-based therapies, the few studies using this approach are insufficient to address the real benefit of such strategies over others currently applied. Cell-based therapies, using cell transplantation, have already shown promising results in humans in what concerns the positive effect of autologous disc or stem cells, namely in pain relief (Elabd et al. 2016; Mochida et al. 2015; Orozco et al. 2011; Pettine et al. 2015; Yoshikawa et al. 2010). *Ex vivo*, MSCs migrating from the CEP toward a nucleotomized disc have also demonstrated beneficial effects in extracellular remodeling (Pereira et al. 2016), although more studies would be needed to confirm the real potential of stem cell recruitment in restoring IVD matrix and function, namely, by applying such strategy *in vivo*. Moreover, a deep understanding of the degenerative process and degenerative stages in which this approach can be valuable is needed. Recent advances in the imaging techniques will also further contribute to the early diagnosis of IVD degeneration, allowing the application of treatments that aim to halt and prevent further degeneration, in the early stages of the degenerative process.

Studies on the stimuli of endogenous stem cell recruitment as a treatment for IVD degeneration are becoming more frequent, and the recent findings on the chemokines released during the degenerative process will further contribute to the full understanding of IVD endogenous regeneration mechanisms through which new therapies can act.

ACKNOWLEDGMENTS

The authors would like to acknowledge funding from European Union funds through "Projetos Estruturados de I&D&I—Norte-01-0145-FEDER-000012, Portugal 2020—FEDER." RMG and CLP would like to acknowledge FCT–Fundação para a Ciência e a Tecnologia for the FCT Investigator Starting grant (IF/00638/2014) and the PhD grant (SFRH/BD/85779/2012), respectively.

REFERENCES

Ahn, S.H., Y.W. Cho, M.W. Ahn et al. 2002. "mRNA expression of cytokines and chemokines in herniated lumbar intervertebral discs." *Spine (Phila Pa 1976)* 27 (9):911–7.

Aiuti, A., I.J. Webb, C. Bleul, T. Springer, and J.C. Gutierrez-Ramos. 1997. "The chemokine SDF-1 is a chemoattractant for human CD34+ hematopoietic progenitor cells and provides a new mechanism to explain the mobilization of CD34+ progenitors to peripheral blood." *J Exp Med* 185 (1):111–20.

Anders, H.J., P. Romagnani, and A. Mantovani. 2014. "Pathomechanisms: Homeostatic chemokines in health, tissue regeneration, and progressive diseases." *Trends Mol Med* 20 (3):154–65. doi: 10.1016/j.molmed.2013.12.002.

Antoniou, J., T. Steffen, F. Nelson et al. 1996. "The human lumbar intervertebral disc: Evidence for changes in the biosynthesis and denaturation of the extracellular matrix with growth, maturation, aging, and degeneration." *J Clin Invest* 98 (4):996–1003. doi: 10.1172/JCI118884.

Bachelerie, F., G.J. Graham, M. Locati et al. 2014. "New nomenclature for atypical chemokine receptors." *Nat Immunol* 15 (3):207–8. doi: 10.1038/ni.2812.

Bara, J.J., R.G. Richards, M. Alini, and M.J. Stoddart. 2014. "Concise review: Bone marrow-derived mesenchymal stem cells change phenotype following in vitro culture:

Implications for basic research and the clinic." *Stem Cells* 32 (7):1713–23. doi: 10.1002/stem.1649.

Boos, N., R. Rieder, V. Schade et al. 1995. "1995 Volvo Award in clinical sciences. The diagnostic accuracy of magnetic resonance imaging, work perception, and psychosocial factors in identifying symptomatic disc herniations." *Spine (Phila Pa 1976)* 20 (24):2613–25.

Buckwalter, J.A. 1995. "Aging and degeneration of the human intervertebral disc." *Spine (Phila Pa 1976)* 20 (11):1307–14.

Butcher, E.C. 1991. "Leukocyte-endothelial cell recognition: Three (or more) steps to specificity and diversity." *Cell* 67 (6):1033–6.

Capoccia, B.J., R.M. Shepherd, and D.C. Link. 2006. "G-CSF and AMD3100 mobilize monocytes into the blood that stimulate angiogenesis *in vivo* through a paracrine mechanism." *Blood* 108 (7):2438–45. doi: 10.1182/blood-2006-04-013755.

Cappello, R., J.L.E. Bird, D. Pfeiffer, M.T. Bayliss, and J. Dudhia. 2006. "Notochordal cell produce and assemble extracellular matrix in a distinct manner, which may be responsible for the maintenance of healthy nucleus pulposus." *Spine* 31 (8):873–82. doi: 10.1097/01.brs.0000209302.00820.fd.

Ceradini, D.J., A.R. Kulkarni, M.J. Callaghan et al. 2004. "Progenitor cell trafficking is regulated by hypoxic gradients through HIF-1 induction of SDF-1." *Nat Med* 10 (8):858–64. doi: http://www.nature.com/nm/journal/v10/n8/suppinfo/nm1075_S1.html.

Cheng, Z., L. Ou, X. Zhou et al. 2008. "Targeted migration of mesenchymal stem cells modified with CXCR4 gene to infarcted myocardium improves cardiac performance." *Mol Ther* 16 (3):571–9. doi: 10.1038/sj.mt.6300374.

Cheung, K.M., J. Karppinen, D. Chan et al. 2009. "Prevalence and pattern of lumbar magnetic resonance imaging changes in a population study of one thousand and forty-three individuals." *Spine (Phila Pa 1976)* 34 (9):934–40. doi: 10.1097/BRS.0b013e3181a01b3f.

Christopher, M.J., F. Liu, M.J. Hilton, F. Long, and D.C. Link. 2009. "Suppression of CXCL12 production by bone marrow osteoblasts is a common and critical pathway for cytokine-induced mobilization." *Blood* 114 (7):1331–9. doi: 10.1182/blood-2008-10-184754.

Crean, J.K., S. Roberts, D.C. Jaffray, S.M. Eisenstein, and V.C. Duance. 1997. "Matrix metalloproteinases in the human intervertebral disc: Role in disc degeneration and scoliosis." *Spine (Phila Pa 1976)* 22 (24):2877–84.

Cross, D.P., and C. Wang. 2011. "Stromal-derived factor-1 alpha-loaded PLGA microspheres for stem cell recruitment." *Pharm Res* 28 (10):2477–89. doi: 10.1007/s11095-011-0474-x.

da Silva Meirelles, L., P.C. Chagastelles, and N.B. Nardi. 2006. "Mesenchymal stem cells reside in virtually all post-natal organs and tissues." *J Cell Sci* 119 (Pt 11):2204–13. doi: 10.1242/jcs.02932.

Devereaux, M.W. 2007. "Anatomy and examination of the spine." *Neurol Clin* 25 (2):331–51. doi: 10.1016/j.ncl.2007.02.003.

Dwyer, R.M., S.M. Potter-Beirne, K.A. Harrington et al. 2007. "Monocyte chemotactic protein-1 secreted by primary breast tumors stimulates migration of mesenchymal stem cells." *Clin Cancer Res* 13 (17):5020–7. doi: 10.1158/1078-0432.CCR-07-0731.

Elabd, C., C.J. Centeno, J.R. Schultz et al. 2016. "Intra-discal injection of autologous, hypoxic cultured bone marrow-derived mesenchymal stem cells in five patients with chronic lower back pain: A long-term safety and feasibility study." *J Transl Med* 14:253. doi: 10.1186/s12967-016-1015-5.

Fong, E.L., C.K. Chan, and S.B. Goodman. 2011. "Stem cell homing in musculoskeletal injury." *Biomaterials* 32 (2):395–409. doi: 10.1016/j.biomaterials.2010.08.101.

Fox, J.M., G. Chamberlain, B.A. Ashton, and J. Middleton. 2007a. "Recent advances into the understanding of mesenchymal stem cell trafficking." *Br J Haematol* 137 (6):491–502. doi: 10.1111/j.1365-2141.2007.06610.x.

Goncalves, R.M., J.C. Antunes, and M.A. Barbosa. 2012. "Mesenchymal stem cell recruitment by stromal derived factor-1-delivery systems based on chitosan/poly (gamma-glutamic acid) polyelectrolyte complexes." *Eur Cell Mater* 23:249–60; discussion 260–1.

Grad, S., C. Bow, J. Karppinen et al. 2016. "Systemic blood plasma CCL5 and CXCL6: Potential biomarkers for human lumbar disc degeneration." *Eur Cell Mater* 31:1–10.

Granero-Molto, F., J.A. Weis, L. Longobardi, and A. Spagnoli. 2008. "Role of mesenchymal stem cells in regenerative medicine: Application to bone and cartilage repair." *Exp Opin Biol Ther* 8 (3):255–68. doi: 10.1517/14712598.8.3.255.

Gruber, H.E., and E.N. Hanley, Jr. 1998. "Analysis of aging and degeneration of the human intervertebral disc. Comparison of surgical specimens with normal controls." *Spine (Phila Pa 1976)* 23 (7):751–7.

He, X., J. Ma, and E. Jabbari. 2010. "Migration of marrow stromal cells in response to sustained release of stromal-derived factor-1alpha from poly (lactide ethylene oxide fumarate) hydrogels." *Int J Pharm* 390 (2):107–16. doi: 10.1016/j.ijpharm.2009.12.063.

Hegewald, A.A., K. Neumann, G. Kalwitz et al. 2012. "The chemokines CXCL10 and XCL1 recruit human annulus fibrosus cells." *Spine (Phila Pa 1976)* 37 (2):101–7. doi: 10.1097/BRS.0b013e318210ed55.

Henriksson, H.B., N. Papadimitriou, S. Tschernitz et al. 2015. "Indications of that migration of stem cells is influenced by the extra cellular matrix architecture in the mammalian intervertebral disk region." *Tissue Cell* 47 (5):439–55. doi: 10.1016/j.tice.2015.08.001.

Henriksson, H., M. Thornemo, C. Karlsson et al. 2009. "Identification of cell proliferation zones, progenitor cells and a potential stem cell niche in the intervertebral disc region: A study in four species." *Spine (Phila Pa 1976)* 34 (21):2278–87. doi: 10.1097/BRS.0b013e3181a95ad2.

Henriksson, H.B., E. Svala, E. Skioldebrand, A. Lindahl, and H. Brisby. 2012. "Support of concept that migrating progenitor cells from stem cell niches contribute to normal regeneration of the adult mammal intervertebral disc: A descriptive study in the New Zealand white rabbit." *Spine* 37 (9):722–32. doi: 10.1097/BRS.0b013e318231c2f7.

Herberts, C.A., M.S. Kwa, and H.P. Hermsen. 2011. "Risk factors in the development of stem cell therapy." *J Transl Med* 9:29. doi: 10.1186/1479-5876-9-29.

Honczarenko, M., Y. Le, M. Swierkowski et al. 2006. "Human bone marrow stromal cells express a distinct set of biologically functional chemokine receptors." *Stem Cells* 24 (4):1030–41. doi: 10.1634/stemcells.2005-0319.

Hunter, C.J., J.R. Matyas, and N.A. Duncan. 2004. "Cytomorphology of notochordal and chondrocytic cells from the nucleus pulposus: A species comparison." *J Anat* 205 (5):357–62. doi: 10.1111/j.0021-8782.2004.00352.x.

Iikura, M., M. Miyamasu, M. Yamaguchi et al. 2001. "Chemokine receptors in human basophils: Inducible expression of functional CXCR4." *J Leukoc Biol* 70 (1):113–20.

Illien-Junger, S., G. Pattappa, M. Peroglio et al. 2012. "Homing of mesenchymal stem cells in induced degenerative intervertebral discs in a whole organ culture system." *Spine (Phila Pa 1976)* 37 (22):1865–73. doi: 10.1097/BRS.0b013e3182544a8a.

Ip, J.E., Y. Wu, J. Huang et al. 2007. "Mesenchymal stem cells use integrin beta1 not CXC chemokine receptor 4 for myocardial migration and engraftment." *Mol Biol Cell* 18 (8):2873–82. doi: 10.1091/mbc.E07-02-0166.

Jensen, M.C., M.N. Brant-Zawadzki, N. Obuchowski et al. 1994. "Magnetic resonance imaging of the lumbar spine in people without back pain." *N Engl J Med* 331 (2):69–73. doi: 10.1056/NEJM199407143310201.

Ji, W., F. Yang, J. Ma et al. 2013. "Incorporation of stromal cell-derived factor-1alpha in PCL/gelatin electrospun membranes for guided bone regeneration." *Biomaterials* 34 (3):735–45. doi: 10.1016/j.biomaterials.2012.10.016.

Jones, C.P., S.C. Pitchford, C.M. Lloyd, and S.M. Rankin. 2009. "CXCR2 mediates the recruitment of endothelial progenitor cells during allergic airways remodeling." *Stem Cells* 27 (12):3074–81. doi: 10.1002/stem.222.

Kanemoto, M., S. Hukuda, Y. Komiya, A. Katsuura, and J. Nishioka. 1996. "Immunohistochemical study of matrix metalloproteinase-3 and tissue inhibitor of metalloproteinase-1 human intervertebral discs." *Spine (Phila Pa 1976)* 21 (1):1–8.

Karp, J.M., and G.S. Leng Teo. 2009. "Mesenchymal stem cell homing: The devil is in the details." *Cell Stem Cell* 4 (3):206–16. doi: 10.1016/j.stem.2009.02.001.

Kawaguchi, S., T. Yamashita, G. Katahira et al. 2002. "Chemokine profile of herniated intervertebral discs infiltrated with monocytes and macrophages." *Spine (Phila Pa 1976)* 27 (14):1511–6.

Kepler, C.K., D.Z. Markova, F. Dibra et al. 2013. "Expression and relationship of proinflammatory chemokine RANTES/CCL5 and cytokine IL-1beta in painful human intervertebral discs." *Spine (Phila Pa 1976)* 38 (11):873–80. doi: 10.1097/BRS.0b013e318285ae08.

Kidd, L.J., A.S. Stephens, J.S. Kuliwaba et al. 2010. "Temporal pattern of gene expression and histology of stress fracture healing." *Bone* 46 (2):369–78. doi: 10.1016/j.bone.2009.10.009.

Kiel, M.J., and S.J. Morrison. 2008. "Uncertainty in the niches that maintain haematopoietic stem cells." *Nat Rev Immunol* 8 (4):290–301. doi: 10.1038/nri2279.

Kim, J.Y., X. Xin, E.K. Moioli et al. 2010. "Regeneration of dental-pulp-like tissue by chemotaxis-induced cell homing." *Tissue Eng Part A* 16 (10):3023–31. doi: 10.1089/ten.TEA.2010.0181.

Kim, Y.H., and Y. Tabata. 2016. "Recruitment of mesenchymal stem cells and macrophages by dual release of stromal cell-derived factor-1 and a macrophage recruitment agent enhances wound closure." *J Biomed Mater Res A* 104 (4):942–56. doi: 10.1002/jbm.a.35635.

Kitaori, T., H. Ito, E.M. Schwarz et al. 2009. "Stromal cell-derived factor 1/CXCR4 signaling is critical for the recruitment of mesenchymal stem cells to the fracture site during skeletal repair in a mouse model." *Arthritis Rheum* 60 (3):813–23. doi: 10.1002/art.24330.

Kuslich, S.D., C.L. Ulstrom, and C.J. Michael. 1991. "The tissue origin of low back pain and sciatica: A report of pain response to tissue stimulation during operations on the lumbar spine using local anesthesia." *Orthop Clin North Am* 22 (2):181–7.

Lapidot, T., A. Dar, and O. Kollet. 2005. "How do stem cells find their way home?" *Blood* 106 (6):1901–10. doi: 10.1182/blood-2005-04-1417.

Le Maitre, C.L., A.J. Freemont, and J.A. Hoyland. 2005. "The role of interleukin-1 in the pathogenesis of human intervertebral disc degeneration." *Arthritis Res Ther* 7 (4):R732–45. doi: 10.1186/ar1732.

Lee, C.H., J.L. Cook, A. Mendelson et al. 2010a. "Regeneration of the articular surface of the rabbit synovial joint by cell homing: A proof of concept study." *Lancet* 376 (9739):440–8. doi: 10.1016/S0140-6736(10)60668-X.

Lee, D.Y., T.J. Cho, H.R. Lee et al. 2010b. "Distraction osteogenesis induces endothelial progenitor cell mobilization without inflammatory response in man." *Bone* 46 (3):673–9. doi: 10.1016/j.bone.2009.10.018.

Li, Y., X. Yu, S. Lin et al. 2007. "Insulin-like growth factor 1 enhances the migratory capacity of mesenchymal stem cells." *Biochem Biophys Res Commun* 356 (3):780–4. doi: 10.1016/j.bbrc.2007.03.049.

Liu, H., M. Li, L. Du, P. Yang, and S. Ge. 2015. "Local administration of stromal cell-derived factor-1 promotes stem cell recruitment and bone regeneration in a rat periodontal bone defect model." *Mater Sci Eng C Mater Biol Appl* 53:83–94. doi: 10.1016/j.msec.2015.04.002.

Liu, S., H. Liang, S.M. Lee, Z. Li, J. Zhang, and Q. Fei. 2017a. "Isolation and identification of stem cells from degenerated human intervertebral discs and their migration characteristics." *Acta Biochim Biophys Sin (Shanghai)* 49 (2):101–9. doi: 10.1093/abbs/gmw121.

Liu, W., D. Liu, J. Zheng et al. 2017b. "Annulus fibrosus cells express and utilize C-C chemokine receptor 5 (CCR5) for migration." *Spine J.* doi: 10.1016/j.spinee.2017.01.010.

Liu, Z.J., Y. Zhuge, and O.C. Velazquez. 2009. "Trafficking and differentiation of mesenchymal stem cells." *J Cell Biochem* 106 (6):984–91. doi: 10.1002/jcb.22091.

Luster, A.D. 1998. "Chemokines—Chemotactic cytokines that mediate inflammation." *N Engl J Med* 338 (7):436–45. doi: 10.1056/NEJM199802123380706.

Ma, H.C., X.L. Shi, H.Z. Ren, X.W. Yuan, and Y.T. Ding. 2014. "Targeted migration of mesenchymal stem cells modified with CXCR4 to acute failing liver improves liver regeneration." *World J Gastroenterol* 20 (40):14884–94. doi: 10.3748/wjg.v20.i40.14884.

MacArthur, J.W., Jr., B.P. Purcell, Y. Shudo et al. 2013. "Sustained release of engineered stromal cell-derived factor 1-alpha from injectable hydrogels effectively recruits endothelial progenitor cells and preserves ventricular function after myocardial infarction." *Circulation* 128 (11 Suppl 1):S79–86. doi: 10.1161/CIRCULATIONAHA.112.000343.

Martin, B.I., R.A. Deyo, S.K. Mirza et al. 2008. "Expenditures and health status among adults with back and neck problems." *JAMA* 299 (6):656–64. doi: 10.1001/jama.299.6.656.

Mendez-Ferrer, S., D. Lucas, M. Battista, and P.S. Frenette. 2008. "Haematopoietic stem cell release is regulated by circadian oscillations." *Nature* 452 (7186):442–7. doi: 10.1038/nature06685.

Mercier, F.E., C. Ragu, and D.T. Scadden. 2011. "The bone marrow at the crossroads of blood and immunity." *Nat Rev Immunol* 12 (1):49–60. doi: 10.1038/nri3132.

MGC Project Team, G. Temple, D.S. Gerhard et al. 2009. "The completion of the Mammalian Gene Collection (MGC)." *Genome Res* 19 (12):2324–33. doi: 10.1101/gr.095976.109.

Mochida, J., D. Sakai, Y. Nakamura et al. 2015. "Intervertebral disc repair with activated nucleus pulposus cell transplantation: A three-year, prospective clinical study of its safety." *Eur Cell Mater* 29:202–12; discussion 212.

Molinos, M., C.R. Almeida, J. Caldeira et al. 2015. "Inflammation in intervertebral disc degeneration and regeneration." *J R Soc Interface* 12 (108):20150429. doi: 10.1098/rsif.2015.0429.

Muller, W.A., S.A. Weigl, X. Deng, and D.M. Phillips. 1993. "PECAM-1 is required for transendothelial migration of leukocytes." *J Exp Med* 178 (2):449–60.

Naderi-Meshkin, H., A.R. Bahrami, H.R. Bidkhori, M. Mirahmadi, and N. Ahmadiankia. 2015. "Strategies to improve homing of mesenchymal stem cells for greater efficacy in stem cell therapy." *Cell Biol Int* 39 (1):23–34. doi: 10.1002/cbin.10378.

Naumann, U., E. Cameroni, M. Pruenster et al. 2010. "CXCR7 functions as a scavenger for CXCL12 and CXCL11." *PLOS One* 5 (2):e9175. doi: 10.1371/journal.pone.0009175.

Nomiyama, H., N. Osada, and O. Yoshie. 2013. "Systematic classification of vertebrate chemokines based on conserved synteny and evolutionary history." *Genes Cells* 18 (1):1–16. doi: 10.1111/gtc.12013.

Orozco, L., R. Soler, C. Morera et al. 2011. "Intervertebral disc repair by autologous mesenchymal bone marrow cells: A pilot study." *Transplantation* 92 (7):822–8. doi: 10.1097/TP.0b013e3182298a15.

Ozaki, Y., M. Nishimura, K. Sekiya et al. 2007. "Comprehensive analysis of chemotactic factors for bone marrow mesenchymal stem cells." *Stem Cell Dev* 16 (1):119–29. doi: 10.1089/scd.2006.0032.

Palframan, R.T., P.D. Collins, T.J. Williams, and S.M. Rankin. 1998. "Eotaxin induces a rapid release of eosinophils and their progenitors from the bone marrow." *Blood* 91 (7):2240–8.

Pattappa, G., Z. Li, M. Peroglio et al. 2012. "Diversity of intervertebral disc cells: Phenotype and function." *J Anat* 221 (6):480–96. doi: 10.1111/j.1469-7580.2012.01521.x.

Pattappa, G., M. Peroglio, D. Sakai et al. 2014. "CCL5/RANTES is a key chemoattractant released by degenerative intervertebral discs in organ culture." *Eur Cell Mater* 27:124–36; discussion 136.

Peled, A., I. Petit, O. Kollet et al. 1999. "Dependence of human stem cell engraftment and repopulation of NOD/SCID mice on CXCR4." *Science* 283 (5403):845–8.

Pelekanos, R.A., M.J. Ting, V.S. Sardesai et al. 2014. "Intracellular trafficking and endocytosis of CXCR4 in fetal mesenchymal stem/stromal cells." *BMC Cell Biol* 15:15. doi: 10.1186/1471-2121-15-15.

Pereira, C.L., R.M. Goncalves, M. Peroglio et al. 2014. "The effect of hyaluronan-based delivery of stromal cell-derived factor-1 on the recruitment of MSCs in degenerating intervertebral discs." *Biomaterials* 35 (28):8144–53. doi: 10.1016/j.biomaterials.2014.06.017.

Pereira, C.L., G.Q. Teixeira, C. Ribeiro-Machado et al. 2016. "Mesenchymal stem/stromal cells seeded on cartilaginous endplates promote intervertebral disc regeneration through extracellular matrix remodeling." *Sci Rep* 6:33836. doi: 10.1038/srep33836.

Pettine, K.A., M.B. Murphy, R.K. Suzuki, and T.T. Sand. 2015. "Percutaneous injection of autologous bone marrow concentrate cells significantly reduces lumbar discogenic pain through 12 months." *Stem Cells* 33 (1):146–56. doi: 10.1002/stem.1845.

Phillips, K.L., N. Chiverton, A.L. Michael et al. 2013. "The cytokine and chemokine expression profile of nucleus pulposus cells: Implications for degeneration and regeneration of the intervertebral disc." *Arthritis Res Ther* 15 (6):R213. doi: 10.1186/ar4408.

Pitchford, S.C., R.C. Furze, C.P. Jones, A.M. Wengner, and S.M. Rankin. 2009. "Differential mobilization of subsets of progenitor cells from the bone marrow." *Cell Stem Cell* 4 (1):62–72. doi: 10.1016/j.stem.2008.10.017.

Pittenger, M.F., A.M. Mackay, S.C. Beck et al. 1999. "Multilineage potential of adult human mesenchymal stem cells." *Science* 284 (5411):143–7.

Ponte, A.L., E. Marais, N. Gallay et al. 2007. "The *in vitro* migration capacity of human bone marrow mesenchymal stem cells: Comparison of chemokine and growth factor chemotactic activities." *Stem Cells* 25 (7):1737–45. doi: 10.1634/stemcells.2007-0054.

Purcell, B.P., J.A. Elser, A. Mu, K.B. Margulies, and J.A. Burdick. 2012. "Synergistic effects of SDF-1alpha chemokine and hyaluronic acid release from degradable hydrogels on directing bone marrow derived cell homing to the myocardium." *Biomaterials* 33 (31):7849–57. doi: 10.1016/j.biomaterials.2012.07.005.

Rabbany, S.Y., J. Pastore, M. Yamamoto et al. 2010. "Continuous delivery of stromal cell-derived factor-1 from alginate scaffolds accelerates wound healing." *Cell Transplant* 19 (4):399–408. doi: 10.3727/096368909X481782.

Rankin, S.M. 2012. "Chemokines and adult bone marrow stem cells." *Immunol Lett* 145 (1–2):47–54. doi: 10.1016/j.imlet.2012.04.009.

Ries, C., V. Egea, M. Karow et al. 2007. "MMP-2, MT1-MMP, and TIMP-2 are essential for the invasive capacity of human mesenchymal stem cells: Differential regulation by inflammatory cytokines." *Blood* 109 (9):4055–63. doi: 10.1182/blood-2006-10-051060.

Risbud, M., A. Guttapalli, T.T. Tsai et al. 2007. "Evidence for skeletal progenitor cells in the degenerate human intervertebral disc." *Spine (Phila Pa 1976)* 32 (23):2537–44. doi: 10.1097/BRS.0b013e318158dea6.

Risbud, M.V., and I.M. Shapiro. 2014. "Role of cytokines in intervertebral disc degeneration: Pain and disc content." *Nat Rev Rheumatol* 10 (1):44–56. doi: 10.1038/nrrheum.2013.160.

Roberts, S., B. Caterson, J. Menage et al. 2000. "Matrix metalloproteinases and aggrecanase: Their role in disorders of the human intervertebral disc." *Spine (Phila Pa 1976)* 25 (23):3005–13.

Roberts, S., H. Evans, J. Trivedi, and J. Menage. 2006. "Histology and pathology of the human intervertebral disc." *J Bone Joint Surg Am* 88 Suppl 2:10–4. doi: 10.2106/JBJS.F.00019.

Ruster, B., S. Gottig, R.J. Ludwig et al. 2006. "Mesenchymal stem cells display coordinated rolling and adhesion behavior on endothelial cells." *Blood* 108 (12):3938–44. doi: 10.1182/blood-2006-05-025098.

Ryser, M.F., F. Ugarte, S. Thieme et al. 2008. "mRNA transfection of CXCR4-GFP fusion—Simply generated by PCR—Results in efficient migration of primary human mesenchymal stem cells." *Tissue Eng Part C Methods* 14 (3):179–84. doi: 10.1089/ten.tec.2007.0359.

Sadelain, M. 2004. "Insertional oncogenesis in gene therapy: How much of a risk?" *Gene Ther* 11 (7):569–73. doi: 10.1038/sj.gt.3302243.

Sakai, D., Y. Nakamura, T. Nakai et al. 2012. "Exhaustion of nucleus pulposus progenitor cells with aging and degeneration of the intervertebral disc." *Nat Commun* 3:1264. doi: 10.1038/ncomms2226.

Sakai, D., K. Nishimura, M. Tanaka et al. 2015. "Migration of bone marrow-derived cells for endogenous repair in a new tail-looping disc degeneration model in the mouse: A pilot study." *Spine J* 15 (6):1356–65. doi: 10.1016/j.spinee.2013.07.491.

Scadden, D.T. 2006. "The stem-cell niche as an entity of action." *Nature* 441 (7097):1075–9. doi: 10.1038/nature04957.

Schenk, S., N. Mal, A. Finan et al. 2007. "Monocyte chemotactic protein-3 is a myocardial mesenchymal stem cell homing factor." *Stem Cells* 25 (1):245–51. doi: 10.1634/stemcells.2006-0293.

Schioppa, T., B. Uranchimeg, A. Saccani et al. 2003. "Regulation of the chemokine receptor CXCR4 by hypoxia." *J Exp Med* 198 (9):1391–402. doi: 10.1084/jem.20030267.

Schmorl, G., and H. Junghanns. 1971. *Human Spine in Health and Disease*. Translated by E.F. Besemann. Edited by Herbert Junghanns and E.F. Besemann. New York: Grune & Stratton.

Seguin, C.A., R.M. Pilliar, P.J. Roughley, and R.A. Kandel. 2005. "Tumor necrosis factor-alpha modulates matrix production and catabolism in nucleus pulposus tissue." *Spine (Phila Pa 1976)* 30 (17):1940–8.

Shamji, M.F., L.A. Setton, W. Jarvis et al. 2010. "Proinflammatory cytokine expression profile in degenerated and herniated human intervertebral disc tissues." *Arthritis Rheum* 62 (7):1974–82. doi: 10.1002/art.27444.

Shen, W., X. Chen, J. Chen et al. 2010. "The effect of incorporation of exogenous stromal cell-derived factor-1 alpha within a knitted silk-collagen sponge scaffold on tendon regeneration." *Biomaterials* 31 (28):7239–49. doi: 10.1016/j.biomaterials.2010.05.040.

Shepherd, R.M., B.J. Capoccia, S.M. Devine, J. Dipersio, K.M. Trinkaus, D. Ingram, and D.C. Link. 2006. "Angiogenic cells can be rapidly mobilized and efficiently harvested from the blood following treatment with AMD3100." *Blood* 108 (12):3662–7. doi: 10.1182/blood-2006-06-030577.

Shi, M., J. Li, L. Liao et al. Zhao. 2007. "Regulation of CXCR4 expression in human mesenchymal stem cells by cytokine treatment: Role in homing efficiency in NOD/SCID mice." *Haematologica* 92 (7):897–904.

Shinmei, M., T. Kikuchi, M. Yamagishi, and Y. Shimomura. 1988. "The role of interleukin-1 on proteoglycan metabolism of rabbit annulus fibrosus cells cultured *in vitro*." *Spine* 13 (11):1284–90. doi: 10.1097/00007632-198811000-00014.

Singh, A., S. Suri, and K. Roy. 2009. "In-situ crosslinking hydrogels for combinatorial delivery of chemokines and siRNA-DNA carrying microparticles to dendritic cells." *Biomaterials* 30 (28):5187–200. doi: 10.1016/j.biomaterials.2009.06.001.

Smith, L.J., N.L. Nerurkar, K.S. Choi, B.D. Harfe, and D.M. Elliott. 2011. "Degeneration and regeneration of the intervertebral disc: Lessons from development." *Dis Model Mech* 4 (1):31–41. doi: 10.1242/dmm.006403.

Sokol, C.L., and A.D. Luster. 2015. "The chemokine system in innate immunity." *Cold Spring Harb Perspect Biol* 7 (5). doi: 10.1101/cshperspect.a016303.

Song, M., H. Jang, J. Lee et al. 2014. "Regeneration of chronic myocardial infarction by injectable hydrogels containing stem cell homing factor SDF-1 and angiogenic peptide Ac-SDKP." *Biomaterials* 35 (8):2436–45. doi: 10.1016/j.biomaterials.2013.12.011.

Springer, T.A. 1994. "Traffic signals for lymphocyte recirculation and leukocyte emigration: The multistep paradigm." *Cell* 76 (2):301–14.

Studer, R.K., N. Vo, G. Sowa, C. Ondeck, and J. Kang. 2011. "Human nucleus pulposus cells react to IL-6: Independent actions and amplification of response to IL-1 and TNF-alpha." *Spine (Phila Pa 1976)* 36 (8):593–9. doi: 10.1097/BRS.0b013e3181da38d5.

Sugiyama, T., H. Kohara, M. Noda, and T. Nagasawa. 2006. "Maintenance of the hematopoietic stem cell pool by CXCL12-CXCR4 chemokine signaling in bone marrow stromal cell niches." *Immunity* 25 (6):977–88. doi: 10.1016/j.immuni.2006.10.016.

Sukegawa, A., N. Iwasaki, Y. Kasahara et al. 2012. "Repair of rabbit osteochondral defects by an acellular technique with an ultrapurified alginate gel containing stromal cell-derived factor-1." *Tissue Eng Part A* 18 (9–10):934–45. doi: 10.1089/ten.TEA.2011.0380.

Surmi, B.K., and A.H. Hasty. 2010. "The role of chemokines in recruitment of immune cells to the artery wall and adipose tissue." *Vascul Pharmacol* 52 (1–2):27–36. doi: 10.1016/j.vph.2009.12.004.

Tam, V., I. Rogers, D. Chan, V.Y. Leung, and K.M. Cheung. 2014. "A comparison of intravenous and intradiscal delivery of multipotential stem cells on the healing of injured intervertebral disk." *J Orthop Res* 32 (6):819–25. doi: 10.1002/jor.22605.

Thieme, S., M. Ryser, M. Gentsch et al. 2009. "Stromal cell-derived factor-1alpha-directed chemoattraction of transiently CXCR4-overexpressing bone marrow stromal cells into functionalized three-dimensional biomimetic scaffolds." *Tissue Eng Part C Methods* 15 (4):687–96. doi: 10.1089/ten.TEC.2008.0556.

Tzeng, Y.S., H. Li, Y.L. Kang et al. 2011. "Loss of Cxcl12/Sdf-1 in adult mice decreases the quiescent state of hematopoietic stem/progenitor cells and alters the pattern of hematopoietic regeneration after myelosuppression." *Blood* 117 (2):429–39. doi: 10.1182/blood-2010-01-266833.

Videman, T., M.C. Battié, L.E. Gibbons et al. 2003. "Associations between back pain history and lumbar MRI findings." *Spine* 28 (6):582–8.

Vos, T., A.D. Flaxman, M. Naghavi et al. 2012. "Years lived with disability (YLDs) for 1160 sequelae of 289 diseases and injuries 1990–2010: A systematic analysis for the Global Burden of Disease Study 2010." *Lancet* 380 (9859):2163–96. doi: 10.1016/S0140-6736(12)61729-2.

Walker, B.F. 2000. "The prevalence of low back pain: A systematic review of the literature from 1966 to 1998." *J Spinal Disord* 13 (3):205–17.

Walker, M.H., and D.G. Anderson. 2004. "Molecular basis of intervertebral disc degeneration." *Spine J* 4 (6 Suppl):158S–166S. doi: 10.1016/j.spinee.2004.07.010.

Wang, J., Y. Tian, K.L. Phillips et al. 2013. "Tumor necrosis factor alpha- and interleukin-1beta-dependent induction of CCL3 expression by nucleus pulposus cells promotes macrophage migration through CCR1." *Arthritis Rheum* 65 (3):832–42. doi: 10.1002/art.37819.

Wang, Y., and D.J. Irvine. 2011. "Engineering chemoattractant gradients using chemokine-releasing polysaccharide microspheres." *Biomaterials* 32 (21):4903–13. doi: 10.1016/j.biomaterials.2011.03.027.

Wei, J.-N., F. Cai, F. Wang et al. 2016. "Transplantation of CXCR4 overexpressed mesenchymal stem cells augments regeneration in degenerated intervertebral discs." *DNA Cell Biol* 35 (5):241–8. doi: 10.1089/dna.2015.3118.

Weiler, C., A.G. Nerlich, R. Schaaf et al. 2010. "Immunohistochemical identification of notochordal markers in cells in the aging human lumbar intervertebral disc." *Eur Spine J* 19 (10):1761–70. doi: 10.1007/s00586-010-1392-z.

Wynn, R.F., C.A. Hart, C. Corradi-Perini et al. 2004. "A small proportion of mesenchymal stem cells strongly expresses functionally active CXCR4 receptor capable of promoting migration to bone marrow." *Blood* 104 (9):2643–5. doi: 10.1182/blood-2004-02-0526.

Yellowley, C. 2013. "CXCL12/CXCR4 signaling and other recruitment and homing pathways in fracture repair." *BoneKEy Reports* 2:300. doi: 10.1038/bonekey.2013.34.

Yoshikawa, T., Y. Ueda, K. Miyazaki, M. Koizumi, and Y. Takakura. 2010. "Disc regeneration therapy using marrow mesenchymal cell transplantation: A report of two case studies." *Spine (Phila Pa 1976)* 35 (11):E475–80. doi: 10.1097/BRS.0b013e3181cd2cf4.

Zhang, D., G.C. Fan, X. Zhou et al. 2008. "Over-expression of CXCR4 on mesenchymal stem cells augments myoangiogenesis in the infarcted myocardium." *J Mol Cell Cardiol* 44 (2):281–92. doi: 10.1016/j.yjmcc.2007.11.010.

Zhang, G., Y. Nakamura, X. Wang et al. 2007. "Controlled release of stromal cell-derived factor-1 alpha in situ increases c-kit+ cell homing to the infarcted heart." *Tissue Eng* 13 (8):2063–71. doi: 10.1089/ten.2006.0013.

Zhang, W., J. Chen, J. Tao et al. 2013a. "The use of type 1 collagen scaffold containing stromal cell-derived factor-1 to create a matrix environment conducive to partial-thickness cartilage defects repair." *Biomaterials* 34 (3):713–23. doi: 10.1016/j.biomaterials.2012.10.027.

Zhang, W., L. Nie, Y. Wang et al. 2013b. "CCL20 secretion from the nucleus pulposus improves the recruitment of CCR6-expressing Th17 cells to degenerated IVD tissues." *PLoS One* 8 (6):e66286. doi: 10.1371/journal.pone.0066286.

Zhang, Y., L. Liu, S. Wang et al. 2016. "Production of CCL20 on nucleus pulposus cells recruits IL-17-producing cells to degenerated IVD tissues in rat models." *J Mol Histol* 47 (1):81–9. doi: 10.1007/s10735-015-9651-2.

Zhao, X., S. Jain, H. Benjamin Larman, S. Gonzalez, and D.J. Irvine. 2005. "Directed cell migration via chemoattractants released from degradable microspheres." *Biomaterials* 26 (24):5048–63. doi: 10.1016/j.biomaterials.2004.12.003.

Zlotnik, A., and O. Yoshie. 2012. "The chemokine superfamily revisited." *Immunity* 36 (5):705–16. doi: 10.1016/j.immuni.2012.05.008.

7 Immunomodulation in Degenerated Intervertebral Disc

Graciosa Q. Teixeira, Mário Adolfo Barbosa, and Raquel M. Gonçalves

CONTENTS

7.1	Introduction	184
7.2	Immunogenic Phenotype of IVD Cell Populations and Induced Immune Cell Response	185
7.3	Key Proinflammatory Molecules in IVD Degeneration and Associated Inflammation	187
	7.3.1 TNF-α	187
	7.3.2 IL-1β	194
	7.3.3 IL-6	195
	7.3.4 Toll-Like Receptors	195
7.4	microRNAs	196
7.5	Immune Cell Activation	197
	7.5.1 T Cells	200
	7.5.2 Macrophages	201
7.6	Other Factors Involved in Innervation, Vascularization, and Pain	202
7.7	Strategies for Immunomodulation of Degenerated IVD	202
	7.7.1 Molecular Therapy: Clinical Trials	203
	7.7.2 Molecular Therapy: *In Vivo* and *Ex Vivo* Studies	210
7.8	Gene Therapy	211
7.9	Cell-Based Therapies	212
	7.9.1 Endogenous Therapies	212
	7.9.2 Exogenous Stem Cell Delivery: Clinical Trials	213
	7.9.3 Exogenous Stem Cell Delivery: *In Vivo* and *In Vitro* Studies	213
7.10	Concluding Remarks and Future Perspectives	214
Acknowledgments		214
References		214

7.1 INTRODUCTION

Low back pain (LBP) has been described to affect approximately two-thirds of the world population at some point in their life (Andersson 1999; Deyo and Weinstein 2001), and it is considered the number one disease regarding global years lived with disability (Vos et al. 2012).

Disc degeneration is linked with aging (Roberts et al. 2006), as recently reviewed by Vo et al. (2016). Nonetheless, it has been also observed in young children (11- to 16-years-old) (Boos et al. 2002). The intervertebral disc's (IVD's) well-defined microstructural organization and biochemical composition are affected by aging molecular mechanisms and can ultimately lead to a cell-mediated structural failure (Iatridis et al. 2009; Vo et al. 2016). With age, variations in abundance and structure of IVD's extracellular matrix (ECM) macromolecules may be a consequence of catabolism and anabolism imbalance (Roughley 2004), but in cases of early degeneration, abnormal age-related changes also occur (Iatridis et al. 2009).

Degenerated IVD pathogenesis might be affected by multiple factors, such as gene polymorphisms, as recently reviewed by Martirosyan et al. (2016), which include genes that mediate apoptosis, contribute to structural proteins, and encode molecules involved in inflammatory pathways.

The aging/degenerative process of IVD is characterized by an initial increase in cell proliferation and formation of cell clusters, as well as alterations in cell cycle and an increase in cell senescence and apoptosis, with increased production of pro-apoptotic (Fas ligand [FasL], caspase-3) proteins, and death (Bertolo et al. 2011; Gruber et al. 2009; Richardson et al. 2007; Roberts et al. 2006).

The nucleus pulposus (NP) changes from gelatinous to a more fibrous structure, cracks and fissures often occur, namely in the AF, and there is a decrease in IVD water content. This is common due to a turnover of ECM components, a shift from collagen type II (COL2) to type I (COL1) production by NP cells, and a decrease in aggrecan (ACAN) synthesis (Bertolo et al. 2011; Richardson et al. 2007). An up-regulation of specific metalloproteinases (MMPs), such as MMP-1, -2, -3, -7, -8, -10, and -13, a disintegrin and MMP with thrombospondin motifs (ADAMTS)-1, -4, -5, -9 and -15, and tissue inhibitors of MMPs (TIMPs)-1 and -2 were observed (Bachmeier et al. 2009; Doita et al. 2001; Le Maitre et al. 2007; Pockert et al. 2009; Vo et al. 2013).

During degeneration, several changes may occur in the capillaries arising from the vertebral bodies, namely, atherosclerosis, reduced capillary density, occlusion of the marrow spaces, and cartilaginous end plate (CEP) obstruction due to calcification/ increased mineralization (Grant et al. 2016; Huang, Urban, and Luk 2014). It has been hypothesized that an increase in free calcium ions (Ca2+) may impair CEP homeostasis, compromising nutrient diffusion and availability to the cells, consequently leading to alterations in cell metabolism and viability (Grant et al. 2016; Huang, Urban, and Luk 2014). It has also been reported that ECM component degradation may promote obstruction of the CEP, contributing to the drastic decrease in oxygen and nutrient diffusion into the disc (Huang, Urban, and Luk 2014; Ogata and Whiteside 1981). Particularly the NP is subjected to high mechanical and osmotic pressures, severe hypoxia, and limited nutrients supply (Mehrkens et al. 2012).

Additionally, blood vessels begin to grow into the disc from the outer areas of the AF (Roberts et al. 2006).

Moreover, a wide number of inflammatory mediators, including prostaglandins, namely, prostaglandin E_2 (PGE_2), interleukins (IL-1, -6, -8, -12, and -17), tumor necrosis factor (TNF)-α, and interferon (IFN)-γ have been described as crucial players in the catabolic processes in human NP and annulus fibrosus (AF), nerve ingrowth, and pain (Cuellar et al. 2010; Le Maitre, Hoyland, and Freemont 2007a; Purmessur et al. 2013; Risbud and Shapiro 2014; Shamji et al. 2010). Furthermore, increased amounts of nitric oxide (NO) have also been detected (Kang et al. 1996; O'Donnell and O'Donnell 1996; Saal et al. 1990). As the latest reviews point out, inflammation is an important contributor to the pathogenesis of IVD degeneration (Gorth, Shapiro, and Risbud et al. 2015; Risbud and Shapiro 2014; Wuertz and Haglund 2013). A balance between inflammatory response and tissue resorption may be achieved by controlling the levels of proinflammatory cytokines known to be involved in enzymatic degrading activity (Le Maitre, Hoyland, and Freemont 2007b).

7.2 IMMUNOGENIC PHENOTYPE OF IVD CELL POPULATIONS AND INDUCED IMMUNE CELL RESPONSE

An association between disc degeneration, herniation, and inflammation has been established over time (Johnson et al. 2015; Molinos et al. 2015). IVD cells can secrete proinflammatory cytokines to induce and enhance inflammation (Le Maitre, Hoyland, and Freemont 2007a), and an inflammatory response occurs not only in the IVD but also in the surrounding tissues (Risbud and Shapiro 2014). Therefore, an in-depth characterization of the synergic interplay between degeneration, inflammation, and pain could promote the development of more advanced and targeted therapies for IVD degeneration and LBP (Molinos et al. 2015; Teixeira et al. 2015, 2016). In this section, we discuss the contributions of different factors to cellular and tissue level changes seen during disc degeneration (schematically summarized in Figure 7.1).

IVD cells express several inflammatory factors already during homeostasis (Molinos et al. 2015). However, an initial insult, related with aging and degeneration, leads to an up-regulation of inflammation mediators such as key proinflammatory cytokines, namely, IL-1β and TNF-α, but also IL-6, IL-17, IFN-γ, and chemokines, among others (Burke et al. 2002a; Huang et al. 2008; Kang et al. 1997; Kokubo et al. 2008; Le Maitre, Freemont, and Hoyland 2005; Le Maitre, Hoyland, and Freemont 2007a; Park, Chang, and Kim 2002; Risbud and Shapiro 2014; Shamji et al. 2010; Specchia et al. 2002; Takahashi et al. 1996; Weiler et al. 2005). These are initiating events of the IVD degenerative cascade (Risbud and Shapiro 2014; Walter et al. 2015). They contribute to the increase in cell senescence, unbalanced anabolism, and catabolism in ECM synthesis (Cuellar et al. 2013; Le Maitre, Freemont, and Hoyland 2005; Purmessur et al. 2013; Seguin et al. 2005; Shamji et al. 2010). Herniated discs are known to induce a specific autoimmune response (Sun et al. 2013). Macrophages, leucocytes, neutrophils, and T cells were found in extruded tissues (Kokubo et al. 2008; Shamji et al. 2010; Risbud and Shapiro 2014), surrounded by granulation

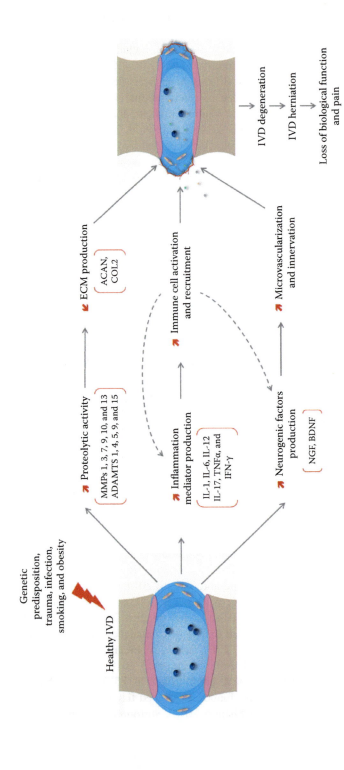

FIGURE 7.1 Role of the cytokines involved in different phases of IVD degeneration and herniation, leading to back and radicular pain. In the first phase of degeneration, IVD cells express several catabolic molecules in the inflammatory environment, promoting ACAN and COL2 degradation, which leads to mechanical instability and ECM breakdown. In many cases, AF tearing and herniation occur. Second, the release of cytokines and chemokines by the IVD cells enhances activation and infiltration of immune cells, which also produce proinflammatory factors by themselves, further amplifying the inflammatory response. Of notice, together with the infiltration of immune cells, there is also microvascularization and innervation by nociceptive nerve fibers arising from the dorsal root ganglion (DRG) and production of neurogenic factors, particularly NGF and BDNF, produced by the herniated disc and immune cells.

tissue, neovascularization, and innervation (Burke et al. 2002b; Freemont et al. 2002a). Vascular and nerve ingrowth into the avascular IVD occurs from the outer layers of the AF into the NP (Sakai and Andersson 2015).

The factors produced by both IVD and immune cells, as well as their effect in degeneration, inflammatory state, and associated pain will be discussed ahead. As Molinos et al. (2015) highlighted, there are numerous inflammatory mediators found in the human IVD, which may be produced by NP, AF, and/or infiltrating inflammatory cells, as summarized in Table 7.1.

7.3 KEY PROINFLAMMATORY MOLECULES IN IVD DEGENERATION AND ASSOCIATED INFLAMMATION

Both IVD cells and leucocytes secrete IL-1β and TNF-α (Le Maitre, Freemont, and Hoyland 2005; Le Maitre, Hoyland, and Freemont 2007a). IL-1 and TNF-α have been identified in painful hernia samples, being associated with the mechanism of sensory nerves ingrowth into the NP (Hayashi et al. 2008), damage of the dorsal root ganglion (DRG) and neuropathic pain (Igarashi et al. 2000; Leung and Cahill 2010; Murata et al. 2006; Olmarker and Larsson 1998). The expression of TNF-α and IL-1β was shown to increase with age and severity of degeneration, as observed by analysis of human hernia samples from donors with different ages, being these cytokines by themselves degeneration precursors (Bachmeier et al. 2007; Johnson et al. 2015; Le Maitre, Freemont, and Hoyland 2005; Le Maitre, Hoyland, and Freemont 2007a; Wang et al. 2014).

7.3.1 TNF-α

TNF-α was shown to be one of the first cytokines highly produced by human IVD cells in both IVD degeneration and herniation scenarios (Dudli, Haschtmann, and Ferguson 2012; Le Maitre, Hoyland, and Freemont 2007a; Ulrich et al. 2007; Weiler et al. 2005). It was shown that by exposing bovine organ cultures to TNF-α, as it may occur from injured surrounding tissues, it can penetrate in healthy intact IVDs, induce the expression of additional proinflammatory cytokines, and alter the tissue mechanical behavior (Millward-Sadler et al. 2009; Walter et al. 2015). NP cells stimulated with TNF-α and IL-1β showed a strong induction of ECM degrading enzyme expression, namely, ADAMTS-4 and -5, MMP-1, -2, -3, and -13 (Jimbo et al. 2005; Krupkova et al. 2016; Le Maitre, Freemont, and Hoyland 2005; Seguin et al. 2005; Shen et al. 2003; Wang et al. 2011d, 2014), and other proinflammatory factors, such as IL-6 or cyclooxygenase (COX)-2 (Fujita et al. 2012; Jimbo et al. 2005), previously identified in human IVD degenerated samples (Bachmeier et al. 2009; Pockert et al. 2009).

TNF-α belongs to a superfamily of ligand/receptor proteins designated TNF/TNFR superfamily proteins. Human TNF is synthesized as a type II transmembrane protein (membrane-bound TNF [mTNF]), forming stable homotrimers. mTNF is processed by TNF-α-converting enzyme (TACE) into soluble TNF (sTNF) (Black et al. 1997; Johnson et al. 2015; Risbud and Shapiro 2014). Both sTNF-α and mTNF-α can bind through the TNF homology domain to the cysteine-rich domains (CRDs) of its receptors (TNFRs), TNFR1 or TNFR2, which act as TNF antagonists (Leung and Cahill 2010). TACE, TNFR1, and TNFR2 are expressed in human NP tissue

TABLE 7.1
Inflammation Mediators Expressed with Degeneration by IVD Cells and Infiltrating Cells in Painful Human IVDs

Mediators	Tissue	Tissue Collection	Disorder	References
TNF-α	AF	Autopsy, biopsy	Degeneration	Dongfeng et al. 2011
	AF + NP	Autopsy, biopsy	Degeneration, herniation (protrusion, extrusion, sequestration)	Bachmeier et al. 2007; Le Maitre, Hoyland, and Freemont 2007a; Weiler et al. 2005
	AF + NP	Autopsy, biopsy	Degeneration, herniation	Le Maitre, Hoyland, and Freemont 2007a; Weiler et al. 2005
	AF + NP + CEP	Biopsy	Herniation, spondylosis	Kokubo et al. 2008
	NP	Autopsy, biopsy	Degeneration	Richardson et al. 2009
	NP	Biopsy	Herniation	Park et al. 2011
	NP	Biopsy	Herniation (protrusion, extrusion, sequestration)	Chen et al. 2017
	IVD	Autopsy, biopsy	Degeneration, herniation	Akyol et al. 2010
	IVD	Biopsy	Degeneration, herniation	Lee et al. 2009b
	IVD	Biopsy	Herniation (protrusion, extrusion, sequestration)	Takahashi et al. 1996
	IVD	Biopsy	Herniation (extrusion, sequestration)	Miyamoto et al. 2000
	IVD	Biopsy	Herniation (subligamentous extensions, transligamentous extensions including sequestration)	Ahn et al. 2002
TNFR1	AF + NP	Autopsy, biopsy	Degeneration, herniation (protrusion, extrusion, sequestration)	Bachmeier et al. 2007; Le Maitre, Hoyland, and Freemont 2007a
TNFR2, TACE	AF + NP	Autopsy, biopsy	Degeneration, herniation (protrusion, extrusion, sequestration)	Bachmeier et al. 2007
IL-1α	AF + NP	Autopsy, biopsy	Degeneration	Le Maitre, Freemont, and Hoyland 2005
	IVD	Biopsy	Herniation (protrusion, extrusion, sequestration)	Takahashi et al. 1996

(Continued)

TABLE 7.1 (CONTINUED)
Inflammation Mediators Expressed with Degeneration by IVD Cells and Infiltrating Cells in Painful Human IVDs

Mediators	Tissue	Tissue Collection	Disorder	References
	IVD	Biopsy	Herniation (subligamentous extensions, transligamentous extensions including sequestration)	Ahn et al. 2002
IL-1β	AF + NP	Autopsy, biopsy	Degeneration, herniation	Le Maitre, Hoyland, and Freemont 2007a
	AF + NP	Autopsy, biopsy	Degeneration	Le Maitre, Freemont, and Hoyland 2005
	NP	Autopsy, biopsy	Degeneration	Richardson et al. 2009
	NP	Autopsy, biopsy	Herniation	Gronblad et al. 1994
	NP	Biopsy	Herniation	Park et al. 2011
	IVD	Autopsy, biopsy	Degeneration, herniation	Akyol et al. 2010
	IVD	Biopsy	Degeneration, herniation	Lee et al. 2009b
	IVD	Biopsy	Herniation (protrusion, extrusion, sequestration)	Takahashi et al. 1996
	IVD	Biopsy	Herniation (extrusion, sequestration)	Miyamoto et al. 2000
IL-1Ra	AF + NP	Autopsy, biopsy	Degeneration	Le Maitre, Freemont, and Hoyland 2005
IL-1R1	AF + NP	Autopsy, biopsy	Degeneration, herniation	Le Maitre, Hoyland, and Freemont 2007a
	AF + NP	Autopsy, biopsy	Degeneration	Le Maitre, Freemont, and Hoyland 2005
IL-2	IVD	Autopsy, biopsy	Degeneration, herniation	Akyol et al. 2010
IL-4	AF + NP	Autopsy, biopsy	Degeneration, herniation	Shamji et al. 2010
	NP	Biopsy	Herniation (subligamentous extrusion and protrusion, sequestration and transligamentous extrusion)	Park, Chang, and Kim 2002
	IVD	Autopsy, biopsy	Degeneration, herniation	Akyol et al. 2010
IL-6	AF + NP	Autopsy, biopsy	Degeneration, herniation	Shamji et al. 2010
	IVD	Biopsy	Degeneration, herniation	Lee et al. 2009b
	IVD	Biopsy	Degeneration, herniation (protrusion, extrusion, sequestration)	Burke et al. 2002a

(*Continued*)

TABLE 7.1 (CONTINUED)
Inflammation Mediators Expressed with Degeneration by IVD Cells and Infiltrating Cells in Painful Human IVDs

Mediators	Tissue	Tissue Collection	Disorder	References
	IVD	Biopsy	Herniation	Kang et al. 1996
	IVD	Biopsy	Herniation (extrusion, sequestration, protrusion)	Takahashi et al. 1996
	Lavage fluid from disc space	Biopsy	Herniation	Gajendran et al. 2011
IL-8	IVD	Biopsy	Degeneration, herniation	Lee et al. 2009b
	IVD	Biopsy	Degeneration, herniation (protrusion, extrusion, sequestration)	Burke et al. 2002a
	IVD	Biopsy	Herniation (subligamentous extensions, transligamentous extensions including sequestration)	Ahn et al. 2002
	IVD	Biopsy	Herniation (protrusion, extrusion, sequestration), scoliosis	Burke et al. 2002b
IL-10	IVD	Autopsy, biopsy	Degeneration, herniation	Akyol et al. 2010
	IVD	Biopsy	Herniation (subligamentous extensions, transligamentous extensions including sequestration)	Ahn et al. 2002
IL-12	AF + NP	Autopsy, biopsy	Degeneration, herniation	Shamji et al. 2010
	NP	Biopsy	Herniation (subligamentous extrusion and protrusion, sequestration and transligamentous extrusion)	Park, Chang, and Kim 2002
	IVD	Autopsy, biopsy	Degeneration, herniation	Akyol et al. 2010
	IVD	Biopsy	Degeneration, herniation	Lee et al. 2009b
IL-16	NP	Autopsy, biopsy	Degeneration, prolapse, herniation (protrusion, extrusion, sequestration)	Phillips et al. 2013, 2015
IL-17	AF + NP	Autopsy, biopsy	Degeneration, herniation	Shamji et al. 2010
	AF + NP	Biopsy	Degeneration, herniation	Gruber et al. 2013

(*Continued*)

TABLE 7.1 (CONTINUED)
Inflammation Mediators Expressed with Degeneration by IVD Cells and Infiltrating Cells in Painful Human IVDs

Mediators	Tissue	Tissue Collection	Disorder	References
IL-20 (and its receptor subunits)	IVD	Biopsy	Herniation (extrusion, sequestration)	Huang et al. 2008
IL-21	NP	Biopsy	Herniation (protrusion, extrusion, sequestration)	Chen et al. 2017
CCL2, CCL7, CXCL8	NP	Autopsy, biopsy	Degeneration, prolapse, herniation (protrusion, extrusion, sequestration)	Phillips et al. 2013
CCR1, CXCR1, CXCR2	NP	Autopsy, biopsy	Degeneration, prolapse, herniation (protrusion, extrusion, sequestration)	Phillips et al. 2015
IFN-γ	AF + NP	Autopsy, biopsy	Degeneration, herniation	Shamji et al. 2010
	NP	Biopsy	Herniation (subligamentous extrusion and protrusion, sequestration and transligamentous extrusion)	Park, Chang, and Kim 2002
	Lavage fluid from disc space	Biopsy	Herniation	Gajendran et al. 2011
	Lavage fluid from disc space	Biopsy	Degeneration, scoliosis	Cuellar et al. 2010
RANTES	AF + NP	Biopsy	Degeneration, herniation	Gruber et al. 2014a
	IVD	Biopsy	Herniation (subligamentous extensions, transligamentous extensions including sequestration)	Ahn et al. 2002
TGF-β	IVD	Biopsy	Degeneration, herniation	Lee et al. 2009b
TGF-β1	IVD	Biopsy	Herniation (subligamentous extensions, transligamentous extensions including sequestration)	Ahn et al. 2002
Substance P	AF + NP + CEP	Biopsy	Herniation, spondylosis	Kokubo et al. 2008
	NP	Autopsy, biopsy	Degeneration	Richardson et al. 2009

(Continued)

TABLE 7.1 (CONTINUED)
Inflammation Mediators Expressed with Degeneration by IVD Cells and Infiltrating Cells in Painful Human IVDs

Mediators	Tissue	Tissue Collection	Disorder	References
MCP-1	IVD	Biopsy	Herniation (protrusion, extrusion, sequestration), scoliosis	Burke et al. 2002b
	Lavage fluid from disc space	Biopsy	Herniation	Gajendran et al. 2011
MIP-1β	Lavage fluid from disc space	Biopsy	Herniation	Gajendran et al. 2011
NGF	AF + NP + CEP	Biopsy	Herniation, spondylosis	Kokubo et al. 2008
	NP	Autopsy, biopsy	Degeneration	Richardson et al. 2009
	IVD	Biopsy	Degeneration, herniation	Lee et al. 2009b
bFGF	AF + NP + CEP	Biopsy	Herniation, spondylosis	Kokubo et al. 2008
VEGF	AF + NP + CEP	Biopsy	Herniation, spondylosis	Kokubo et al. 2008
	IVD	Biopsy	Degeneration, herniation	Lee et al. 2009b
GDF-5	AF	Biopsy	Degeneration, herniation	Gruber et al. 2014b
GM-CSF	IVD	Biopsy	Herniation: extrusion, sequestration, protrusion	Takahashi et al. 1996
MMPs	AF + NP	Biopsy	Degeneration, herniation (protrusion, extrusion, sequestration)	Bachmeier et al. 2009
	AF + NP + CEP	Biopsy	Herniation, spondylosis	Kokubo et al. 2008
	NP	Autopsy, biopsy	Degeneration	Richardson et al. 2009
	NP	Biopsy	Herniation (protrusion, subligamentous extrusion, transligamentous extrusion, sequestration)	Matsui et al. 1998
	IVD	Biopsy	Herniation	Kang et al. 1996
FasL	NP	Biopsy	Herniation (subligamentous extrusion and protrusion, sequestration and transligamentous extrusion)	Park, Chang, and Kim 2001

(Continued)

TABLE 7.1 (CONTINUED)
Inflammation Mediators Expressed with Degeneration by IVD Cells and Infiltrating Cells in Painful Human IVDs

Mediators	Tissue	Tissue Collection	Disorder	References
FasR	NP	Biopsy	Herniation (subligamentous extrusion and protrusion, sequestration and transligamentous extrusion)	Park et al. 2001
CDMP	AF + NP	Autopsy, biopsy	Degeneration	Le Maitre, Freemont, and Hoyland 2009
COX-2	IVD	Biopsy	Herniation (extrusion, sequestration)	Miyamoto et al. 2000
PGE$_2$	NP	Biopsy	Herniation (protrusion, extrusion, sequestration)	O'Donnell and O'Donnell 1996
	IVD	Biopsy	Degeneration, herniation (protrusion, extrusion, sequestration)	Burke et al. 2002a
	IVD	Biopsy	Herniation	Kang et al. 1996
	IVD	Biopsy	Herniation (extrusion, sequestration)	Miyamoto et al. 2000
NO	IVD	Biopsy	Herniation	Kang et al. 1996
ADAMTS-1, -4, -5, -9, -15	AF + NP	Autopsy, biopsy	Degeneration	Pockert et al. 2009
ADAMTS-7	NP	Biopsy	Herniation (protrusion, extrusion, sequestration)	Chen et al. 2017
TIMP-1, TIMP-2	AF + NP	Biopsy	Degeneration, herniation (protrusion, extrusion, sequestration)	Bachmeier et al. 2009
TIMP-3	AF + NP	Autopsy, biopsy	Degeneration	Pockert et al. 2009
PLA$_2$	AF + NP	Autopsy, biopsy	Herniation, spondylosis, spondylolisthesis (among others)	Miyahara et al. 1996

Source: Adapted from Wuertz, K., Haglund, L., *Global Spine J.*, 3, 175–184, 2013; and Molinos, M., Almeida, C.R., Caldeira, J., Cunha, C., Gonçalves, R.M., Barbosa, M.A., *J. R. Soc. Interface*, 12, 20141191, 2015.

Note: CCR1, C-C chemokine receptor 1; CDMP, carfilage-derived morphogenic protein; CXCR1, C-X-C motif chemokine receptor 1; FasR, Fas receptor.

(Johnson et al. 2015). Binding promotes the recruitment of several factors such as TNFR1-associated death domain protein, receptor-interacting protein 1 (RIP1), TNF-receptor-associated factor 2, and baculoviral IAP repeat containing 1 and 2, resulting in the formation of complex I signaling (Johnson et al. 2015). Downstream signaling is mediated by nuclear factor κ-light-chain-enhancer of activated B cell (NF-κB) and mitogen-activated protein kinase (MAPK) pathways (Risbud and Shapiro 2014; Silke 2011).

NF-κB controls the expression of several inflammatory and catabolic genes, playing an important role in the regulation of inflammatory response (Risbud and Shapiro 2014). NF-κB is one of the most important regulators of the synthesis of cytokines, such as TNF-α, IL-1β, IL-6, and IL-8, as of the expression of COX-2 (Tak and Firestein 2001). It is a direct modulator of HIF-1α expression, which is an important transcription factor in cells under hypoxia and vital to chondrocyte survival (Dudli, Haschtmann, and Ferguson 2012). NF-κB activation may also be involved in cell apoptosis (Tak and Firestein 2001). Regarding the MAPK pathways, they not only control inflammation but also have several other functions, such as cell growth and differentiation, among others, depending on the cell type (Li et al. 2015b).

Moreover, it was shown that TNF-α can activate the Wnt/β-catenin signaling pathway in NP cells, increasing the expression of MMP-13 (Ye et al. 2011), and that the Wnt/β-catenin signaling can also induce TNF-α expression in NP cells (Hiyama et al. 2013). It is hypothesized that this may lead to a prodegenerative feed-forward loop between the two signaling pathways (Hiyama et al. 2013).

7.3.2 IL-1β

Regarding the IL-1 family, among nine other cytokines are IL-1α and IL-1β. Although TNF-α seems to be the first cytokines produced in a degeneration scenario by human IVD cells, IL-1 appears to be the predominant cytokine (Dudli, Haschtmann, and Ferguson 2012; Le Maitre, Freemont, and Hoyland 2005; Le Maitre, Hoyland, and Freemont 2007a; Weiler et al. 2005). Both proteins are encoded by two separate genes and synthesized as pro-peptide precursors (pro-IL-1α and pro-IL-1β) and then activated through intracellular proteolytic cleavage (IL-1α is cleaved by calpain, and IL-1β, by caspase-1), forming membrane-bound mIL-1α and mIL-1β (Gabay, Lamacchia, and Palmer 2010; Johnson et al. 2015; Risbud and Shapiro 2014). Although pro-IL-1β requires extracellular activation by neutrophil proteases, membrane-associated pro-IL-1α is biologically active and can exert both intracellular and extracellular effects (Gabay, Lamacchia, and Palmer 2010). Pro-IL-1α can signal adjacent cells through the IL-1 receptor type 1 (IL-1R1), which was identified by Le Maitre, Freemont, and Hoyland (2005) in nondegenerate and degenerate human IVDs. Moreover, pro-IL-1α retains a nuclear localization sequence, working as transcriptional modulator (Johnson et al. 2015; Risbud and Shapiro 2014). Pro-IL-1α, mIL-1α, and mIL-1β can bind to IL-1R1, recruit the IL-1 receptor accessory protein (IL-1RAcP), and create a complex, which then recruits two adaptor proteins, the myeloid differentiation primary response gene 88 and the IL-1 receptor-activated protein kinase (Johnson et al. 2015; Risbud and Shapiro 2014). This leads to downstream activation of numerous signaling proteins, such as c-Jun N-terminal kinase (JNK), p38, and MAPK, and transcription factors, like NF-κB and activating protein (AP)-1, controlling the expression of several inflammatory and catabolic genes (Johnson et al. 2015; Risbud and Shapiro 2014).

In organ culture models, stimulation with TNF-α and IL-1β down-regulated the expression of ECM components; increased the expression of ECM degrading enzymes, proinflammatory cytokines and PGE_2, and pain-associated molecule nerve growth factor (NGF) (Abe et al. 2007; Krupkova et al. 2016; Markova et al. 2013; Ponnappan et al. 2011; Purmessur et al. 2013; Teixeira et al. 2015; Walter et al. 2015, 2016); and compromised disc biomechanics (Walter et al. 2015). *In vitro*, human disc cells, upon stimulation with IL-1β and TNF-α, produced high levels of regulated upon activation, normal T-cell expressed, and secreted (RANTES, also named CC chemokine ligand [CCL] 5), which was also observed in lumbar disc AF tissue with higher degree of degeneration (Gruber et al. 2014a). Additionally, TNF-α and IL-1β treatment of NP cells also seems to mediate IVD cell proliferation, affecting the NOTCH signaling pathway (Wang et al. 2013).

7.3.3 IL-6

IL-6 is also a cytokine with impact on promoting IL-1- and TNF-α-mediated catabolism in IVD cells (Risbud and Shapiro 2014). Similarly to the effect of TNF-α (Murata et al. 2008), IL-6 was also shown to induce DRG neuron apoptosis (Murata et al. 2011) and to contribute to neuropathic pain (Wei et al. 2013). Secreted by T cells, macrophages, and IVD cells (Rand et al. 1997), IL-6 has been characterized as a proinflammatory cytokine in the context of IVD degeneration, but it is also involved in regenerative or anti-inflammatory events (Scheller et al. 2011). IL-6 forms monomers and dimers, and it can signal through a type I cytokine receptor complex, which includes the ligand-binding IL-6Rα chain and the membrane glycoprotein gp130, a receptor and signal-transducing subunit, leading to the activation of intracellular signaling cascades via gp130 (Rose-John et al. 2007; Scheller et al. 2011). This pathway is limited to cells that express IL-6R on their surface (Rose-John et al. 2007). It signals through Janus kinase/signal transducers and activators of transcription (JAK/STAT), MAPK, and phosphoinositide-3 kinase (PI3K) signal transduction pathways (Scheller et al. 2011), and promoting functions include B- and T-cell growth and differentiation, as well as acute-phase protein induction, among others (Risbud and Shapiro 2014). On the other hand, soluble IL-6R (sIL-6R) can be formed by proteolytic cleavage of the mIL-6R protein or translation from alternatively spliced mRNA (Rose-John et al. 2007). sIL-6R amplifies IL-6-mediated signaling by the activation of cells that express the signal transducer protein gp130 but lack trans-membrane IL-6R, working as paracrine factor (Risbud and Shapiro 2014; Scheller et al. 2011).

7.3.4 TOLL-LIKE RECEPTORS

Toll-like receptors (TLRs) are plasma- and endolysosomal-bound pattern recognition receptors implicated in innate immunity and inflammation (De Nardo 2015; Klawitter et al. 2014). TLRs are usually expressed by immune cells, namely, dendritic cells, macrophages, neutrophils, monocytes, and T and B cells, but can also be expressed by other cell types such as synovial fibroblasts, chondrocytes, and IVD cells (De Nardo 2015; Klawitter et al. 2014). Klawitter et al. (2014) detected the expression of TLRs 1, 3, 5, 6, 9, and 10 in human cells isolated from degenerated discs and observed that TLR 1, 2, 4, and 6 expression was dependent on the IVD's degree of degeneration. While TLRs 1, 2, 4, 5, and 6 are located on the cell surface, TLRs 3, 7, 8, and 9 are in the endosomal/lysosomal compartment (Klawitter et al. 2014). Namely TLRs 2 and 4 have been described to be expressed by human (Ellman et al. 2012; Gawri et al. 2014; Klawitter et al. 2012a, 2012b, 2014) and bovine (Rajan et al. 2013) IVD cells. TLR2 and TLR4 are known to mediate the innate immunity, being highly specific in their pathogen recognition. They activate NF-κB and JNK/p38 signaling pathways, leading to increased expression of TNF-α, IL-1α, IL-1β, IL-6, IL-8, COX-2, IκBα (an inhibitor of NF-κB transcription factor), MMP-1, MMP-13, monocyte chemoattractant protein (MCP)-1, macrophage inflammatory protein (MIP)-2, and mitogen-activated protein kinase phosphatase-1 (Gabay, Lamacchia, and Palmer 2010; Quero et al. 2013; Schaefer et al. 2005).

Furthermore, as Johnson et al. (2015) discussed, several studies have shown that ECM degradation products may act as signaling molecules, as TLR endogenous ligands, playing a relevant role in the enhancement of the inflammatory state. For instance, proteolytically cleaved biglycan activates proinflammatory cascades through binding to TLR2 and TLR4 in macrophages (Schaefer et al. 2005), hyaluronic acid fragments activate the TLR2 signaling pathway in resident IVD cells (Quero et al. 2013), VCAN aggregates activate TLR2 in carcinoma (Kim et al. 2009b), and fibronectin fragments work as endogenous ligands for TLR4 (Okamura et al. 2001). Moreover, it was observed that excessive mechanical loading of IVD cells may up-regulate TLR2 and TLR4 expression (Gawri et al. 2014). Also, a significant increase in TLR2 mRNA expression and production by stimulating human disc cells with IL-1β or TNF-α was seen, which was linked to the NF-κB pathway activation (Klawitter et al. 2014).

7.4 microRNAs

The role of microRNAs (miRNAs) and their potential as biomarkers for early diagnosis of IVD degeneration has lately drawn great attention (Li et al. 2015c; Zhou et al. 2017). To date, the precise role of miRNAs in the pathogenesis of degeneration is not yet elucidated (Liu et al. 2014; Zhou et al. 2017).

microRNAs are small noncoding RNA molecules with about 18 to 22 nucleotides (Li et al. 2015c), transcribed from their respective gene loci as primary miRNAs (pri-microRNAs) (Papagiannakopoulos and Kosik 2008), followed by a series of maturation steps (Sato et al. 2011). pri-miRNAs can be transcribed from specific miRNA-encoding regions of the genome or derive from mRNA intronic sequences (Li et al. 2015c). miRNAs work by selectively binding to the 3'-untranslated region of their target mRNAs through complementary base pairing, leading to mRNA degradation or suppression of protein translation (Wang et al. 2011b; Ying, Chang, and Lin et al. 2013).

As components of several gene regulatory networks, miRNAs are involved in cell proliferation, differentiation, and apoptosis (Cao et al. 2014; Luo, Nie, and Zhang 2013; Mathieu and Ruohola-Baker 2013); tissue development (Bae et al. 2012;

Joglekar, Joglekar, and Hardikar 2009; Khoshgoo et al. 2013; Ying, Chang, and Lin 2013); homeostasis; metabolism; and tumorigenesis (Majid et al. 2012; Xie et al. 2013). Defective expression or alterations in miRNA combination with their target genes can contribute, for instance, to different cancers, including gastrointestinal (Bandres et al. 2009), osteosarcoma (Duan et al. 2011), and hepatocellular carcinoma (Furuta et al. 2010); autoimmune diseases, such as rheumatoid arthritis and osteoarthritis (Buckland 2010); and IVD degeneration (Tsirimonaki et al. 2013; Wang et al. 2011c; Zhao et al. 2014). Bioinformatics analyses are commonly used to investigate miRNA target genes and predict possible signaling pathways (Zhou et al. 2017). Several authors identified miRNAs that were differentially expressed by human NP cells in degenerative samples, compared to controls (Ji et al. 2016; Li et al. 2016; Wang et al. 2011c; Xu et al. 2016; Zhao et al. 2014). miRNAs involved in the mechanisms associated with disc degeneration have been recently revised by Li et al. (2015c) and Zhou et al. (2017) and are summarized in Table 7.2.

7.5 IMMUNE CELL ACTIVATION

The IVD has been defined as an immune-privileged organ (Sun et al. 2013; Wang et al. 2007). A study by Sheikh et al. (2009) did not observe immune response to a xenograft of mouse cells in an immunocompetent rabbit model, which suggests the hypothesis of the existence of immune-privileged sites within the IVD. The immunological privilege was shown to be maintained by FasL (predominantly expressed in activated T lymphocytes and stromal cells of immune-privileged sites) and the physiological barrier together in rat (Takada et al. 2002) and rabbit (Wang et al. 2007, 2011c) models (Kaneyama et al. 2008). In human samples, FasL expression was observed to decrease with degeneration (Kaneyama et al. 2008). FasL belongs to the TNF family, and when binding to its receptor Fas, Fas–FasL pathway activation induces cell apoptosis of T lymphocytes (Bellgrau et al. 1995; Greil, Egle, and Villunger 1998; Griffith et al. 1995) and of IVD cells (Park, Chang, and Kim 2001; Park et al. 2001; Wang et al. 2011a) and contributes to proinflammatory cytokine production (Yamamoto et al. 2013).

AF tear and NP leakage are recognizable to the immune system as a foreign body (Sun et al. 2013). This may induce antigen capture and activation of B cells with the production of autoantibodies and $CD8^+$ cytotoxic T (T_C) cells (Sun et al. 2013). Antibodies/immunoglobulins have been detected in human herniated IVD tissue (Marshall, Trethewie, and Curtain 1977; Pennington, McCarron, and Laros 1988; Shamji et al. 2010; Szymczak-Workman, Workman, and Vignali 2009; Takahashi et al. 1996). The immune system downstream cascades, promote migration and infiltration, in the region, of specific and nonspecific immune cells, which, together with the cytokines that they and IVD cells secrete, intensify the inflammatory response and cause pain (Risbud and Shapiro 2014). Takahashi and colleagues (1996) identified that most of the cytokine-producing cells, in protrusions, are not only IVD cells but also histiocytes, fibroblasts, or endothelial cells in extruded and sequestrated tissues.

Risbud and Shapiro (2014) reviewed the role of different immune cells infiltrating into the IVD, commonly in herniation and back and radicular pain scenarios, which is schematically presented in Figure 7.2 and described in the following sections.

TABLE 7.2
miRNAs Reported to be Involved in Human Degenerative NP

miRNA	Expression	Target	Function	References
Apoptosis Mediators				
miR-27a	↑	PIK3CD	Regulates the PI3K/Akt signaling pathway	Liu et al. 2013a
miR-155	↓	FADD, caspase-3	Involved in the FasL-Fas signaling pathway	Wang et al. 2011c
miR-494	↑	JunD	Mediates TNF-α-induced cell apoptosis	Wang et al. 2015
Cell Proliferation Mediators				
miR-10b	↑	HOXD10	Targets the RhoC-Akt signaling pathway	Yu et al. 2013
miR-15a	↑	MAP3K9	Inhibits NP cells proliferation and induced cells apoptosis by targeting MAP3K9. Involved in MAPKs signal pathway	Cai et al. 2017
miR-21	↑	PTEN	Targets the PTEN/Akt signaling pathway	Liu et al. 2014
miR-27b	↓	MMP13	Induces type II collagen loss by directly targeting MMP-13	Li et al. 2016
miR-184	↑	GAS1	Negatively regulates the GAS1/Akt signaling pathway	Li et al. 2017b
Degeneration and Inflammation Mediators				
miR-7	↑	GDF-5	Mediates IL-1β-induced ECM degradation	Liu et al. 2016a
miR-15b	↑	SMAD3	Mediates IL-1β-induced ECM degradation	Kang et al. 2017
miR-34a	↑	GDF-5	Mediates IL-1β-induced ECM degradation	Liu et al. 2016b
miR-93	↓	MMP-3	Positively regulates COL2 loss by directly targeting MMP-3	Jing and Jiang 2015
miR-98	↓	STAT3	Promotes ECM degradation by targeting IL-6/STAT3 signaling pathway	Ji et al. 2016a
miR-100	↑	FGFR1, FGFR3	Activates MMP-13 through suppression of FGFR3 via imbalance of FGFR1 and FGFR3 levels	Yan et al. 2015
miR-133a	↓	MMP-9	Mediates COL2 loss by directly targeting MMP-9	Xu et al. 2016

(Continued)

TABLE 7.2 (CONTINUED)
miRNAs Reported to be Involved in Human Degenerative NP

miRNA	Expression	Target	Function	References
		Degeneration and Inflammation Mediators		
miR-146a	↓	FADD, IL-1β, IL-6, TNF, MMP-16	Involved in IL-1 induced IVD degeneration and inflammation	Gu et al. 2015
miR-193a-3p	↓	MMP-14	Positively regulates COL2 expression by directly targeting MMP-14	Ji et al. 2016b
miR-377	↓	ADAMTS5	Negatively regulates ACAN degradation by ADAMTS5	Tsirimonaki et al. 2013

Source: Adapted from Li, Z., Yu, X., Shen, J., Chan, M.T., Wu, W.K., *Cell. Prolif.*, 48, 278–283, 2015; Zhou, X., Chen, L., Grad, S. et al., *J. Tissue Eng. Regen. Med.*, 2017.
Note: ↓ Down-regulated. ↑ Up-regulated.

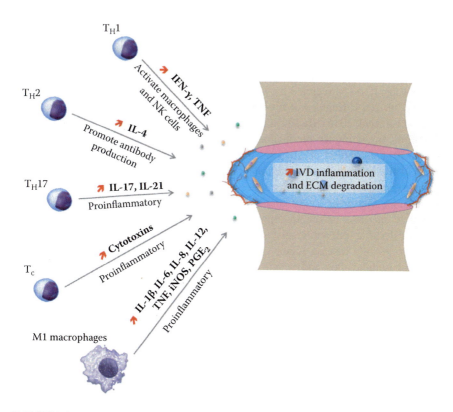

FIGURE 7.2 Role of the different classes of immune cells in amplifying the inflammatory response by disc cells during IVD degeneration.

7.5.1 T Cells

The presence of inflammatory cells, predominantly macrophages, but also mast cells, subtypes of CD4$^+$ T helper (T$_H$) cells, and neutrophils, was observed in painful herniated lumbar discs (Burke et al. 2002a; Doita et al. 1996; Gronblad et al. 1994; Habtemariam et al. 1998; Matsui et al. 1998; Peng et al. 2006; Risbud and Shapiro 2014; Shamji et al. 2010), with significant vascular invasion in noncontained/extruded tissues (Kokubo et al. 2008).

IL-12 and IFN-γ were shown to be highly expressed in herniated disc fragments, compared with bulging discs, which may suggest activation of T$_H$1 CD4$^+$ lymphocytes upon NP exposure to systemic circulation (Cuellar et al. 2010; Park, Chang, and Kim 2002; Shamji et al. 2010). IL-12, known to be produced mainly by macrophages, leads T$_H$1 cells to produce high amounts of IFN-γ, as well as TNF (Risbud and Shapiro 2014; Trinchieri 1994). Of notice, IFN-γ was found to be most commonly elevated in LBP symptomatic patients but absent in asymptomatic controls (Cuellar et al. 2010).

On the other hand, increased levels of IL-4 were found in herniated IVD tissue (Shamji et al. 2010), which suggests the involvement of T$_H$2 CD4$^+$ cells (Risbud and Shapiro 2014). Moreover, higher levels of IL-4 had already been detected in contained discs, when compared to noncontained ones (Park, Chang, and Kim 2002).

The presence of IL-17 was also implicated in IVD degeneration, being identified with CD4$^+$IL-17A$^+$ and CD4$^+$CCR6$^+$ IL-17-producing cells and high levels of IL-17 in degenerated and herniated tissues, in contrast with low levels observed in control tissues obtained from autopsies (Liu, Hou, and Liu 2016; Shamji et al. 2010; Zhang et al. 2013a). IL-17 is secreted by T$_H$17 cells, neutrophils, mast cells (Gaffen 2011; Gruber et al. 2013; Kenna and Brown 2013), as well as by IVD resident cells (Liu, Hou, and Liu et al. 2016). IL-17 is known to induce the activation and mobilization of neutrophils, triggering the production of chemokines and proinflammatory cytokines (Gaffen 2011; Gruber et al. 2013; Kenna and Brown 2013). IL-17 supplementation promoted the production of inflammatory mediators, such as NO, PEG$_2$, and IL-6, and the expression of intercellular adhesion molecule (ICAM)-1 by IVD cells (Gabr et al. 2011). Moreover, costimulation with IL-12 and IFN-γ or TNF-α showed a synergistic increase in the inflammatory mediators and ICAM-1, suggesting an impact of IL-17 at different levels, and an important role of T$_H$17 lymphocytes in the pathology of IVD disease (Gabr et al. 2011). Moreover, IVD cells might recruit additional lymphocytes and immune cells to the IVD (Gabr et al. 2011). IL-17 receptors may signal through JAK/STAT1, MAPK, or NF-κB pathways, correlated with IFN-γ and TNF-α signaling pathways (Albanesi, Cavani, and Girolomoni 1999; Miljkovic and Trajkovic 2004; Weaver et al. 2007).

IL-21, also known to be a cytokine secreted by T$_H$17 cells (Liu et al. 2012; Wei et al. 2007), was recently found in human degenerated IVD (Chen et al. 2017). IL-21 production was shown to contribute to the enhancement of IVD degeneration by stimulation of TNF-α through the JAK/STAT signaling pathway (Chen et al. 2017). It has also been previously shown that IL-21 produced by T$_H$17 cells leads to IL-17 production in a STAT3-dependent manner to promote/sustain T$_H$17 lineage commitment (Wei et al. 2007).

7.5.2 MACROPHAGES

In herniated tissues, it was shown that aside from degenerated IVD cells, also invading monocytes or macrophages (CD68$^+$ cells) may secrete cytokines in the IVD tissue (Kokubo et al. 2008; Peng et al. 2006; Shamji et al. 2010; Wuertz and Haglund 2013). Coculture studies showed that the interaction between IVD cells and macrophages may lead to the production of IL-6, IL-8, inducible NO synthase (iNOS), and PGE$_2$ (Hamamoto et al. 2012; Kim et al. 2008, 2009a, 2012; Takada et al. 2004, 2012; Yamamoto et al. 2013). After tissue injury or infection, monocytes can be recruited to the site as effectors and differentiate into macrophages and dendritic cells (Shi and Pamer 2011). Macrophages are important innate immunity participants, with heterogeneous functions dependent on the microenvironmental cues. Inflammatory macrophages (M1) are described as the "classically activated" subset (Mantovani et al. 2004; Ogle et al. 2016). M1-activated macrophages are part of polarized T$_H$1 response (i.e., stimulation with IFN-γ, lipopolysaccharide, and/or inflammatory cytokines, such as TNF-α), producing numerous inflammatory cytokines (IL-1β, TNF-α, and IL-6), reactive oxygen species, and growth factors, such as vascular endothelial growth factor (VEGF) (Gordon 2003; Mantovani et al. 2004; Mills et al. 2000; Spiller et al. 2014). On the other hand, macrophages can also be polarized towards an anti-inflammatory phenotype (M2), which can further be subdivided into M2a, M2b, and M2c, based on activation signals, cell surface receptors, and functional diversity (Mantovani et al. 2004). Naive macrophages can be polarized *in vitro* by stimulation with IL-4 and/or IL-13 to M2a, with TLR or IL-1R ligands to M2b, or with IL-10 to an M2c phenotype (Mantovani et al. 2004). While M2a macrophages contribute to wound healing, M2b and M2c promote the resolution of inflammation through secretion of IL-10 (Mosser and Edwards 2008). Nonetheless, since macrophage polarization may depend, among other cues, on the amounts of factors present in the area where they migrate to (Mantovani, Garlanda, and Locati 2009), findings from Shamji et al. (2010) from herniated human disc fragments point out to immune lymphocyte activation of the T$_H$1 lineage; hence, macrophages that migrate to herniated IVD tissues will most probably polarize towards an M1 phenotype.

Moreover, infiltrating macrophages, fibroblasts, and endothelial cells, together with native IVD cells, were shown to spontaneously produce MCP-1 and MIP-1α, which, together with IL-8, work as chemotactic molecules for macrophages and other immune cells (Burke et al. 2002b; Gronblad et al. 1994). Several studies hypothesize that the mechanism of spontaneous disc herniation regression may include tissue retraction and dehydration; inflammatory response; and the recruitment, infiltration, and activity of phagocytic cells, among which are neutrophils, monocytes, macrophages, and mast cells (Burke et al. 2002b; Haro et al. 1997; Ikeda et al. 1996; Ito et al. 1996; Kim et al. 2013). Peng and colleagues (2006) detected high numbers of macrophages and mast cells in painful IVDs. Macrophages and mast cells were similarly distributed around blood vessels and among collagenous fibers of scar/granulation tissue, while being absent in nondegenerated controls or aging discs (Peng et al. 2006). Mast cells are highly specialized mononuclear cells, which contribute to disc tissue inflammation, neovascularization, fibrosis, degradation, and secretion of NGF, with a possible causative role in chronic LBP (Freemont et al. 2002b;

Peng et al. 2006). Nonetheless, for instance, Nerlich et al. (2002) also observed that nonherniated NP tissue collected during surgery also presented a high number of resident $CD68^+$ cells. Moreover, Jones et al. (2008) identified, *in vitro*, that IVD cells can undergo phagocytosis, by ingesting latex beads, indicating that endogenous inflammatory-like cells are comprised in the IVD.

Additionally, alterations at systemic level have also been reported, namely, a significant increase in $CD3^+$, $CD4^+$, and $CD4^+/CD8^+$ lymphocytes in the peripheral blood of patients with lumbar disc herniation and with (Ma et al. 2010) or without (Tian et al. 2009) AF rupture. A positive correlation between the percentage of $CD4^+$ T lymphocytes or the ratio $CD4^+/CD8^+$ and pain was also observed (Ma et al. 2010; Tian et al. 2009).

7.6 OTHER FACTORS INVOLVED IN INNERVATION, VASCULARIZATION, AND PAIN

In human extruded or sequestrated discs, other factors have been identified, namely, anti-IL-1, lymphocyte function-associated antigen-1, granulocyte-macrophage colony-stimulating factor (GM-CSF), basic fibroblast growth factor (bFGF) and VEGF, which suggests an active role of those factors in angiogenesis and neovascularization associated with IVD degeneration (Doita et al. 1996; Tolonen et al. 1995). Peng et al. (2006) reported strong expressions of bFGF, transforming growth factor (TGF)-β1, and their receptors, as well as cell proliferation, in granulation tissue from painful lumbar IVDs.

Furthermore, substance P and neurotrophins such as NGF and brain-derived neurotrophic factor (BDNF) have been implicated in the mechanisms associated with an enhancement of innervation and neuropathic pain in some cases of IVD degeneration (Freemont et al. 2002a, 2002b; Ponnappan et al. 2011; Purmessur, Freemont, and Hoyland, 2008; Purmessur et al. 2013). A study by Freemont and colleagues observed production of NGF in painful IVDs with ingrowth of blood vessels and nociceptive nerve fibers. Of notice, NGF expression was not identified in nonpainful or control IVDs (Freemont et al. 2002a).

The production of neurotrophins induces DRG-pain-associated cation channel depolarization (Risbud and Shapiro 2014). The increased expression of transient receptor potential cation channel, subfamily V, member 1 (Trpv1), and the acid-sensing ion channel 3 (ASIC3) induces discogenic pain and further cytokine-mediated disc degeneration (Ohtori et al. 2006; Risbud and Shapiro 2014; Zhang, Huang, and McNaughton 2005).

7.7 STRATEGIES FOR IMMUNOMODULATION OF DEGENERATED IVD

Some regenerative medicine- and tissue engineering-based strategies for degenerated IVD have considered the interplay between IVD degeneration, immune cell response, and inflammation when focused on promoting the production of healthy ECM by native IVD cells, while reducing discogenic pain (Molinos et al. 2015).

Well-balanced approaches targeting not only regeneration but also the modulation of inflammation mediators have been presented as the most promising therapies in reducing IVD-associated pain (Molinos et al. 2015). These include biological approaches (using different molecules such as growth factors), gene therapy, and cell therapies, ranging from autologous/exogenous cell transplantation to endogenous cell stimulation and recruitment, which are under different development levels (clinical trials, *in vivo* trials, and *ex vivo* and *in vitro* studies) and have been reviewed over time (Hughes et al. 2012; Molinos et al. 2015; Richardson et al. 2016; Sakai and Andersson 2015).

7.7.1 Molecular Therapy: Clinical Trials

The modulation and balance of anabolic and anticatabolic responses of IVD cells addressing the aberrant cytokine-rich/proinflammatory degenerative IVD environment are the main target of the molecular therapies proposed so far (Vadala et al. 2015). An overview of these therapies is summarized in Table 7.3. Cocktails or single drug administrations of steroids and anesthetics through epidural delivery or nerve root infiltration, as well as oxygen–ozone (O_2–O_3) gas infiltrations, are routine treatments of discogenic diseases (Bonetti et al. 2005; Burgher et al. 2011). Recently, epidural injection of clonidine, an alpha-2 adrenergic receptor agonist, has also shown potential in patients' pain improvement (Burgher et al. 2011). Clonidine has been previously shown to have anti-inflammatory effects in preclinical studies of nerve injury and may also indirectly influence pain (Romero-Sandoval, McCall, and Eisenach 2005).

Nonetheless, several clinical trials have been evaluating the safety and efficacy of single-dose injections into the NP: for chronic LBP and degenerative disc disease, clinical trials are currently focusing on intradiscal injection of jellified ethanol (NCT02343484), hydrolyzed polyacrylonitrile-based hydrogels (NCT02763956), autologous platelet-rich plasma (PRP), combined with NSAID oral medication (NCT02983747), or recombinant human growth differentiation factor (GDF)-5 (NCT00813813, NCT01124006, and NCT01158924), a member of the TGF-β superfamily and the bone morphogenetic protein (BMP) subfamily, which is known to influence the growth and differentiation of various tissues, including the IVD (Feng et al. 2015). Furthermore, in patients with inflammatory discopathy, intradiscal injection of steroids (NCT00804531 and NCT01694134) has been compared. Drugs with TGF-β antagonist active ingredients (NCT02320019) have also been tested.

In patients suffering from IVD herniation, intradiscal injection of condoliase, a glycosaminoglycan-decomposing enzyme, does degrade the herniated tissue, with high substrate specificity for chondroitin sulfate, dermatan sulfate, and hyaluronic acid (NCT01282606); a recombinant human MMP (NCT01978912); or a fibrin sealant (Yin et al. 2014) have been proposed as potential alternatives. While the MMP and the condoliase studies are currently still in phases II and III of clinical trial, respectively, their selective activity on hernia, leading to its degradation and regression without risk of side effects, is expected to be low (NCT01978912 and NCT01282606). On the other hand, the fibrin sealant was considered to improve pain and function in selected patients with discogenic pain, although neurological

TABLE 7.3
Bioactive Molecules to Target IVD Degeneration, Inflammation, and Discogenic Pain

Therapy	Administration Route	Condition/Model	Posttreatment Follow-Up	Outcomes	References
			Clinical Trials		
Local anesthetic	Transforaminal epidural injection	Disc herniation and radiculitis	Up to 2 years	Significant improvement in all participants who received local anesthetic alone and who received local anesthetic and steroid.	Manchikanti et al. 2014
Oxygen–ozone (O_2–O_3)	Paravertebral injections	LBP due to lumbar disc herniation	Up to 6 months	Minimally invasive; seemed to be safe and effective in reducing root inflammation with a corresponding reduction of pain; reduced disability and intake of analgesic drugs.	Melchionda et al. 2012; Paoloni et al. 2009
	Intradiscal and intraforaminal injection	LBP pain and radicular pain	Up to 1 year	O_2–O_3 nucleolysis provided pain relief in most patients who failed to respond to conservative therapy; no significant differences between O_2–O_3 injection only and combined with steroid.	Zhang et al. 2013b
Pamidronate	Intravenous infusion	Erosive degenerative disc disease; patients fail to respond to NSAIDs	Up to 1 year	Significant improvements in pain and in mean disability scores. The pain no longer showed an inflammatory pattern in 9 of 10 patients.	Poujol et al. 2007
Steroid and O_2–O_3	Intraforaminal and intradiscal injections	Radicular pain related to acute lumbar disk herniation	Up to 6 months	Intraforaminal and intradiscal injections of steroid and O_2–O_3 were more effective than injections of a steroid alone.	Gallucci et al. 2007

(Continued)

TABLE 7.3 (CONTINUED)
Bioactive Molecules to Target IVD Degeneration, Inflammation, and Discogenic Pain

Therapy	Administration Route	Condition/Model	Posttreatment Follow-Up	Outcomes	References
			Clinical Trials		
Steroid	Periradicular infiltration	Radicular pain	Up to 1 year	No additional benefit when compared to local anesthetic injection alone; corticosteroids did not avoid subsequent interventions such as additional root blocks or surgery.	Tafazal et al. 2009
	Transforaminal epidural injection	Acute lumbosacral radiculopathy related to disk herniation	Up to 1 month	Radicular pain improved rapidly with clonidine or triamcinolone. Injections led to greater functional improvement, without differences in analgesia.	Burgher et al. 2011
	Intradiscal injection	Discogenic LBP	Up to 6 months	Potential as short-term alternative for LBP patients unwilling to accept surgery when conservative treatments have failed.	Cao et al. 2011
	Oral administration	Acute sciatica due to herniated disc	Up to 13 months	Modest improvement in function and no improvement in pain after a short course of oral steroids administration, compared with placebo.	Goldberg et al. 2015
	Epidural injection	Sciatica caused by lumbosacral disc prolapse	Up to 3 months	Short-term management of painful sciatica, but no additional long-term improvement over placebo.	Nandi and Chowdhery 2017
TNF-α blocker—adalimumab	Subcutaneous injection	Lumbar disc herniation, sciatica	Up to 6 months, 3 years	Small decrease in leg pain; significantly fewer surgical procedures	Genevay et al. 2010, 2012

(Continued)

TABLE 7.3 (CONTINUED)
Bioactive Molecules to Target IVD Degeneration, Inflammation, and Discogenic Pain

Therapy	Administration Route	Condition/Model	Posttreatment Follow-Up	Outcomes	References
			Clinical Trials		
TNF-α blocker— etanercept	Perispinal administration	Degenerative disc disease, disc herniation, sciatica	Up to 1 month, 230 days	Significant clinical improvement in selected patients with chronic, treatment-resistant disc-related pain.	Tobinick and Britschgi-Davoodifar 2003; Tobinick and Davoodifar 2004
	Subcutaneous injection	Sciatica	Up to 6 weeks	Patients with severe sciatica had sustained improvement after a short treatment with etanercept, compared with standard care plus a short course of methylprednisolone.	Genevay, Stingelin, and Gabay 2004
	Transforaminal epidural injection	Persistent lumbosacral radicular pain secondary to lumbar disc herniation	Up to 26 weeks	Clinically significant reductions in mean daily worst leg pain and worst back pain compared to placebo.	Freeman et al. 2013
	Single intradiscal injection	Discogenic LBP	Up to 2 months	Discogenic LBP alleviation.	Sainoh et al. 2016
TNF-α blocker— infliximab	Intravenous infusion	Sciatica, disc herniation	Up to 6 months	Infliximab was superior in terms of leg pain and back-related disability decrease compared to control but did not appear to interfere with disc herniation resorption.	Autio et al. 2006; Karppinen et al. 2003
	Intravenous infusion	Acute/subacute sciatica secondary to herniated disc	Up to 1 year	Short-term pain reduction, but long-term results did not show differences between infliximab and placebo.	Korhonen et al. 2006

(Continued)

TABLE 7.3 (CONTINUED)
Bioactive Molecules to Target IVD Degeneration, Inflammation, and Discogenic Pain

Therapy	Administration Route	Condition/Model	Posttreatment Follow-Up	Outcomes	References
In Vivo Studies					
BMP-7	Intradiscal injection	Rabbit, *in vivo* disc degeneration	Up to 2 months	Increased disc height up to the 8-week time point and increased NP proteoglycan content at 2 weeks.	An et al. 2005
	Intradiscal injection	Rabbit, *in vivo* disc degeneration	Up to 4 months	A single BMP-7 injection dramatically reversed the decrease in disc height induced by chondroitinase ABC chemonucleolysis.	Imai et al. 2007
BMP-13	Intradiscal injection	Sheep, *in vivo* disc degeneration	Up to 4 months	BMP-13 injected at the time of injury reversed or arrested loss of matrix proteins.	Wei et al. 2009
Corticosteroid	Corticosteroid-loaded ceramic capsule placed adjacent to the punctured disc	Rat, *in vivo* disc degeneration	4 weeks	Continuous sustained release of corticosterone tricalcium phosphate from ceramic capsules could slow the process of degeneration within the traumatized disc in the rat model.	Ragab et al. 2009
COX-2 inhibitor	Epidural injection	Rat, *in vivo* disc degeneration	Up to 1 week	Decrease in mechanical hyperalgesia 1 hour, 3 and 7 days after the epidural injection of COX-2 inhibitor.	Kawakami et al. 2002
Epoxyeicosatrienoic acids	Intradiscal injection	Rat, *in vivo* disc degeneration	1 month	Enhanced the survival of NP cells and inhibited IVD degeneration.	Li et al. 2017b
GDF-5, TGF-β1	Intradiscal injection	Mouse, *in vivo* disc degeneration	Up to 4 weeks	Early intervention avoided or slowed the degenerative process.	Walsh et al. 2004

(Continued)

TABLE 7.3 (CONTINUED)
Bioactive Molecules to Target IVD Degeneration, Inflammation, and Discogenic Pain

In Vivo Studies

Therapy	Administration Route	Condition/Model	Posttreatment Follow-Up	Outcomes	References
IκB kinase-β inhibitor	Intradiscal injection	Rat, *in vivo* disc degeneration	Up to 2 weeks	Injury-induced up-regulation of inflammatory cytokines within the IVD and increased levels of neuropeptides within DRG neurons could be suppressed by inhibiting IκB kinase-β.	Kobori et al. 2014
p38 MAP kinase inhibitor	Intradiscal injection	Rat, *in vivo* disc degeneration	Up to 2 weeks	A direct single application of p38 inhibitor did not suppress calcitonin gene-related peptide expression in DRGs innervating punctured discs.	Hayashi et al. 2009
Phosphodiesterase-2A inhibitor	Intrathecal administration	Rat, *in vivo* noncompressive lumbar disc herniation	Up to 1 week	Alleviates radicular inflammation and mechanical allodynia.	Wang et al. 2017
Platelet-rich plasma (PRP)	Intradiscal injection	Rabbit, *in vivo* disc degeneration	8 weeks	Suppression of degeneration progress.	Nagae et al. 2007
	Injection into and around the IVD	Rat, *in vivo* disc degeneration	Up to 6 weeks	PRP-treated groups retained more normal morphologic features, contained fewer inflammatory cells and showed higher hydration on MRI.	Gullung et al. 2011
Resveratrol	Local application	Rodent, *in vivo* disc degeneration	Up to 2 weeks	Significant pain behavior reduction (it was also seen *in vitro*, in human NP tissue, that resveratrol exhibited an anti-inflammatory and anticatabolic effect).	Wuertz et al. 2011

(Continued)

TABLE 7.3 (CONTINUED)
Bioactive Molecules to Target IVD Degeneration, Inflammation, and Discogenic Pain

Therapy	Administration Route	Condition/Model	Posttreatment Follow-Up	Outcomes	References
In Vivo Studies					
Simvastatin	Intradiscal injection	Rat, *in vivo* disc degeneration	Up to 4 weeks	A single injection of simvastatin loaded in a gel had the potential to retard or regenerate the degenerative disc.	Zhang et al. 2009
Thalidomide	Injection in the epineurium (distal to the NP)	Rat, *in vivo* disc degeneration	Up to 28 days	Significantly inhibited radiculopathic pain *in vivo* (and the expression of proinflammatory mediators and MMPs *in vitro*).	Song et al. 2016
Ex Vivo Studies					
Crocin	Culture medium supplementation	Rat, *ex vivo* disc degeneration	1 week	Effectively suppressed the degeneration-related inflammation and catabolism in rat IVDs, suggesting a potential use as a therapeutic strategy in the treatment of LBP.	Li et al. 2015c
Diclofenac	Intradiscal injection	Bovine, *ex vivo* disc degeneration	Up to 8 days	Df decreased the expression of proinflammatory factors; Df-loaded nanoparticles promoted an up-regulation of extracellular matrix proteins, namely, COL2 and ACAN.	Teixeira et al. 2015, 2016
Epigallocatechin 3-gallate	Culture medium supplementation	Bovine, *ex vivo* disc degeneration	Up to 21 days	The anti-inflammatory and anticatabolic compound epigallocatechin 3-gallate down-regulated the expression of inflammatory and catabolic genes in the NP.	Krupkova et al. 2016

Source: Adapted from Molinos, M., Almeida, C.R., Caldeira, J., Cunha, C., Gonçalves, R.M., Barbosa, M.A., *J. R. Soc. Interface*, 12, 20141191, 2015.

assessments, X-ray, and magnetic resonance imaging (MRI) showed no significant changes (Yin et al. 2014).

O_2–O_3, although promoting tissue stabilization through nucleolysis, is being considered by some clinicians as a successful pain relief approach in some herniated disc patients who failed to respond to conservative therapy (Melchionda et al. 2012; Paoloni et al. 2009; Zhang et al. 2013b), although this is not consensual since many specialists point the absence of functional results to the patients.

TNF-α inhibition and antagonism of TNF-α receptors were shown early on to reduce pain-related symptoms in a chronic constriction injury of rat nerve (Sommer, Marziniak, and Myers 1998). In human trials, a short course of TNF-α inhibitors, such as infliximab, adalimumab, and etanercept, has shown clinical improvement in reducing initial pain in patients with acute or severe sciatica (Genevay et al. 2010, 2012; Goupille et al. 2007; Karppinen et al. 2003; Tobinick and Davoodifar 2004) and the number of patients undergoing surgical procedures. Adalimumab subcutaneous injection, although only showing after 3 years a small decrease in leg pain, significantly reduced the need for back surgery (Genevay et al. 2012). In the case of sciatica, it was shown that intravenous or subcutaneous injection of anti-TNF therapy is short lived and, although lower, might have an associated risk of infection (Goupille et al. 2007). Nonetheless, a single etanercept intradiscal injection was recently shown to alleviate discogenic pain up to 2 months (Sainoh et al. 2016).

7.7.2 MOLECULAR THERAPY: *IN VIVO* AND *EX VIVO* STUDIES

Growth factors have shown overall to enhance ECM production and to stimulate IVD cell proliferation (Masuda 2008). PRP injections into IVD injury models, in rat (Gullung et al. 2011), and rabbit (Nagae et al. 2007) pointed out a maintenance of tissue features, with fewer inflammatory cells, higher fluid content correlated with a more intense signal on MRI (Gullung et al. 2011), and therefore, a delay in the progression of degeneration (Nagae et al. 2007). BMP-7 (An et al. 2005; Imai et al. 2007; Masuda et al. 2006), BMP-13 (Wei et al. 2009), TGF-β1 (Matta et al. 2017; Walsh et al. 2004), and GDF-5 (Chujo et al. 2006; Walsh et al. 2004) have been shown to promote matrix synthesis *in vivo*. While very important, these studies are frequently limited in understanding the effect of the factors injected in native tissue production, disregarding inflammation and pain outputs. Exogenous growth factors were shown to promote matrix synthesis; however, they have the disadvantage of a short biological half-life, ranging from hours to days, and a high cost (Richardson et al. 2016; Winn, Uludag, and Hollinger 1999). Moreover, other works also raise questions about supra-physiologic doses administration for effectiveness and undesired blood vessel ingrowth into the IVD (Zhang et al. 2009).

In vitro tests have also shown great potential of other factors. IL-1Ra released from poly(lactic-co-glycolic acid) microspheres attenuated IL-1β-mediated NP degradation up to 20 days in bovine NP cultures (Gorth et al. 2012). Fullerol nanoparticles were shown to suppress the catabolic activity and adipogenesis of vertebral bone marrow stromal cells under inflammatory stimulus (Liu et al. 2013b). Cobalt protoporphyrin IX treatment of human NP cells from patients with IVD degeneration induced hemeoxygenase-1 expression, which seemed to reverse the effect of IL-1β on the expression of catabolic markers and matrix MMPs (Hu et al. 2016). Natural compounds such as curcumin (Klawitter et al. 2012a) and triptolide (Klawitter et al. 2012b) also exhibited anti-inflammatory, anticatabolic, and antioxidant activity in disc cells.

Other molecules have been successfully tested *ex vivo*: crocin, a bioactive component of saffron (Li et al. 2015a), diclofenac, an NSAID (Teixeira et al. 2015, 2016), and epigallocatechin 3-gallate (Krupkova et al. 2016). All these molecules have shown potential to suppress the degeneration-related inflammation and catabolism in degenerated IVD tissue, suggesting that they can be potentially used as therapeutic drugs in the treatment of LBP.

7.8 GENE THERAPY

Degenerative disc disease is a chronic condition. Therefore, high and long-lasting local levels of different molecules are necessary for a continuous effect of the regenerative therapies (Vadala et al. 2015). Gene therapy has gained significant attention since it promises more prolonged effects in the treatment of IVD degeneration and mediation of inflammation and provides the possibility to locally modulate the expression of a specific gene and the consequent production of its protein (Vadala, Sowa, and Kang 2007; Vadala et al. 2015).

IL-1Ra transfected cells have been suggested as a therapy to inhibit IVD matrix degradation (Le Maitre, Freemont, and Hoyland 2006; Le Maitre, Hoyland, and Freemont 2007b; Muller-Ladner et al. 1997). TGF-β1 transfection of IVD cells through an adenoviral vector was shown to enhance cell activity and proteoglycan synthesis in a rabbit model *in vivo* (Nishida et al. 1999) and in human NP and AF cells *in vitro* (Tan, Hu, and Tan 2003). Also, the transfection of BMP-2, insulin-like growth factor-1 (Li, Yoon, and Hutton 2004), and their combination with TGF-β1 also promoted an increase in proteoglycan synthesis, namely, the combined therapy showed a more promising effect in comparison with the individual transfection treatment (Moon et al. 2008). Moreover, rabbit intradiscal injection of adeno-associated virus serotype 2 vector carrying genes for BMP-2 and tissue inhibitor of metalloproteinases (TIMP)-1 demonstrated an IVD degeneration delay by 12 weeks (Leckie et al. 2012).

Although several therapeutic genes have been successfully identified, the safety of the delivery systems, associated morbidity, and cell irreversible alterations may limit the use of gene transfer vectors in clinics (Molinos et al. 2015; Woods et al. 2011).

Also with great novel therapeutic targeting potential is the mRNA expression of cytokines and chemokines in degenerated IVDs (Ahn et al. 2002). For example, the inhibition of miR-494 protected NP cells from TNF-α-induced apoptosis by targeting JunD (Wang et al. 2015) and the inhibition of miR-34a in NP cells prevented IL-1β-induced ECM degradation by increasing GDF-5 expression (Liu et al. 2016b).

Nonetheless, there is still a long way for the new therapies to go through. Extensive processes of *in vivo* tests and clinical trials are essential to guarantee their safety and long-term effectiveness before widespread use.

7.9 CELL-BASED THERAPIES

Cell-based therapies aim to colonize the IVD with cells capable of differentiating and stimulating endogenous IVD cells' function (Sakai and Andersson 2015). Different cell types have been transplanted over time. NP cells alone (Huang et al. 2011; Nishimura and Mochida 1998; Watanabe et al. 2003) or in combination with AF cells (Ganey et al. 2003; Gruber et al. 2002), elastic cartilage-derived chondrocytes (Gorensek et al. 2004), articular chondrocytes (Acosta et al. 2011), or MSCs have been widely reviewed in the literature (Molinos et al. 2015; Richardson et al. 2016; Sakai and Andersson 2015; Vadala et al. 2016), reporting that cells remain viable thought the studies' time course and that a delayed IVD degeneration is observed. Moreover, a clinical trial using autologous cultured disc-derived chondrocyte transplantation, after discectomy, significantly reduced LBP and allowed retention of hydration in adjacent IVD segments at 2 years when compared to operated patents without cell intervention (Meisel et al. 2007). Allogenic juvenile chondrocytes (NC01771471) and autologous disc chondrocytes (NCT01640457) are currently being tested in phase II clinical trials.

7.9.1 ENDOGENOUS THERAPIES

Progenitor cell populations, as previously discussed, have been pointed out to be present within animal and human IVDs (Brisby et al. 2013; Henriksson et al. 2009; Risbud et al. 2007; Sakai et al. 2012). IVD-derived stem cells were shown to differentiate into chondrogenic and neurogenic lineages, suggesting potential for IVD regeneration (Erwin et al. 2013). They were also shown to play a protective role by modulating IVD inflammatory environment since, for instance, rabbit notochordal cells reduced the expression levels of IL-6, IL-8, and iNOS by human macrophage-exposed AF pellets (Kim et al. 2012). Despite these promising results, Sakai et al. (2012) observed that a population of progenitor cells identified within the human IVD decreases with both age and degeneration, indicating that the isolation of sufficient cell numbers in the NP may be an obstacle when thinking of a clinical application.

Endogenous progenitor cell recruitment/homing to the degenerated disc has also been proposed as an alternative therapeutic approach (Grad et al. 2015). MSC migration was enhanced by degenerative cues and chemoattractor-delivery systems *ex vivo*, in bovine organ culture models (Illien-Junger et al. 2012; Pereira et al. 2014), and *in vivo*, in a mouse tail-looping disc degeneration model (Sakai et al. 2015). *In vivo*, cell homing by the degenerated environment alone is challenging, since it might be widely determined by the degree of neovascularization of the degenerated tissue and of the potential of circulating or bone-marrow-derived MSCs to migrate into the NP (Grad et al. 2015; Sakai et al. 2015). Nonetheless, these results provide important data for the development of novel molecular therapies (Sakai et al. 2015), as discussed in the previous section.

On the other hand, MSC transplantation potential has been linked to their ability to differentiate into an NP cell phenotype, possibly acquiring NP cell-like function, producing IVD native ECM components, or promoting stimulation of endogenous IVD cells, thus enabling anticatabolic and anti-inflammatory effects, as reviewed by

Sakai and Andersson (2015). Moreover, MSCs are also described to have an immunomodulatory role (Prockop and Oh 2012; Yoo et al. 2009).

7.9.2 Exogenous Stem Cell Delivery: Clinical Trials

MSC-based therapies have been tested in a few clinical scenarios of degenerative disc disease and LBP (Orozco et al. 2011; Pettine et al. 2015; Yoshikawa et al. 2010). Yoshikawa and colleagues (2010) reported two case studies in which patients underwent hernia fenestration surgery and degenerated IVD percutaneous engraftment of a collagen sponge containing autologous MSCs. Two years after surgery, an increase in MRI signal intensity of IVDs with cell grafts was observed, suggesting higher hydration. Disc instability and pain symptoms also seemed to have improved (Yoshikawa et al. 2010). Orozco et al. (2011) also showed the safety and feasibility of autologous bone-marrow-derived MSC intradiscal injection. Patients exhibited rapid improvement of pain and disability (85% of maximum in 3 months) that approached 71% of optimal efficacy, described to be comparable with the results of procedures such as spinal fusion or total disc replacement. Although disc height was not recovered, water content was significantly elevated at 12 months (Orozco et al. 2011). Moreover, it was recently reported that percutaneous injection of autologous bone marrow concentrate cells significantly reduced lumbar discogenic pain over 12 months (Pettine et al. 2015).

These results encouraged other trials that are currently ongoing addressing the use of allogenic (NCT02097862) or autologous (NCT02338271 and NCT02529566) cell transplantation and implantation of cell-seeded scaffolds in degenerated IVD (NCT01290367, NCT01513694, NCT01643681, and NCT02412735).

7.9.3 Exogenous Stem Cell Delivery: In Vivo and In Vitro Studies

Sakai and Andersson (2015) reviewed several preclinical studies investigating transplantation of stem cells derived from bone marrow, adipose, synovial, and umbilical cord tissues, as well as from CEP, AF, and NP for IVD regeneration. Overall, an improvement in MRI signaling, disc height maintenance, or up-regulation of IVD ECM component expression was reported (Sakai and Andersson 2015).

The immunomodulatory role of MSCs has been previously addressed in several contexts. *In vitro*, coculture of human adipose-derived MSCs and osteoarthritic chondrocytes induced down-regulation of inflammatory factors such as IL-6, IL-8, IL-1β, MCP-1, MIP-1α, and RANTES expression by MSCs (Manferdini et al. 2013). Proinflammatory cytokines, NO, and other damage-associated molecules from injured tissues have also been shown to activate MSCs to secrete PGE_2, which binds to macrophages and polarizes them to an M2 phenotype that secretes IL-10 (Nemeth et al. 2009).

Ex vivo, synovial explants exposed to MSC-conditioned medium showed down-regulation of IL-1β and MMP-1 and -13 and up-regulation of suppressor of cytokine signaling 1 (van Buul et al. 2012). In cartilage, the expression of IL-1Ra was up-regulated, while ADAMTS-5 and COL2 were down-regulated. MSC-conditioned medium reduced NO production in cartilage explants, and the presence of the NF-κB inhibitor, IκBα, was increased in synoviocytes and chondrocytes treated with MSC-conditioned medium (van Buul et al. 2012). MSCs administered systemically were shown to secrete anti-inflammatory TNF-α stimulated gene/protein-6 in myocardial infarction in mice (Lee et al. 2009a) and in rat injured cornea (Roddy et al. 2011).

In the IVD context, FasL protein (found in other immune privileged sites) was shown to be expressed in the NP region after MSC intradiscal administration into beagle nucleotomized IVDs, indicating that MSCs either differentiated into cells expressing FasL or stimulated the few remaining NP cells to express it (Hiyama et al. 2008). Moreover, IL-1Ra was shown to mediate the anti-inflammatory and antifibrotic effects of MSCs in a mouse model of lung injury (Ortiz et al. 2007). However, MSCs' mechanism of action in the IVD and their impact on inflammation mediators are often disregarded in the multiple studies across the literature (Molinos et al. 2015).

7.10 CONCLUDING REMARKS AND FUTURE PERSPECTIVES

This chapter covered numerous works on immunomodulatory and therapeutic approaches that have potential to promote a proregenerative milieu in the IVD. Despite the inherent variability and contradictions arising from different studies, as suggested in the review work of Molinos et al. (2015), integrated strategies contemplating the different features of IVD degeneration may contribute to a better translation of *ex vivo* and *in vivo* results and therapeutics to humans.

ACKNOWLEDGMENTS

The authors would like to acknowledge to European Union funds through "Projetos Estruturados de I&D&I—Norte-01-0145-FEDER-000012, Portugal 2020—FEDER." RMG and GQT would like to acknowledge FCT—Fundação para a Ciência e a Tecnologia for the FCT Investigator Starting Grant (IF/00638/2014) and the PhD grant (SFRH/BD/88429/2012), respectively, and EUROSPINE TRF (2017_5).

REFERENCES

Abe, Y., K. Akeda, H.S. An et al. 2007. "Proinflammatory cytokines stimulate the expression of nerve growth factor by human intervertebral disc cells." *Spine (Phila Pa 1976)* 32:635–42.

Acosta, F.L. Jr., L. Metz, H.D. Adkisson et al. 2011. "Porcine intervertebral disc repair using allogenic juvenile articular chondrocytes or mesenchymal stem cells." *Tissue Eng Part A* 17:3045–55.

Ahn, S.H., Y.W. Cho, M.W. Ahn et al. 2002. "mRNA expression of cytokines and chemokines in herniated lumbar intervertebral discs." *Spine (Phila Pa 1976)* 27:911–7.

Akyol, S., B.S. Eraslan, H. Etyemez, T. Tanriverdi, and M. Hanci. 2010. "Catabolic cytokine expressions in patients with degenerative disc disease." *Turk Neurosurg* 20:492–9.

Albanesi, C., A. Cavani, and G. Girolomoni. 1999. "IL-17 is produced by nickel-specific T lymphocytes and regulates ICAM-1 expression and chemokine production in human

keratinocytes: Synergistic or antagonist effects with IFN-gamma and TNF-alpha." *J Immunol* 162:494–502.
An, H.S., K. Takegami, H. Kamada et al. 2005. "Intradiscal administration of osteogenic protein-1 increases intervertebral disc height and proteoglycan content in the nucleus pulposus in normal adolescent rabbits." *Spine (Phila Pa 1976)* 30:25–32.
Andersson, G.B. 1999. "Epidemiological features of chronic low back pain." *Lancet* 354: 581–5.
Autio, R.A., J. Karppinen, J. Niinimäki et al. 2006. "The effect of infliximab, a monoclonal antibody against TNF-alpha, on disc herniation resorption: A randomized controlled study." *Spine (Phila Pa 1976)* 31:2641–5.
Bachmeier, B.E., A. Nerlich, N. Mittermaier et al. 2009. "Matrix metalloproteinase expression levels suggest distinct enzyme roles during lumbar disc herniation and degeneration." *Eur Spine J* 18:1573–86.
Bachmeier, B.E., A.G. Nerlich, C. Weiler et al. 2007. "Analysis of tissue distribution of TNF-alpha, TNF-alpha-receptors, and the activating TNF-alpha-converting enzyme suggests activation of the TNF-alpha system in the aging intervertebral disc." *Ann N Y Acad Sci* 1096:44–54.
Bae, Y., T. Yang, H.C. Zeng et al. 2012. "miRNA-34c regulates Notch signaling during bone development." *Hum Mol Genet* 21:2991–3000.
Bandres, E., N. Bitarte, F. Arias et al. 2009. "microRNA-451 regulates macrophage migration inhibitory factor production and proliferation of gastrointestinal cancer cells." *Clin Cancer Res* 15:2281–90.
Bellgrau, D., D. Gold, H. Selawry et al. 1995. "A role for CD95 ligand in preventing graft rejection." *Nature* 377:630–2.
Bertolo, A., T. Thiede, N. Aebli et al. 2011. "Human mesenchymal stem cell co-culture modulates the immunological properties of human intervertebral disc tissue fragments *in vitro*." *Eur Spine J* 20:592–603.
Black, R.A., C.T. Rauch, C.J. Kozlosky et al. 1997. "A metalloproteinase disintegrin that releases tumour-necrosis factor-alpha from cells." *Nature* 385:729–33.
Bonetti, M., A. Fontana, B. Cotticelli et al. 2005. "Intraforaminal O(2)-O(3) versus peri-radicular steroidal infiltrations in low back pain: Randomized controlled study." *AJNR Am J Neuroradiol* 26:996–1000.
Boos, N., S. Weissbach, H. Rohrbach et al. 2002. "Classification of age-related changes in lumbar intervertebral discs: 2002 Volvo Award in basic science." *Spine (Phila Pa 1976)* 27:2631–44.
Brisby, H., N. Papadimitriou, C. Brantsing et al. 2013. "The presence of local mesenchymal progenitor cells in human degenerated intervertebral discs and possibilities to influence these *in vitro*: A descriptive study in humans." *Stem Cells Dev* 22:804–14.
Buckland, J. 2010. "Biomarkers: microRNAs under the spotlight in inflammatory arthritis." *Nat Rev Rheumatol* 6:436.
Burgher, A.H., B.C. Hoelzer, D.R. Schroeder, G.A. Wilson, and M.A. Huntoon. 2011. "Transforaminal epidural clonidine versus corticosteroid for acute lumbosacral radiculopathy due to intervertebral disk herniation." *Spine (Phila Pa 1976)* 36: E293–300.
Burke, J.G., R.W. Watson, D. McCormack et al. 2002a. "Intervertebral discs which cause low back pain secrete high levels of proinflammatory mediators." *J Bone Joint Surg Br* 84:196–201.
Burke, J.G., R.W. Watson, D. McCormack et al. 2002b. "Spontaneous production of monocyte chemoattractant protein-1 and interleukin-8 by the human lumbar intervertebral disc." *Spine (Phila Pa 1976)* 27:1402–7.
Cai, P., T. Yang, X. Jiang et al. "Role of miR-15a in intervertebral disc degeneration through targeting MAP3K9." *Biomed Pharmacother* 87:568–74.

Cao, H., X. Hu, Q. Zhang et al. 2014. "Upregulation of let-7a inhibits vascular smooth muscle cell proliferation *in vitro* and in vein graft intimal hyperplasia in rats." *J Surg Res* 192:223–33.

Cao, P., L. Jiang, C. Zhuang et al. 2011. "Intradiscal injection therapy for degenerative chronic discogenic low back pain with end plate Modic changes." *Spine J* 11:100–6.

Chen, B., Y. Liu, Y. Zhang et al. 2017. "IL-21 is positively associated with intervertebral disc degeneration by interaction with TNF-alpha through the JAK-STAT signaling pathway." *Inflammation* 40:612–22.

Chujo, T., H.S. An, K. Akeda et al. 2006. "Effects of growth differentiation factor-5 on the intervertebral disc—*In vitro* bovine study and *in vivo* rabbit disc degeneration model study." *Spine (Phila Pa 1976)* 31:2909–17.

Cuellar, J.M., P.M. Borges, V.G., Cuellar et al. 2013. "Cytokine expression in the epidural space: A model of noncompressive disc herniation-induced inflammation." *Spine (Phila Pa 1976)* 38:17–23.

Cuellar, J.M., S.R. Golish, M.W. Reuter et al. 2010. "Cytokine evaluation in individuals with low back pain using discographic lavage." *Spine J* 10:212–8.

De Nardo, D. 2015. "Toll-like receptors: Activation, signalling and transcriptional modulation." *Cytokine* 74:181–9.

Deyo, R.A. and J.N. Weinstein. 2001. "Low back pain." *N Engl J Med* 344:363–70.

Doita, M., T. Kanatani, T. Ozaki et al. 2001. "Influence of macrophage infiltration of herniated disc tissue on the production of matrix metalloproteinases leading to disc resorption." *Spine (Phila Pa 1976)* 26:1522–27.

Doita, M., T. Kanatani, T. Harada, and K. Mizuno. 1996. "Immunohistologic study of the ruptured intervertebral disc of the lumbar spine." *Spine (Phila Pa 1976)* 21:235–41.

Dongfeng, R., S. Hou, W. Wu et al. 2011. "The expression of tumor necrosis factor-alpha and CD68 in high-intensity zone of lumbar intervertebral disc on magnetic resonance image in the patients with low back pain." *Spine (Phila Pa 1976)* 36:E429–33.

Duan, Z., E. Choy, D. Harmon et al. 2011. "MicroRNA-199a-3p is downregulated in human osteosarcoma and regulates cell proliferation and migration." *Mol Cancer Ther* 10:1337–45.

Dudli, S., D. Haschtmann, and S.J. Ferguson. 2012. "Fracture of the vertebral endplates, but not equienergetic impact load, promotes disc degeneration *in vitro*." *J Orthop Res* 30:809–16.

Ellman, M.B., J.S. Kim, H.S. An et al. 2012. "Toll-like receptor adaptor signaling molecule MyD88 on intervertebral disk homeostasis: *In vitro, ex vivo* studies." *Gene* 505:283–90.

Erwin, W.M., D. Islam, E. Eftekarpour, R.D. Inman, M.Z. Karim, and M.G. Fehlings. 2013. "Intervertebral disc-derived stem cells: Implications for regenerative medicine and neural repair." *Spine (Phila Pa 1976)* 38:211–6.

Feng, C., H. Liu, Y. Yang, B. Huang and Y. Zhou. 2015. "Growth and differentiation factor-5 contributes to the structural and functional maintenance of the intervertebral disc." *Cell Physiol Biochem* 35:1–16.

Freeman, B.J., G.L. Ludbrook, S. Hall et al. 2013. "Randomized, double-blind, placebo-controlled, trial of transforaminal epidural etanercept for the treatment of symptomatic lumbar disc herniation." *Spine (Phila Pa 1976)* 38:1986–94.

Freemont, A.J., A. Watkins, C. Le Maitre et al. 2002a. "Nerve growth factor expression and innervation of the painful intervertebral disc." *J Pathol* 197:286–92.

Freemont, A.J., M. Jeziorska, J.A. Hoyland, P., Rooney, and S. Kumar. 2002b. "Mast cells in the pathogenesis of chronic back pain: A hypothesis." *J Pathol* 197:281–5.

Fujita, N., S.S. Gogate, K. Chiba et al. 2012. "Prolyl hydroxylase 3 (PHD3) modulates catabolic effects of tumor necrosis factor-α (TNF-α) on cells of the nucleus pulposus through co-activation of nuclear factor κB (NF-κB)/p65 signaling." *J Biol Chem* 287:39942–53.

Gabay, C., C. Lamacchia, and G. Palmer. 2010. "IL-1 pathways in inflammation and human diseases." *Nat Rev Rheumatol* 6:232–41.

Gabr, M.A., L. Jing, A.R. Helbling et al. 2011. "Interleukin-17 synergizes with IFNγ or TNFα to promote inflammatory mediator release and intercellular adhesion molecule-1 (ICAM-1) expression in human intervertebral disc cells." *J Orthop Res* 29:1–7.

Gaffen, S.L. 2011. "Recent advances in the IL-17 cytokine family." *Curr Opin Immunol* 23:613–9.

Gajendran, V.K., M.W. Reuter, S.R. Golish, L.S. Hanna, and G.J. Scuderi. 2011. "Is the fibronectin-aggrecan complex present in cervical disk disease?" *PM R* 3:1030–4.

Gallucci, M., N. Limbucci, L. Zugaro et al. 2007. "Sciatica: Treatment with intradiscal and intraforaminal injections of steroid and oxygen-ozone versus steroid only." *Radiology* 242(3):907–13.

Ganey, T., J. Libera, V. Moos et al. 2003. "Disc chondrocyte transplantation in a canine model: A treatment for degenerated or damaged intervertebral disc." *Spine (Phila Pa 1976)* 28:2609–20.

Gawri, R., D.H. Rosenzweig, E., Krock et al. 2014. "High mechanical strain of primary intervertebral disc cells promotes secretion of inflammatory factors associated with disc degeneration and pain." *Arthritis Res Ther* 16:R21.

Genevay, S., S. Viatte, A. Finckh et al. 2010. "Adalimumab in severe and acute sciatica: A multicenter, randomized, double-blind, placebo-controlled trial." *Arthritis Rheum* 62:2339–46.

Genevay, S., A. Finckh, P. Zufferey et al. 2012. "Adalimumab in acute sciatica reduces the long-term need for surgery: A 3-year follow-up of a randomised double-blind placebo-controlled trial." *Ann Rheum Dis* 71:560–2.

Genevay, S., S. Stingelin, and C. Gabay. 2004. "Efficacy of etanercept in the treatment of acute, severe sciatica: A pilot study." *Ann Rheum Dis* 63:1120–3.

Goldberg, H., W. Firtch, M. Tyburski et al. 2015. "Oral steroids for acute radiculopathy due to a herniated lumbar disk: A randomized clinical trial." *JAMA* 313:1915–23.

Gordon, S. 2003. "Alternative activation of macrophages." *Nat Rev Immunol* 3:23–35.

Gorensek, M., C. Jaksimović, N. Kregar-Velikonja et al. 2004. "Nucleus pulposus repair with cultured autologous elastic cartilage derived chondrocytes." *Cell Mol Biol Lett* 9:363–73.

Gorth, D.J., R.L. Mauck, J.A. Chiaro et al. 2012. "IL-1ra delivered from poly (lactic-co-glycolic acid) microspheres attenuates IL-1beta-mediated degradation of nucleus pulposus *in vitro*." *Arthritis Res Ther* 14:R179.

Gorth, D.J., I.M. Shapiro, and M.V. Risbud. 2015. "Discovery of the drivers of inflammation induced chronic low back pain: From bacteria to diabetes." *Discov Med* 20:177–84.

Goupille, P., D. Mulleman, G. Paintaud, H. Watier, and J.P. Valat. 2007. "Can sciatica induced by disc herniation be treated with tumor necrosis factor alpha blockade?" *Arthritis Rheum* 56:3887–95.

Grad, S., M. Peroglio, Z. Li, and M. Alini. 2015. "Endogenous cell homing for intervertebral disk regeneration." *J Am Acad Orthop Surg* 23:264–6.

Grant, M.P., L.M. Epure, R. Bokhari et al. 2016. "Human cartilaginous endplate degeneration is induced by calcium and the extracellular calcium-sensing receptor in the intervertebral disc." *Eur Cell Mater* 32:137–51.

Greil, R., A. Egle, and A. Villunger. 1998. "On the role and significance of Fas (Apo-1/CD95) ligand (FasL) expression in immune privileged tissues and cancer cells using multiple myeloma as a model." *Leuk Lymphoma* 31:477–90.

Griffith, T.S., T. Brunner, S.M., Fletcher, D.R., Green, and T.A. Ferguson. 1995. "Fas ligand-induced apoptosis as a mechanism of immune privilege." *Science* 270:1189–92.

Gronblad, M., J. Virri, J. Tolonen et al. 1994. "A controlled immunohistochemical study of inflammatory cells in disc herniation tissue." *Spine (Phila Pa 1976)* 19:2744–51.

Gruber, H.E., G.L. Hoelscher, J.A. Ingram et al. 2014a. "Production and expression of RANTES (CCL5) by human disc cells and modulation by IL-1-beta and TNF-alpha in 3D culture." *Exp Mol Pathol* 96:133–8.

Gruber, H.E., G.L. Hoelscher, J.A. Ingram, H.J. Norton, and E.N. Hanley Jr. 2013. "Increased IL-17 expression in degenerated human discs and increased production in cultured annulus cells exposed to IL-1ss and TNF-alpha." *Biotech Histochem* 88:302–10.

Gruber, H.E., G.L. Hoelscher, J.A. Ingram, S. Bethea, and E.N. Hanley Jr. 2014b. "Growth and differentiation factor-5 (GDF-5) in the human intervertebral annulus cells and its modulation by IL-1ss and TNF-alpha *in vitro*." *Exp Mol Pathol* 96:225–9.

Gruber, H.E., J.A. Ingram, D.E. Davis, and E.N. Hanley Jr. 2009. "Increased cell senescence is associated with decreased cell proliferation *in vivo* in the degenerating human annulus." *Spine J* 9:210–5.

Gruber, H.E., T. Johnson, H.J. Norton, and E.N. Hanley Jr. 2002. "The sand rat model for disc degeneration: Radiologic characterization of age-related changes: Cross-sectional and prospective analyses." *Spine (Phila Pa 1976)* 27:230–4.

Gu, S.-X., X. Li, J.L. Hamilton et al. 2015. "MicroRNA-146a reduces IL-1 dependent inflammatory responses in the intervertebral disc." *Gene* 555:80–7.

Gullung, G.B., J.W. Woodall, M.A. Tucci, J. James, D.A. Black, and R.A. McGuire. 2011. "Platelet-rich plasma effects on degenerative disc disease: Analysis of histology and imaging in an animal model." *Evid Based Spine Care J* 2:13–8.

Habtemariam, A., M. Gronblad, J. Virri, S. Seitsalo, and E. Karaharju. 1998. "A comparative immunohistochemical study of inflammatory cells in acute-stage and chronic-stage disc herniations." *Spine (Phila Pa 1976)* 23:2159–66.

Hamamoto, H., H. Miyamoto, M. Doita et al. 2012. "Capability of nondegenerated and degenerated discs in producing inflammatory agents with or without macrophage interaction." *Spine (Phila Pa 1976)* 37:161–7.

Haro, H., H. Komori, A. Okawa et al. 1997. "Sequential dynamics of monocyte chemotactic protein-1 expression in herniated nucleus pulposus resorption." *J Orthop Res* 15:734–41.

Hayashi, S., A. Taira, G. Inoue et al. 2008. "TNF-alpha in nucleus pulposus induces sensory nerve growth: A study of the mechanism of discogenic low back pain using TNF-alpha-deficient mice." *Spine (Phila Pa 1976)* 33:1542–6.

Hayashi, Y., S. Ohtori, M. Yamashita et al. 2009. "Direct single injection of p38 mitogen-activated protein kinase inhibitor does not affect calcitonin gene-related peptide expression in dorsal root ganglion neurons innervating punctured discs in rats." *Spine (Phila Pa 1976)* 34:2843–7.

Henriksson, H., M. Thornemo, C. Karlsson et al. 2009. "Identification of cell proliferation zones, progenitor cells and a potential stem cell niche in the intervertebral disc region: A study in four species." *Spine (Phila Pa 1976)* 34:2278–87.

Hiyama, A., J. Mochida, T. Iwashina et al. 2008. "Transplantation of mesenchymal stem cells in a canine disc degeneration model." *J Orthop Res* 26:589–600.

Hiyama, A., K. Yokoyama, T. Nukaga, D. Sakai, and J. Mochida. 2013. "A complex interaction between Wnt signaling and TNF-alpha in nucleus pulposus cells." *Arthritis Res Ther* 15:R189.

Hu, B., C. Shi, C. Xu et al. 2016. "Heme oxygenase-1 attenuates IL-1beta induced alteration of anabolic and catabolic activities in intervertebral disc degeneration." *Sci Rep* 6:21190.

Huang, B., Y. Zhuang, C.Q. Li, L.T. Liu, and Y. Zhou. 2011. "Regeneration of the intervertebral disc with nucleus pulposus cell-seeded collagen II/hyaluronan/chondroitin-6-sulfate tri-copolymer constructs in a rabbit disc degeneration model." *Spine (Phila Pa 1976)* 36:2252–9.

Huang, K.Y., R.M. Lin, W.Y. Chen et al. 2008. "IL-20 may contribute to the pathogenesis of human intervertebral disc herniation." *Spine (Phila Pa 1976)* 33:2034–40.

Huang, Y.C., J.P. Urban, and K.D. Luk. 2014. "Intervertebral disc regeneration: Do nutrients lead the way?" *Nat Rev Rheumatol* 10:561–6.

Hughes, S.P., A.J. Freemont, D.W. Hukins, A.H. McGregor, and S. Roberts. 2012. "The pathogenesis of degeneration of the intervertebral disc and emerging therapies in the management of back pain." *J Bone Joint Surg Br* 94:1298–304.

Iatridis, J.C., A.J. Michalek, D. Purmessur, and C.L. Korecki. 2009. "Localized intervertebral disc injury leads to organ level changes in structure, cellularity, and biosynthesis." *Cell Mol Bioeng* 2:437–47.

Igarashi, T., S. Kikuchi, V. Shubayev, and R.R. Myers. 2000. "2000 Volvo Award winner in basic science studies: Exogenous tumor necrosis factor-alpha mimics nucleus pulposus-induced neuropathology. Molecular, histologic, and behavioral comparisons in rats." *Spine (Phila Pa 1976)* 25:2975–80.

Ikeda, T., T. Nakamura, T. Kikuchi et al. 1996. "Pathomechanism of spontaneous regression of the herniated lumbar disc: Histologic and immunohistochemical study." *J Spinal Disord* 9:136–40.

Illien-Junger, S., G. Pattappa, M. Peroglio et al. 2012. "Homing of mesenchymal stem cells in induced degenerative intervertebral discs in a whole organ culture system." *Spine (Phila Pa 1976)* 37:1865–73.

Imai, Y., M. Okuma, H.S. An et al. 2007. "Restoration of disc height loss by recombinant human osteogenic protein-1 injection into intervertebral discs undergoing degeneration induced by an intradiscal injection of chondroitinase ABC." *Spine (Phila Pa 1976)* 32:1197–205.

Ito, T., M. Yamada, F. Ikuta et al. 1996. "Histologic evidence of absorption of sequestration-type herniated disc." *Spine (Phila Pa 1976)* 21:230–4.

Ji, M.L., J. Lu, P.L. Shi et al. 2016a. "Dysregulated miR-98 contributes to extracellular matrix degradation by targeting IL-6/STAT3 signaling pathway in human intervertebral disc degeneration." *J Bone Miner Res* 31:900–9.

Ji, M.L., X.J. Zhang, P.L. Shi et al. 2016b. "Downregulation of microRNA-193a-3p is involved in invertebral disc degeneration by targeting MMP14." *J Mol Med (Berl)* 94:457–68.

Jimbo, K., J.S. Park, K. Yokosuka, K. Sato, and K. Nagata. 2005. "Positive feedback loop of interleukin-1beta upregulating production of inflammatory mediators in human intervertebral disc cells *in vitro*." *J Neurosurg Spine* 2:589–95.

Jing, W., and W. Jiang. 2015. "MicroRNA-93 regulates collagen loss by targeting MMP3 in human nucleus pulposus cells." *Cell Prolif* 48:284–92.

Joglekar, M.V., V.M. Joglekar, and A.A. Hardikar. 2009. "Expression of islet-specific microRNAs during human pancreatic development." *Gene Expr Patterns* 9:109–13.

Johnson, Z.I., Z.R. Schoepflin, H. Choi, I.M. Shapiro, and M.V. Risbud. 2015. "Disc in flames: Roles of TNF-alpha and IL-1beta in intervertebral disc degeneration." *Eur Cell Mater* 30:104–17.

Jones, P., L. Gardner, J. Menage, G.T. Williams, and S. Roberts. 2008. "Intervertebral disc cells as competent phagocytes *in vitro*: Implications for cell death in disc degeneration." *Arthritis Res Ther*. 10:R86.

Kaneyama, S., K. Nishida, T. Takada et al. 2008. "Fas ligand expression on human nucleus pulposus cells decreases with disc degeneration processes." *J Orthop Sci* 13:130–5.

Kang, J.D., H.I. Georgescu, L. McIntyre-Larkin et al. 1996. "Herniated lumbar intervertebral discs spontaneously produce matrix metalloproteinases, nitric oxide, interleukin-6, and prostaglandin E2." *Spine (Phila Pa 1976)* 21:271–7.

Kang, J.D., M. Stefanovic-Racic, L.A. McIntyre, H.I. Georgescu, and C.H. Evans. 1997. "Toward a biochemical understanding of human intervertebral disc degeneration and herniation. Contributions of nitric oxide, interleukins, prostaglandin E2, and matrix metalloproteinases." *Spine (Phila Pa 1976)* 22:1065–73.

Kang, L., C. Yang, H. Yin et al. 2017. "MicroRNA-15b silencing inhibits IL-1β-induced extracellular matrix degradation by targeting SMAD3 in human nucleus pulposus cells." *Biotechnol Lett* 39:623–32.

Karppinen, J., T. Korhonen, A. Malmivaara et al. 2003. "Tumor necrosis factor-alpha monoclonal antibody, infliximab, used to manage severe sciatica." *Spine (Phila Pa 1976)* 28:750–4.

Kawakami, M., T. Matsumoto, H. Hashizume, K. Kuribayashi, and T. Tamaki. 2002. "Epidural injection of cyclooxygenase-2 inhibitor attenuates pain-related behavior following application of nucleus pulposus to the nerve root in the rat." *J Orthop Res* 20:376–81.

Kenna, T.J., and M.A. Brown. 2013. "The role of IL-17-secreting mast cells in inflammatory joint disease." *Nat Rev Rheumatol* 9:375–9.

Khoshgoo, N., R. Kholdebarin, B.M. Iwasiow, and R. Keijzer. 2013. "MicroRNAs and lung development." *Pediatr Pulmonol* 48:317–23.

Kim, J.H., B.M. Deasy, H.Y. Seo et al. 2009a. "Differentiation of intervertebral notochordal cells through live automated cell imaging system in vitro." *Spine (Phila Pa 1976)* 34:2486–93.

Kim, J.H., H.J. Moon, J.H. Lee et al. 2012. "Rabbit notochordal cells modulate the expression of inflammatory mediators by human annulus fibrosus cells cocultured with activated macrophage-like THP-1 cells." *Spine (Phila Pa 1976)* 37:1856–64.

Kim, J.H., R.K. Studer, G.A. Sowa, N.V. Vo, and J.D. Kang. 2008. "Activated macrophage-like THP-1 cells modulate annulus fibrosus cell production of inflammatory mediators in response to cytokines." *Spine (Phila Pa 1976)* 33:2253–9.

Kim, S.G., J.C. Yang, T.W. Kim, and K.H. Park. 2013. "Spontaneous regression of extruded lumbar disc herniation: Three cases report." *Korean J Spine* 10:78–81.

Kim, S., H. Takahashi, W.W. Lin et al. 2009b. "Carcinoma-produced factors activate myeloid cells through TLR2 to stimulate metastasis." *Nature* 457:102–6.

Klawitter, M., L. Quero, J. Klasen et al. 2012a. "Curcuma DMSO extracts and curcumin exhibit an anti-inflammatory and anti-catabolic effect on human intervertebral disc cells, possibly by influencing TLR2 expression and JNK activity." *J Inflamm (Lond)* 9:29.

Klawitter, M., L. Quero, J. Klasen et al. 2012b. "Triptolide exhibits anti-inflammatory, anti-catabolic as well as anabolic effects and suppresses TLR expression and MAPK activity in IL-1beta treated human intervertebral disc cells." *Eur Spine J* 21 (Suppl 6): S850–9.

Klawitter, M., M. Hakozaki, H. Kobayashi et al. 2014. "Expression and regulation of toll-like receptors (TLRs) in human intervertebral disc cells." *Eur Spine J* 23:1878–91.

Kobori, S., M. Miyagi, S. Orita et al. 2014. "Inhibiting IκB kinase-β downregulates inflammatory cytokines in injured discs and neuropeptides in dorsal root ganglia innervating injured discs in rats." *Spine (Phila Pa 1976)* 39:1171–7.

Kokubo, Y., K. Uchida, S. Kobayashi et al. 2008. "Herniated and spondylotic intervertebral discs of the human cervical spine: Histological and immunohistological findings in 500 en bloc surgical samples. Laboratory investigation." *J Neurosurg Spine* 9:285–95.

Korhonen, T., J. Karppinen, L. Paimela et al. 2006. "The treatment of disc-herniation-induced sciatica with infliximab: One-year follow-up results of FIRST II, a randomized controlled trial." *Spine (Phila Pa 1976)* 31:2759–66.

Krupkova, O., M. Hlavna, J. Amir Tahmasseb et al. 2016. "An inflammatory nucleus pulposus tissue culture model to test molecular regenerative therapies: Validation with epigallocatechin 3-gallate." *Int J Mol Sci* 17:E1640.

Le Maitre, C.L., A.J. Freemont, and J.A. Hoyland. 2005. "The role of interleukin-1 in the pathogenesis of human intervertebral disc degeneration." *Arthritis Res Ther* 7: R732–45.

Le Maitre, C.L., A.J. Freemont, and J A. Hoyland. 2006. "A preliminary *in vitro* study into the use of IL-1Ra gene therapy for the inhibition of intervertebral disc degeneration." *Int J Exp Pathol* 87:17–28.

Le Maitre, C.L., A.J. Freemont, and J.A. Hoyland. 2009. "Expression of cartilage-derived morphogenetic protein in human intervertebral discs and its effect on matrix synthesis in degenerate human nucleus pulposus cells." *Arthritis Res Ther* 11:R137.

Le Maitre, C.L., J.A. Hoyland, and A.J. Freemont. 2007a. "Catabolic cytokine expression in degenerate and herniated human intervertebral discs: IL-1beta and TNFalpha expression profile." *Arthritis Res Ther* 9:R77.

Le Maitre, C.L., J.A. Hoyland, and A.J. Freemont. 2007b. "Interleukin-1 receptor antagonist delivered directly and by gene therapy inhibits matrix degradation in the intact degenerate human intervertebral disc: An in situ zymographic and gene therapy study." *Arthritis Res Ther* 9:R83.

Le Maitre, C.L., A. Pockert, D.J. Buttle, A.J. Freemont, and J.A. Hoyland. 2007. "Matrix synthesis and degradation in human intervertebral disc degeneration." *Biochem Soc Trans* 35:652–5.

Leckie, S.K., B.P. Bechara, R.A. Hartman et al. 2012. "Injection of AAV2-BMP2 and AAV2-TIMP1 into the nucleus pulposus slows the course of intervertebral disc degeneration in an *in vivo* rabbit model." *Spine J* 12:7–20.

Lee, R.H., A.A. Pulin, M.J. Seo et al. 2009a. "Intravenous hMSCs improve myocardial infarction in mice because cells embolized in lung are activated to secrete the anti-inflammatory protein TSG-6." *Cell Stem Cell* 5:54–63.

Lee, S., C.S. Moon, D. Sul et al. 2009b. "Comparison of growth factor and cytokine expression in patients with degenerated disc disease and herniated nucleus pulposus." *Clin Biochem* 42:1504–11.

Leung, L., and C.M. Cahill. 2010. "TNF-α and neuropathic pain—A review." *J Neuroinflammation* 7:27.

Li, H.R., Q. Cui, Z.Y. Dong et al. 2016. "Downregulation of miR-27b is involved in loss of type II collagen by directly targeting matrix metalloproteinase 13 (MMP13) in human intervertebral disc degeneration." *Spine (Phila Pa 1976)* 41:E116–23.

Li, J., H. Guan, H. Liu et al. 2017a. "Epoxyeicosanoids prevent intervertebral disc degeneration *in vitro* and *in vivo*." *Oncotarget* 8:3781–97.

Li, J., S.T. Yoon, and W.C. Hutton. 2004. "Effect of bone morphogenetic protein-2 (BMP-2) on matrix production, other BMPs, and BMP receptors in rat intervertebral disc cells." *J Spinal Disord Tech* 17:423–8.

Li, K., Y. Li, Z. Ma, and J. Zhao. 2015a. "Crocin exerts anti-inflammatory and anti-catabolic effects on rat intervertebral discs by suppressing the activation of JNK." *Int J Mol Med* 36:1291–9.

Li, W., P. Wang, Z. Zhang et al. 2017b. "MiR-184 Regulates Proliferation in Nucleus Pulposus Cells by Targeting GAS1." *World Neurosurg* 97:710–715.e1.

Li, Y., K. Li, Y. Hu, B. Xu, and J. Zhao. 2015b. "Piperine mediates LPS induced inflammatory and catabolic effects in rat intervertebral disc." *Int J Clin Exp Pathol* 8:6203–13.

Li, Z., X. Yu, J. Shen, M.T. Chan, and W.K. Wu. 2015c. "MicroRNA in intervertebral disc degeneration." *Cell Prolif* 48:278–83.

Liu, G., P. Cao, H. Chen et al. 2013a. "miR-27a regulates apoptosis in nucleus pulposus cells by targeting PI3K." *PLoS One* 8:e75251.

Liu, H., X. Huang, X. Liu et al. 2014. "miR-21 promotes human nucleus pulposus cell proliferation through PTEN/AKT signaling." *Int J Mol Sci* 15:4007–18.

Liu, Q., L. Jin, F.H. Shen, G. Balian, and X.J. Li. 2013b. "Fullerol nanoparticles suppress inflammatory response and adipogenesis of vertebral bone marrow stromal cells— A potential novel treatment for intervertebral disc degeneration." *Spine J* 13:1571–80.

Liu, R., Q. Wu, D. Su et al. 2012. "A regulatory effect of IL-21 on T follicular helper-like cell and B cell in rheumatoid arthritis." *Arthritis Res Ther* 14:R255.

Liu, W., Y. Zhang, P. Xia et al. 2016a. "MicroRNA-7 regulates IL-1beta-induced extracellular matrix degeneration by targeting GDF5 in human nucleus pulposus cells." *Biomed Pharmacother* 83:1414–21.

Liu, W., Y. Zhang, X. Feng et al. 2016b. "Inhibition of microRNA-34a prevents IL-1beta-induced extracellular matrix degradation in nucleus pulposus by increasing GDF5 expression." *Exp Biol Med (Maywood)* 241:1924–32.

Liu, X.-G., H.-W, Hou, and Y.-L. Liu. 2016. "Expression levels of IL-17 and TNF-α in degenerated lumbar intervertebral discs and their correlation." *Exp Ther Med* 11: 2333–40.

Luo, W., Q. Nie, and X. Zhang. 2013. "MicroRNAs involved in skeletal muscle differentiation." *J Genet Genomics* 40:107–16.

Ma, X.L., P. Tian, T. Wang, and J.X. Ma. 2010. "A study of the relationship between type of lumbar disc herniation, straight leg raising test and peripheral T lymphocytes." *Orthop Surg* 2:52–7.

Majid, S., A.A. Dar, S. Saini et al. 2012. "MicroRNA-1280 inhibits invasion and metastasis by targeting ROCK1 in bladder cancer." *PLoS One* 7:e46743.

Manchikanti, L., K.A. Cash, V. Pampati, and F.J. Falco. 2014. "Transforaminal epidural injections in chronic lumbar disc herniation: A randomized, double-blind, active-control trial." *Pain Physician* 17:E489–501.

Manferdini, C., M. Maumus, E. Gabusi et al. 2013. "Adipose-derived mesenchymal stem cells exert antiinflammatory effects on chondrocytes and synoviocytes from osteoarthritis patients through prostaglandin E2." *Arthritis Rheum* 65:1271–81.

Mantovani, A., C. Garlanda, and M. Locati. 2009. "Macrophage diversity and polarization in atherosclerosis: A question of balance." *Arterioscler Thromb Vasc Biol* 29:1419–23.

Mantovani, A., A. Sica, S. Sozzani et al. 2004. "The chemokine system in diverse forms of macrophage activation and polarization." *Trends Immunol* 25:677–86.

Markova, D.Z., C.K. Kepler, S. Addya et al. 2013. "An organ culture system to model early degenerative changes of the intervertebral disc II: Profiling global gene expression changes." *Arthritis Res Ther* 15:R121.

Marshall, L.L., E.R. Trethewie, and C.C. Curtain. 1977. "Chemical radiculitis. A clinical, physiological and immunological study." *Clin Orthop Relat Res* (129):61–7.

Martirosyan, N.L., A.A. Patel, A. Carotenuto et al. 2016. "Genetic alterations in intervertebral disc disease." *Front Surg* 3:59.

Masuda, K. 2008. "Biological repair of the degenerated intervertebral disc by the injection of growth factors." *Eur Spine J* 17:441–51.

Masuda, K., Y. Imai, M. Okuma et al. 2006. "Osteogenic protein-1 injection into a degenerated disc induces the restoration of disc height and structural changes in the rabbit anular puncture model." *Spine (Phila Pa 1976)* 31:742–54.

Mathieu, J., and H. Ruohola-Baker. 2013. "Regulation of stem cell populations by microRNAs." *Adv Exp Med Biol* 786:329–51.

Matsui, Y., M. Maeda, W. Nakagami, and H. Iwata. 1998. "The involvement of matrix metalloproteinases and inflammation in lumbar disc herniation." *Spine (Phila Pa 1976)* 23:863–9.

Matta, A., M.Z. Karim, D.E. Isenman, and W.M. Erwin. 2017. "Molecular therapy for degenerative disc disease: Clues from secretome analysis of the notochordal cell-rich nucleus pulposus." *Sci Rep* 7:45623.

Mehrkens, A., A.M. Müller, V. Valderrabano, S. Schären, and P. Vavken. 2012. "Tissue engineering approaches to degenerative disc disease—A meta-analysis of controlled animal trials." *Osteoarthritis Cartilage* 20:1316–25.

Meisel, H.J., V. Siodla, T. Ganey et al. 2007. "Clinical experience in cell-based therapeutics: Disc chondrocyte transplantation A treatment for degenerated or damaged intervertebral disc." *Biomol Eng* 24:5–21.

Melchionda, D., P. Milillo, G. Manente, L. Stoppino, and L. Macarini. 2012. "Treatment of radiculopathies: A study of efficacy and tollerability of paravertebral oxygen-ozone injections compared with pharmacological anti-inflammatory treatment." *J Biol Regul Homeost Agents* 26:467–74.

Miljkovic, D., and V. Trajkovic. 2004. "Inducible nitric oxide synthase activation by interleukin-17." *Cytokine Growth Factor Rev* 15:21–32.

Mills, C.D., K. Kincaid, J.M. Alt, M.J., Heilman, and A.M. Hill. 2000. "M-1/M-2 macrophages and the Th1/Th2 paradigm." *J Immunol* 164:6166–73.

Millward-Sadler, S.J., P.W. Costello, A.J. Freemont, and J.A. Hoyland. 2009. "Regulation of catabolic gene expression in normal and degenerate human intervertebral disc cells: Implications for the pathogenesis of intervertebral disc degeneration." *Arthritis Res Ther* 11:R65.

Miyahara, K., T. Ishida, S. Hukuda et al. 1996. "Human group II phospholipase A2 in normal and diseased intervertebral discs." *Biochim Biophys Acta* 1316:183–90.

Miyamoto, H., R. Saura, T. Harada, M. Doita, and K. Mizuno. 2000. "The role of cyclooxygenase-2 and inflammatory cytokines in pain induction of herniated lumbar intervertebral disc." *Kobe J Med Sci* 46:13–28.

Molinos, M., C.R. Almeida, J. Caldeira et al. 2015. "Inflammation in intervertebral disc degeneration and regeneration." *J R Soc Interface* 12:20141191.

Moon, S.H., K. Nishida, L.G. Gilbertson et al. 2008. "Biologic response of human intervertebral disc cells to gene therapy cocktail." *Spine (Phila Pa 1976)* 33:1850–5.

Mosser, D.M., and J.P. Edwards. 2008. "Exploring the full spectrum of macrophage activation." *Nat Rev Immunol* 8:958–69.

Muller-Ladner, U., C.R. Roberts, B.N. Franklin et al. 1997. "Human IL-1Ra gene transfer into human synovial fibroblasts is chondroprotective." *J Immunol* 158:3492–8.

Murata, Y., A. Onda, B. Rydevik et al. 2006. "Changes in pain behavior and histologic changes caused by application of tumor necrosis factor-alpha to the dorsal root ganglion in rats." *Spine (Phila Pa 1976)* 31:530–5.

Murata, Y., B. Rydevik, U., Nannmark et al. 2011. "Local application of interleukin-6 to the dorsal root ganglion induces tumor necrosis factor-alpha in the dorsal root ganglion and results in apoptosis of the dorsal root ganglion cells." *Spine (Phila Pa 1976)* 36: 926–32.

Murata, Y., U. Nannmark, B., Rydevik, K., Takahashi, and K. Olmarker. 2008. "The role of tumor necrosis factor-alpha in apoptosis of dorsal root ganglion cells induced by herniated nucleus pulposus in rats." *Spine (Phila Pa 1976)* 33:155–62.

Nagae, M., T. Ikeda, Y. Mikami et al. 2007. "Intervertebral disc regeneration using plateletrich plasma and biodegradable gelatin hydrogel microspheres." *Tissue Eng* 13:147–58.

Nandi, J., and A. Chowdhery. 2017. "A randomized controlled clinical trial to determine the effectiveness of caudal epidural steroid injection in lumbosacral sciatica." *J Clin Diagn Res* 11:Rc04–8.

Nemeth, K., A. Leelahavanichkul, P.S. Yuen et al. 2009. "Bone marrow stromal cells attenuate sepsis via prostaglandin E(2)-dependent reprogramming of host macrophages to increase their interleukin-10 production." *Nat Med* 15:42–9.

Nerlich, A.G., C. Weiler, J. Zipperer, M. Narozny, and N. Boos. 2002. "Immunolocalization of phagocytic cells in normal and degenerated intervertebral discs." *Spine (Phila Pa 1976)* 27:2484–90.

Nishida, K., J.D. Kang, L.G. Gilbertson et al. 1999. "Modulation of the biologic activity of the rabbit intervertebral disc by gene therapy: An *in vivo* study of adenovirus-mediated

transfer of the human transforming growth factor beta 1 encoding gene." *Spine (Phila Pa 1976)* 24:2419–25.
Nishimura, K., and J. Mochida. 1998. "Percutaneous reinsertion of the nucleus pulposus. An experimental study." *Spine (Phila Pa 1976)* 23:1531–9.
O'Donnell, J.L., and A.L. O'Donnell. 1996. "Prostaglandin E2 content in herniated lumbar disc disease." *Spine (Phila Pa 1976)* 21:1653–6.
Ogata, K., and L.A. Whiteside. 1981. "1980 Volvo Award winner in basic science. Nutritional pathways of the intervertebral disc. An experimental study using hydrogen washout technique." *Spine (Phila Pa 1976)* 6:211–6.
Ogle, M.E., C.E. Segar, S. Sridhar, and E.A. Botchwey. 2016. "Monocytes and macrophages in tissue repair: Implications for immunoregenerative biomaterial design." *Exp Biol Med* 241:1084–97.
Ohtori, S., G. Inoue, T. Koshi et al. 2006. "Up-regulation of acid-sensing ion channel 3 in dorsal root ganglion neurons following application of nucleus pulposus on nerve root in rats." *Spine (Phila Pa 1976)* 31:2048–52.
Okamura, Y., M. Watari, E.S. Jerud et al. 2001. "The extra domain A of fibronectin activates Toll-like receptor 4." *J Biol Chem* 276:10229–33.
Olmarker, K., and K. Larsson. 1998. Tumor necrosis factor alpha and nucleus-pulposus-induced nerve root injury. *Spine (Phila Pa 1976)* 23:2538–44.
Orozco, L., R. Soler, C. Morera et al. 2011. Intervertebral disc repair by autologous mesenchymal bone marrow cells: A pilot study. *Transplantation* 92:822–8.
Ortiz, L.A., M. Dutreil, C. Fattman et al. 2007. "Interleukin 1 receptor antagonist mediates the antiinflammatory and antifibrotic effect of mesenchymal stem cells during lung injury." *Proc Natl Acad Sci U S A* 104:11002–7.
Paoloni, M., L. Di Sante, A. Cacchio et al. 2009. "Intramuscular oxygen-ozone therapy in the treatment of acute back pain with lumbar disc herniation: A multicenter, randomized, double-blind, clinical trial of active and simulated lumbar paravertebral injection." *Spine (Phila Pa 1976)* 34:1337–44.
Papagiannakopoulos, T., and K.S. Kosik. 2008. "MicroRNAs: Regulators of oncogenesis and stemness." *BMC Med* 6:15.
Park, J.B., H. Chang, and K.W. Kim. 2001. "Expression of Fas ligand and apoptosis of disc cells in herniated lumbar disc tissue." *Spine (Phila Pa 1976)* 26:618–21.
Park, J.B., H. Chang, and Y.S. Kim. 2002. "The pattern of interleukin-12 and T-helper types 1 and 2 cytokine expression in herniated lumbar disc tissue." *Spine (Phila Pa 1976)* 27:2125–8.
Park, J.B., K.W. Kim, C.W. Han, and H. Chang. 2001. "Expression of Fas receptor on disc cells in herniated lumbar disc tissue." *Spine (Phila Pa 1976)* 26:142–6.
Park, J.Y., S.U. Kuh, H.S. Park, and K.S. Kim. 2011. "Comparative expression of matrix-associated genes and inflammatory cytokines-associated genes according to disc degeneration: Analysis of living human nucleus pulposus." *J Spinal Disord Tech* 24:352–7.
Peng, B., J. Hao, S. Hou et al. 2006. "Possible pathogenesis of painful intervertebral disc degeneration." *Spine (Phila Pa 1976)* 31:560–6.
Pennington, J.B., R.F. McCarron, and G.S. Laros. 1988. "Identification of IgG in the canine intervertebral disc." *Spine (Phila Pa 1976)* 13:909–12.
Pereira, C.L., R.M. Gonçalves, M. Peroglio et al. 2014. "The effect of hyaluronan-based delivery of stromal cell-derived factor-1 on the recruitment of MSCs in degenerating intervertebral discs." *Biomaterials* 35:8144–53.
Pettine, K.A., M.B. Murphy, R.K. Suzuki, and T.T. Sand. 2015. "Percutaneous injection of autologous bone marrow concentrate cells significantly reduces lumbar discogenic pain through 12 months." *Stem Cells* 33:146–56.

Phillips, K.L., N. Chiverton, A.L. Michael et al. 2013. "The cytokine and chemokine expression profile of nucleus pulposus cells: Implications for degeneration and regeneration of the intervertebral disc." *Arthritis Res Ther* 15:R213.

Phillips, K.L., K. Cullen, N. Chiverton et al. 2015. "Potential roles of cytokines and chemokines in human intervertebral disc degeneration: Interleukin-1 is a master regulator of catabolic processes." *Osteoarthritis Cartilage* 23:1165–77.

Pockert, A.J., S.M. Richardson, C.L. Le Maitre et al. 2009. "Modified expression of the ADAMTS enzymes and tissue inhibitor of metalloproteinases 3 during human intervertebral disc degeneration." *Arthritis Rheum* 60:482–91.

Ponnappan, R.K., D.Z. Markova, P.J. Antonio et al. 2011. "An organ culture system to model early degenerative changes of the intervertebral disc." *Arthritis Res Ther* 13:R171.

Poujol, D., J.M. Ristori, J.J. Dubost, and M. Soubrier. 2007. "Efficacy of pamidronate in erosive degenerative disk disease: A pilot study." *Joint Bone Spine* 74:663–4.

Prockop, D.J., and J.Y. Oh. 2012. "Mesenchymal stem/stromal cells (MSCs): Role as guardians of inflammation." *Mol Ther* 20:14–20.

Purmessur, D., A.J. Freemont, and J.A. Hoyland. 2008. "Expression and regulation of neurotrophins in the nondegenerate and degenerate human intervertebral disc." *Arthritis Res Ther* 10:R99.

Purmessur, D., B.A. Walter, P.J. Roughley et al. 2013. "A role for TNFα in intervertebral disc degeneration: A non-recoverable catabolic shift." *Biochem Biophys Res Commun* 433:151–6.

Quero, L., M. Klawitter, A. Schmaus et al. 2013. "Hyaluronic acid fragments enhance the inflammatory and catabolic response in human intervertebral disc cells through modulation of toll-like receptor 2 signalling pathways." *Arthritis Res Ther* 15:R94.

Ragab, A.A., J.W. Woodall Jr., M.A. Tucci et al. 2009. "A preliminary report on the effects of sustained administration of corticosteroid on traumatized disc using the adult male rat model." *J Spinal Disord Tech* 22:473–8.

Rajan, N.E., O. Bloom, R. Maidhof et al. 2013. "Toll-like receptor 4 (TLR4) expression and stimulation in a model of intervertebral disc inflammation and degeneration." *Spine (Phila Pa 1976)* 38:1343–51.

Rand, N., F. Reichert, Y. Floman, and S. Rotshenker. 1997. "Murine nucleus pulposus-derived cells secrete interleukins-1-beta, -6, and -10 and granulocyte-macrophage colony-stimulating factor in cell culture." *Spine (Phila Pa 1976)* 22:2598–602.

Richardson, S.M., A. Mobasheri, A.J. Freemont, and J.A. Hoyland. 2007. "Intervertebral disc biology, degeneration and novel tissue engineering and regenerative medicine therapies." *Histol Histopathol* 22:1033–41.

Richardson, S.M., G. Kalamegam, P.N. Pushparaj et al. 2016. "Mesenchymal stem cells in regenerative medicine: Focus on articular cartilage and intervertebral disc regeneration." *Methods* 99:69–80.

Richardson, S.M., P. Doyle, B.M. Minogue, K. Gnanalingham, and J.A. Hoyland. 2009. "Increased expression of matrix metalloproteinase-10, nerve growth factor and substance P in the painful degenerate intervertebral disc." *Arthritis Res Ther* 11:R126.

Risbud, M.V., A. Guttapalli, T.T. Tsai et al. 2007. "Evidence for skeletal progenitor cells in the degenerate human intervertebral disc." *Spine (Phila Pa 1976)* 32:2537–44.

Risbud, M.V., and I.M. Shapiro. 2014. "Role of cytokines in intervertebral disc degeneration: Pain and disc content." *Nat Rev Rheumatol* 10:44–56.

Roberts, S., H. Evans, J. Trivedi, and J. Menage. 2006. "Histology and pathology of the human intervertebral disc." *J Bone Joint Surg Am* 88 Suppl 2:10–4.

Roddy, G.W., J.Y. Oh, R.H. Lee et al. 2011. "Action at a distance: Systemically administered adult stem/progenitor cells (MSCs) reduce inflammatory damage to the cornea without engraftment and primarily by secretion of TNF-alpha stimulated gene/protein 6." *Stem Cells* 29:1572–79.

Romero-Sandoval, E.A., C. McCall, and J.C. Eisenach. 2005. "Alpha2-adrenoceptor stimulation transforms immune responses in neuritis and blocks neuritis-induced pain." *J Neurosci* 25:8988–94.

Rose-John, S., G.H. Waetzig, J. Scheller, J. Grotzinger, and D. Seegert. 2007. "The IL-6/sIL-6R complex as a novel target for therapeutic approaches." *Expert Opin Ther Targets* 11:613–24.

Roughley, P.J. 2004. "Biology of intervertebral disc aging and degeneration: Involvement of the extracellular matrix." *Spine (Phila Pa 1976)* 29:2691–9.

Saal, J.S., R.C. Franson, R. Dobrow et al. 1990. "High levels of inflammatory phospholipase A2 activity in lumbar disc herniations." *Spine (Phila Pa 1976)* 15:674–8.

Sainoh, T., S. Orita, M. Miyagi et al. 2016. "Single intradiscal administration of the tumor necrosis factor-alpha inhibitor, etanercept, for patients with discogenic low back pain." *Pain Med* 17:40–5.

Sakai, D., and G.B.J. Andersson. 2015. "Stem cell therapy for intervertebral disc regeneration: Obstacles and solutions." *Nat Rev Rheumatol* 11:243–56.

Sakai, D., Y. Nakamura, T. Nakai et al. 2012. "Exhaustion of nucleus pulposus progenitor cells with aging and degeneration of the intervertebral disc." *Nat Commun* 3:1264.

Sakai, D., K. Nishimura, M. Tanaka et al. 2015. "Migration of bone marrow-derived cells for endogenous repair in a new tail-looping disc degeneration model in the mouse: A pilot study." *Spine J* 15:1356–65.

Sato, F., S. Tsuchiya, S.J., Meltzer, and K. Shimizu. 2011. "MicroRNAs and epigenetics." *FEBS J* 278:1598–609.

Schaefer, L., A. Babelova, E. Kiss et al. 2005. "The matrix component biglycan is proinflammatory and signals through Toll-like receptors 4 and 2 in macrophages." *J Clin Invest* 115:2223–33.

Scheller, J., A. Chalaris, D. Schmidt-Arras, and S. Rose-John. 2011. "The pro- and anti-inflammatory properties of the cytokine interleukin-6." *Biochim Biophys Acta* 1813:878–8.

Seguin, C.A., R.M. Pilliar, P.J. Roughley, and R.A. Kandel. 2005. "Tumor necrosis factor-alpha modulates matrix production and catabolism in nucleus pulposus tissue." *Spine (Phila Pa 1976)* 30:1940–8.

Shamji, M.F., L.A. Setton, W. Jarvis et al. 2010. "Proinflammatory cytokine expression profile in degenerated and herniated human intervertebral disc tissues." *Arthritis Rheum* 62:1974–82.

Shi, C., and E.G. Pamer. 2011. "Monocyte recruitment during infection and inflammation". *Nat Rev Immunol.* 11:762–64 Springer-Verlag Wien.

Sheikh, H., K. Zakharian, R.P. De La Torre et al. 2009. "*In vivo* intervertebral disc regeneration using stem cell-derived chondroprogenitors." *J Neurosurg Spine* 10:265–72.

Shen, B., J. Melrose, P. Ghosh, and F. Taylor. 2003. "Induction of matrix metalloproteinase-2 and -3 activity in ovine nucleus pulposus cells grown in three-dimensional agarose gel culture by interleukin-1beta: A potential pathway of disc degeneration." *Eur Spine J* 12:66–75.

Silke, J. 2011. "The regulation of TNF signalling: What a tangled web we weave." *Curr Opin Immunol* 23:620–6.

Sommer, C., M. Marziniak, and R.R. Myers. 1998. "The effect of thalidomide treatment on vascular pathology and hyperalgesia caused by chronic constriction injury of rat nerve." *Pain* 74:83–91.

Song, T., X. Ma, K. Gu et al. 2016. "Thalidomide represses inflammatory response and reduces radiculopathic pain by inhibiting IRAK-1 and NF-kappaB/p38/JNK signaling." *J Neuroimmunol* 290:1–8.

Specchia, N., A. Pagnotta, A. Toesca, and F. Greco. 2002. "Cytokines and growth factors in the protruded intervertebral disc of the lumbar spine." *Eur Spine J* 11:145–51.

Spiller, K.L., R.R. Anfang, K.J. Spiller et al. 2014. "The role of macrophage phenotype in vascularization of tissue engineering scaffolds." *Biomaterials* 35:4477–88.

Sun, Z., M. Zhang, X.H. Zhao et al. 2013. "Immune cascades in human intervertebral disc: The pros and cons." *Int J Clin Exp Pathol* 6:1009–14.

Szymczak-Workman, A.L., C.J. Workman, and D.A. Vignali. 2009. "Cutting edge: Regulatory T cells do not require stimulation through their TCR to suppress." *J Immunol* 182:5 188–92.

Tafazal, S., L. Ng, N. Chaudhary, and P. Sell. 2009. "Corticosteroids in peri-radicular infiltration for radicular pain: A randomised double blind controlled trial. One year results and subgroup analysis." *Eur Spine J* 18:1220–5.

Tak, P.P., and G.S. Firestein. 2001. "NF-κB:A key role in inflammatory diseases." *J Clin Invest* 107:7–11.

Takada, T., K. Nishida, K. Maeno et al. 2012. "Intervertebral disc and macrophage interaction induces mechanical hyperalgesia and cytokine production in a herniated disc model in rats." *Arthritis Rheum* 64: 2601–10.

Takada, T., K. Nishida, M. Doita, and M. Kurosaka. 2002. "Fas ligand exists on intervertebral disc cells: A potential molecular mechanism for immune privilege of the disc." *Spine (Phila Pa 1976)* 27:1526–30.

Takada, T., K. Nishida, M. Doita, H. Miyamoto, and M. Kurosaka. 2004. "Interleukin-6 production is upregulated by interaction between disc tissue and macrophages." *Spine (Phila Pa 1976)* 29:1089–93.

Takahashi, H., T. Suguro, Y. Okazima et al. 1996. "Inflammatory cytokines in the herniated disc of the lumbar spine." *Spine (Phila Pa 1976)* 21:218–24.

Tan, Y., Y. Hu, and J. Tan. 2003. "Extracellular matrix synthesis and ultrastructural changes of degenerative disc cells transfected by Ad/CMV-hTGF-beta 1." *Chin Med J* 116: 1399–403.

Teixeira, G.Q., A. Boldt, I. Nagl et al. 2015. "A degenerative/pro-inflammatory intervertebral disc organ culture: An *ex vivo* model for anti-inflammatory drug and cell therapy." *Tissue Eng Part C Methods* 22:8–19.

Teixeira, G.Q., C.L. Pereira, F. Castro et al. 2016. "Anti-inflammatory Chitosan/Poly-gamma-glutamic acid nanoparticles control inflammation while remodeling extracellular matrix in degenerated intervertebral disc." *Acta Biomater* 42:168–79.

Tian, P., X.L. Ma, T. Wang, J.X., Ma, and X. Yang. 2009. "Correlation between radicalgia and counts of T lymphocyte subsets in the peripheral blood of patients with lumbar disc herniation." *Orthop Surg* 1:317–21.

Tobinick, E.L., and S. Britschgi-Davoodifar. 2003. "Perispinal TNF-alpha inhibition for discogenic pain." *Swiss Med Wkly* 133:170–7.

Tobinick, E., and S. Davoodifar. 2004. "Efficacy of etanercept delivered by perispinal administration for chronic back and/or neck disc-related pain: A study of clinical observations in 143 patients." *Curr Med Res Opin* 20:1075–85.

Tolonen, J., M. Grönblad, J. Virri et al. 1995. "Basic fibroblast growth factor immunoreactivity in blood vessels and cells of disc herniations." *Spine (Phila Pa 1976)* 20:271–6.

Trinchieri, G. 1994. "Interleukin-12: A cytokine produced by antigen-presenting cells with immunoregulatory functions in the generation of T-helper cells type 1 and cytotoxic lymphocytes." *Blood* 84:4008–27.

Tsirimonaki, E., C. Fedonidis, S.G. Pneumaticos et al. 2013. "PKCepsilon signalling activates ERK1/2, and regulates aggrecan, ADAMTS5, and miR377 gene expression in human nucleus pulposus cells." *PLoS One* 8:e82045.

Ulrich, J.A., E.C. Liebenberg, D.U. Thuillier, and J.C. Lotz. 2007. "ISSLS prize winner: Repeated disc injury causes persistent inflammation." *Spine (Phila Pa 1976)* 32:2812–9.

Vadala, G., F. Russo, A. Di Martino, and V. Denaro. 2015. "Intervertebral disc regeneration: From the degenerative cascade to molecular therapy and tissue engineering." *J Tissue Eng Regen Med* 9:679–90.

Vadala, G., G.A. Sowa, and J.D. Kang. 2007. "Gene therapy for disc degeneration." *Expert Opin Biol Ther* 7:185–96.

van Buul, G.M., E. Villafuertes, P.K. Bos et al. 2012. "Mesenchymal stem cells secrete factors that inhibit inflammatory processes in short-term osteoarthritic synovium and cartilage explant culture." *Osteoarthritis Cartilage* 20:1186–96.

Vo, N.V., R.A. Hartman, P.R. Patil et al. 2016. "Molecular mechanisms of biological aging in intervertebral discs." *J Orthop Res* 34:1289–306.

Vo, N.V., R.A. Hartman, T. Yurube et al. 2013. "Expression and regulation of metalloproteinases and their inhibitors in intervertebral disc aging and degeneration." *Spine J* 13:331–41.

Vos, T., A.D. Flaxman, M. Naghavi et al. 2012. "Years lived with disability (YLDs) for 1160 sequelae of 289 diseases and injuries 1990–2010: A systematic analysis for the Global Burden of Disease Study 2010." *Lancet* 380:2163–96.

Walsh, A.J., D.S. Bradford, and J.C. Lotz. 2004. "*In vivo* growth factor treatment of degenerated intervertebral discs." *Spine (Phila Pa 1976)* 29:156–63.

Walter, B.A., M. Likhitpanichkul, S. Illien-Junger et al. 2015. "TNFα transport induced by dynamic loading alters biomechanics of intact intervertebral discs." *PLoS One* 10: e0118358.

Walter, B.A., D. Purmessur, A. Moon et al. 2016. "Reduced tissue osmolarity increases TRPV4 expression and pro-inflammatory cytokines in intervertebral disc cells." *Eur Cell Mater* 32:123–36.

Wang, F., J.M. Jiang, C.H. Deng et al. 2011a. "Expression of Fas receptor and apoptosis in vertebral endplates with degenerative disc diseases categorized as Modic type I or II." *Injury* 42:790–5.

Wang, F., G. Niu, X. Chen, and F. Cao. 2011b. "Molecular imaging of microRNAs." *Eur J Nucl Med Mol Imaging* 38:1572–9.

Wang, H.Q., X.D. Yu, Z.H. Liu et al. 2011c. "Deregulated miR-155 promotes Fas-mediated apoptosis in human intervertebral disc degeneration by targeting FADD and caspase-3." *J Pathol* 225:232–42.

Wang, H., Y. Tian, J. Wang et al. 2013. "Inflammatory cytokines induce NOTCH signaling in nucleus pulposus cells: Implications in intervertebral disc degeneration." *J Biol Chem* 288:16761–74.

Wang, J.N., X.J. Zhao Z.H. Liu et al. Fu. 2017. "Selective phosphodiesterase-2A inhibitor alleviates radicular inflammation and mechanical allodynia in non-compressive lumbar disc herniation rats." *Eur Spine J* 26:1961–8.

Wang, J., T. Tang, H. Yang et al. 2007. "The expression of Fas ligand on normal and stabbed-disc cells in a rabbit model of intervertebral disc degeneration: A possible pathogenesis." *J Neurosurg Spine* 6:425–30.

Wang, J., D. Markova, D.G. Anderson et al. 2011d. "TNF-alpha and IL-1beta promote a disintegrin-like and metalloprotease with thrombospondin type I motif-5-mediated aggrecan degradation through syndecan-4 in intervertebral disc." *J Biol Chem* 286:39738–49.

Wang, T., P. Li, X. Ma et al. 2015. "MicroRNA-494 inhibition protects nucleus pulposus cells from TNF-alpha-induced apoptosis by targeting JunD." *Biochimie* 115:1–7.

Wang, X., H. Wang, H. Yang et al. 2014. "Tumor necrosis factor-α- and interleukin-1β-dependent matrix metalloproteinase-3 expression in nucleus pulposus cells requires cooperative signaling via syndecan 4 and mitogen-activated protein kinase-NF-κB axis: Implications in inflammatory disc disease." *Am J Pathol* 184:2560–72.

Watanabe, K., J. Mochida, T. Nomura et al. 2003. "Effect of reinsertion of activated nucleus pulposus on disc degeneration: An experimental study on various types of collagen in degenerative discs." *Connect Tissue Res* 44:104–8.

Weaver, C.T., R.D. Hatton, P.R. Mangan, and L.E. Harrington. 2007. "IL-17 family cytokines and the expanding diversity of effector T cell lineages." *Annu Rev Immunol* 25:821–52.

Wei, A., L.A. Williams, D. Bhargav et al. 2009. "BMP13 prevents the effects of annular injury in an ovine model." *Int J Biol Sci* 5:388–96.

Wei, L., A. Laurence, K.M. Elias, and J.J. O'Shea. 2007. "IL-21 is produced by Th17 cells and drives IL-17 production in a STAT3-dependent manner." *J Biol Chem* 282:34605–10.

Wei, X.H., X.D. Na, G.J. Liao et al. 2013. "The up-regulation of IL-6 in DRG and spinal dorsal horn contributes to neuropathic pain following L5 ventral root transection." *Exp Neurol* 241:159–68.

Weiler, C., A.G. Nerlich, B.E. Bachmeier, and N. Boos. 2005. "Expression and distribution of tumor necrosis factor alpha in human lumbar intervertebral discs: A study in surgical specimen and autopsy controls." *Spine (Phila Pa 1976)* 30:44–54.

Winn, S.R., H. Uludag, and J.O. Hollinger. 1999. "Carrier systems for bone morphogenetic proteins." *Clin Orthop Relat Res* (367 Suppl):S95–106.

Woods, B.I., N. Vo, G. Sowa, and J.D. Kang. 2011. "Gene therapy for intervertebral disk degeneration." *Orthop Clin North Am* 42:563–574.

Wuertz, K., and L. Haglund. 2013. "Inflammatory mediators in intervertebral disk degeneration and discogenic pain." *Global Spine J* 3:175–84.

Wuertz, K., L. Quero, M. Sekiguchi et al. 2011. "The red wine polyphenol resveratrol shows promising potential for the treatment of nucleus pulposus-mediated pain *in vitro* and *in vivo*." *Spine (Phila Pa 1976)* 36:E1373–84.

Xie, W., Z. Li, M. Li, N. Xu, and Y. Zhang. 2013. "miR-181a and inflammation: miRNA homeostasis response to inflammatory stimuli *in vivo*." *Biochem Biophys Res Commun* 430:647–52.

Xu, Y.Q., Z.H. Zhang, Y.F. Zheng, and S.Q. Feng. 2016. "Dysregulated miR-133a mediates loss of type II collagen by directly targeting matrix metalloproteinase 9 (MMP9) in human intervertebral disc degeneration." *Spine (Phila Pa 1976)* 41:E717–24.

Yamamoto, J., K. Maeno, T. Takada et al. 2013. "Fas ligand plays an important role for the production of pro-inflammatory cytokines in intervertebral disc nucleus pulposus cells." *J Orthop Res* 31:608–15.

Yan, N., S. Yu, H. Zhang, and T. Hou. 2015. "Lumbar disc degeneration is facilitated by miR-100-mediated FGFR3 suppression." *Cell Physiol Biochem* 36:2229–36.

Ye, S., J. Wang, S. Yang et al. 2011. "Specific inhibitory protein Dkk-1 blocking Wnt/beta-catenin signaling pathway improve protectives effect on the extracellular matrix." *J Huazhong Univ Sci Technol Med Sci* 31:657–62.

Yin, W., K. Pauza, W.J. Olan, J.F. Doerzbacher, and K.J. Thorne. 2014. "Intradiscal injection of fibrin sealant for the treatment of symptomatic lumbar internal disc disruption: Results of a prospective multicenter pilot study with 24-month follow-up." *Pain Med* 15:16–31.

Ying, S.Y., D.C. Chang, and S.L. Lin (2013). "The MicroRNA." *Methods Mol Biol* 936: 1–19.

Yoo, K.H., I.K. Jang, M.W. Lee et al. 2009. "Comparison of immunomodulatory properties of mesenchymal stem cells derived from adult human tissues." *Cell Immunol* 259:150–6.

Yoshikawa, T., Y. Ueda, K. Miyazaki, M. Koizumi, and Y. Takakura. 2010. "Disc regeneration therapy using marrow mesenchymal cell transplantation: A report of two case studies." *Spine (Phila Pa 1976)* 35:E475–80.

Yu, X., Z. Li, J. Shen et al. 2013. "MicroRNA-10b promotes nucleus pulposus cell proliferation through RhoC-Akt pathway by targeting HOXD10 in intervetebral disc degeneration." *PLoS One* 8:e83080.

Zhang, H., L. Wang, J.B. Park et al. 2009. "Intradiscal injection of simvastatin retards progression of intervertebral disc degeneration induced by stab injury." *Arthritis Res Ther* 11:R172.

Zhang, W., L. Nie, Y. Wang et al. 2013a. "CCL20 Secretion from the nucleus pulposus improves the recruitment of CCR6-expressing Th17 cells to degenerated IVD tissues." *PLoS ONE* 8:e66286.

Zhang, X., J. Huang, and P.A. McNaughton. 2005. "NGF rapidly increases membrane expression of TRPV1 heat-gated ion channels." *EMBO J* 24:4211–23.

Zhang, Y., Y. Ma, J. Jiang, T. Ding, and J. Wang. 2013b. "Treatment of the lumbar disc herniation with intradiscal and intraforaminal injection of oxygen-ozone." *J Back Musculoskelet Rehabil* 26:317–22.

Zhao, B.O., Q. Yu, H. Li, X. Guo, and X. He. 2014. "Characterization of microRNA expression profiles in patients with intervertebral disc degeneration." *Int J Mol Med* 33:43–50.

Zhou, X., L. Chen, S. Grad et al. 2017. "The roles and perspectives of microRNAs as biomarkers for intervertebral disc degeneration." *J Tissue Eng Regen Med* 11:3481–7. doi: 10.1002/term.2261.

8 Gene Delivery for Intervertebral Disc

Gianluca Vadalà, Luca Ambrosio, and Vincenzo Denaro

CONTENTS

8.1	Introduction	231
8.2	Intervertebral Disc Degeneration	232
8.3	Modulating Disc Cell Activity	233
8.4	Gene Therapy for IDD	234
8.5	Gene Delivery Systems and Strategies	235
	8.5.1 Viral Systems	236
	8.5.1.1 Retrovirus	236
	8.5.1.2 Adenovirus	236
	8.5.1.3 Adeno-Associated Virus	237
	8.5.2 Nonviral Systems	237
8.6	RNA Interference	238
8.7	Proof of Principle of a Gene Therapy for IDD	239
	8.7.1 Anabolic Factors	240
	8.7.2 Anticatabolic Factors	241
	8.7.3 Transcription Factors	242
	8.7.4 Cell Survival and Apoptosis	243
	8.7.5 Multiple Targets	244
8.8	*In Vivo* Efficacy and Feasibility	245
8.9	Safety Concerns	246
8.10	Regulating Transgene Expression	246
8.11	Conclusion	247
References		248

8.1 INTRODUCTION

Low back pain (LBP) is an extremely common condition affecting approximately 632 million people worldwide (Vos et al. 2012). It is estimated to be the main cause of disability in people under 45 years of age and the second most common reason for a symptomatic physician visit in the United States (Hart, Deyo, and Cherkin 1995). This results in significant social and economical burdens, with a lifetime prevalence of up to 85% (Lively 2002) and national economic losses higher than $90 billion per year (Luo et al. 2004).

In most cases, intrinsic LBP appears to be caused by intervertebral disc degeneration (IDD).

IDD is a chronic process that is mainly characterized by progressive loss of proteoglycan and, consequently, of water content inside the nucleus pulposus (NP) matrix. This process leads to decreased disc height and degeneration of surrounding structures, thus becoming clinically evident by causing LBP, disc herniation, spinal stenosis, segmental instability, radiculopathy, myelopathy, and other secondary disorders.

Current treatment options for IDD range from nonsurgical procedures, such as bed rest, analgesic and myorelaxant medications, and physical therapy, to interventional procedures (including ablation techniques and epidural steroid injections) and surgical approaches, such as discectomy, laminectomy, spinal fusion, and total disc replacement. However, none of these options have shown full efficacy in stopping IDD or preventing relapses because their effect is not exerted by triggering the underlying pathophysiological process but by slowing it down or solely relieving clinical symptoms. Recent advances in molecular and cellular biology have allowed scientists to identify several pathways involved in IDD, thus raising the possibility to target the single genes involved in order to modulate the process and, possibly, arrest or reverse the degenerative cascade.

8.2 INTERVERTEBRAL DISC DEGENERATION

The intervertebral disc (IVD) is a fibrocartilaginous structure that connects two contiguous vertebrae and provides shock absorption and mechanical load support in both static and dynamic conditions (i.e., flexion, extension, bending, rotation, and weight bearing) (Humzah and Soames 1988).

The IVD mainly consists in three highly specialized tissues: the NP, the annulus fibrosus (AF), and the cartilaginous end plate (CEP), which constitutes the interface between the disc and the adjacent vertebral bodies.

The NP is composed of an amorphous, gelatinous matrix primarily containing proteoglycans and type II collagen fibers in a random orientation, among which sparse and scattered chondrocyte-like cells can be found. These cells, whose number dramatically decreases with age, are committed to maintain the extracellular matrix homeostasis (Trout et al. 1982). Proteoglycans are high-molecular-weight components of the NP matrix consisting of a core protein with several covalently attached glycosaminoglycans (GAGs), repetitive disaccharides (mostly chondroitin sulfate and keratan sulfate) that are highly polar and able to attract water. The major noncollagenous protein in the NP matrix is aggrecan, a large proteoglycan that interacts with hyaluronan-forming stable ternary aggregates in the extracellular environment, providing the NP itself with a number of peculiar chemical and biomechanical features (Sivan, Wachtel, and Roughley 2014).

The AF is constituted by a fibrocartilaginous ring that surrounds the NP and forms the external part of the IVD: it is primarily composed of parallel type I collagen fibers, oriented radially and perpendicularly in different lamellae, with an interlamellar matrix being interposed (Cassidy, Hiltner, and Baer 1989). Among fibers, fibroblast-like cells, which are responsible for type I collagen, proteoglycans, and non-collagenous proteins synthesis, can be detected (Bruehlmann et al. 2002).

IDD is a chronic, progressive, and age-dependent process, characterized by gradual loss of proteoglycan and, consequently, of free water content in the NP matrix, thus leading to decreased disc height and deterioration of its biomechanical properties.

To date, the etiology and pathophysiology of this process remain unclear (Roughley, Melching 2006). Although many environmental risk factors have been identified, including smoking (Nasto et al. 2014), physical inactivity, being overweight (Samartzis et al. 2012), and hormonal imbalances (Skrzypiec et al. 2007), genetic influences have been shown to play a significant role (Battie et al. 2004).

Several gene polymorphisms have been associated with IDD, such as the ones encoding for vitamin D receptors (Videman et al. 1998), type IX collagen (Annunen et al. 1999; Paassilta et al. 2001), aggrecan, and metalloproteinase-3 (MMP-3) (Kawaguchi et al. 1999; Takahashi et al. 2001).

However, as stated previously, the progressive decrease in aggrecan content within the NP seems to be the characteristic event mainly leading to IDD. This phenomenon is hypothesized to be due to either a reduction in aggrecan synthesis, an increase in its breakdown, or a combination of both. The high aggrecan levels in the healthy IVD provide it with an elevated water content, which constitutes approximately 85%–90% of the entire NP matrix. Such a hydration rate is responsible for maintaining disc height and the load-bearing capacity of the IVD (Urban and McMullin 1985).

As IDD gradually establishes, the drop in aggrecan and water content leads to a decrease in disc height and in a disruption of the normal distribution of forces across the spine. This leads to AF fissuring, disc herniation, facet hypertrophy, osteophyte formation, and subchondral sclerosis (Urban and McMullin 1988). When this latter occurs in the CEP, nutrient uptake from the already scarcely vascularized endplate becomes even more compromised, thus boosting the degenerative process. Nonetheless, the inherent acidic environment, intrinsic avascularity, low nourishment, and anatomical isolation that characterize the IVD may probably be additional issues regarding the apparent inability of the disc for self-restoring the proteoglycan content and respond to degenerative triggers (Urban, Smith, and Fairbank 2004).

Accordingly, increasing the disc capacity to counteract these detrimental changes by stimulating proteoglycan synthesis may reestablish the normal water content inside the NP, thus ameliorating its original biomechanical properties.

In this regard, several genes involved in matrix synthesis and catabolism regulation have been identified as possible targets of a gene therapy for IDD (Masuda, Oegema, and An 2004).

8.3 MODULATING DISC CELL ACTIVITY

The concept of treating IDD, as well as other musculoskeletal disorders, by directly modulating cell activity originated from the application of growth factors, small peptides involved in cell proliferation, differentiation, migration, and matrix synthesis.

Thompson, Oegema, and Bradford (1991) were the first to demonstrate that the exogenous administration of transforming growth factor (TGF)-β1 to NP cells resulted in an increased proteoglycan synthesis *in vitro*.

This finding was further corroborated by Takegami et al. (2002), who showed that OP-1 (osteogenic protein-1, also known as bone morphogenetic protein [BMP]-7) might stimulate AF and NP cells to enhance proteoglycan production in static conditions and after the administration of interleukin (IL)-1, a proinflammatory cytokine.

Li, Yoon, and Hutton (2004) proved that the exogenous application of BMP-2 increased type II collagen and aggrecan content and intrinsic OP-1 levels in rat IVD cells *in vitro*, while Kim et al. (2003) showed that the same growth factor, in a human recombinant form, increased aggrecan and type I and II collagen synthesis by human IVD cells *in vitro*, while they did detect a higher expression of osteocalcin. This latter, as a marker of bone formation, is often associated with the administration of BMP-2 in other tissues, which is known to be an osteoinductive agent: low osteocalcin levels in this study may indicate that this growth factor has the potential to lead to a chondrocytic phenotype in IVD cells (Kim et al. 2003).

On the other hand, Osada et al. (1996) showed that exogenous IGF-1 was capable of increasing proteoglycan content and NP cell proliferation *in vitro*.

Recently, Ellman et al. tested a combined peptide therapy using lactoferricin B (which was shown to exert anti-inflammatory, anticatabolic, and proanabolic effects on NP cells) and BMP-7 on bovine disc cells *in vitro*, demonstrating that the two factors together had a synergistic effect that resulted in higher aggrecan content when compared to the administration of a single factor alone (Ellman et al. 2013).

Nonetheless, these growth factors have been proved to be efficacious *in vivo* as well: An et al. (2005) injected healthy rabbit discs with OP-1 and found a higher proteoglycan and water content when compared to noninjected controls at 2 weeks posttreatment, while (Masuda et al. (2006) administered the same growth factor in a rabbit model of IDD induced by annular puncture and showed that OP-1 was able to restore disc height and structural alterations in degenerated discs.

Although these studies clearly demonstrate the intrinsic ability of growth factors to modulate NP cell biological activity, one major concern is related to the short half-life of this peptide, which ranges from hours to days, thus not allowing control of a chronic and progressive condition as IDD, unless growth factor levels are kept elevated for a significant period of time (Winn, Uludag, and Hollinger 1999).

Having the capacity to modulate the expression of target genes for a prolonged time, gene therapy has been excitingly considered to treat IDD and reverse the degenerative process.

8.4 GENE THERAPY FOR IDD

Gene therapy consists in the transfer of exogenous genetic material (DNA or RNA) into the genome of target cells so as to modulate gene expression by enhancing the synthesis of beneficial and/or missing proteins or by inhibiting the synthesis of detrimental products. Differently from the administration of growth factors, this approach, by permanently modifying the host cell genome, provides a long-term synthesis of the desired product.

The original concept was formulated to treat inheritable monogenic diseases, such as cystic fibrosis, spinal muscular atrophy, and Duchenne's muscular dystrophy, by effectively replacing the defective genes with exogenous functional sequences.

To date, this approach has been applied to several polygenic and multifactorial disorders, including cancer, heart diseases, arthritis, and IDD (Evans and Robbins 1995).

The synthesis of exogenous genetic sequences requires an expensive, long time-taking, and demanding process: first, a reverse transcriptase enzyme is utilized to generate a complementary DNA (cDNA) using an RNA as a template, most often the messenger RNA (mRNA) encoded by the target gene. Then, the cDNA is cloned in a plasmid and put under the regulation of a promoter sequence that modulates the expression of the cDNA itself. Eventually, the plasmid is introduced into a vector, which drives the transfer of the exogenous gene inside the host cell. Once the cDNA is internalized by the host cell, it is transferred to the nucleus, where it is transcribed into an mRNA, which is then translated by ribosomes, thus allowing for the synthesis of the desired protein (Woods et al. 2011).

8.5 GENE DELIVERY SYSTEMS AND STRATEGIES

Given that naked DNA is not taken up by cells, vectors are needed to store, transport, and deliver plasmid DNA to target cells in order to guarantee a sustained transgene expression.

Apart from selecting the appropriate gene and vector, it is crucial to choose the ideal delivery strategy depending on the characteristics of target organs and cells, of the underlying disease, of the vector, and, in addition, well considering safety issues. There are two main strategies for gene delivery: the *in vivo* and the *ex vivo* strategy.

The former consists of the transfer of vectors loaded with the appropriate gene directly into targeted host tissues and cells, while the latter involves isolation of host cells, which are then cultured and transfected with the vector *in vitro* and eventually reimplanted into the host.

From a theoretical point of view, the *ex vivo* strategy has the advantage of allowing for controlling the transfection process *in vitro* and for monitoring the presence of any abnormal cellular reaction before reimplanting modified cells back into the host, thus being safer. However, the need for both a harvest and a reimplantation site increases surgical morbidity (Evans and Robbins 1995), while other major concerns regard the harvesting procedure and cell viability *in vitro*. This is particularly relevant when applying an *ex vivo* gene therapy to IDD: NP cells are physiologically exposed to hostile conditions, such as low oxygen tension, acidic pH, and low nutrient availability, which are difficult to be replicated *in vitro*. In addition, the harvesting and reimplantation procedure would require a double access to the disc, which might boost IDD itself. These reasons have convinced most authors that an *in vivo* approach would be more suitable for an IDD gene therapy.

Gene delivery systems can be divided into two main categories: viral systems (which include genome-incorporating virus, e.g., retrovirus and lentivirus, and non-genome incorporating, e.g., herpes virus, adenovirus, and adeno-associated virus [AAV]) and nonviral systems (Vadala, Sowa, and Kang et al. 2007).

8.5.1 Viral Systems

Viral vectors are able to infect host cells and transfer their genome inside them. While they appear to be significantly effective as vectors for gene therapy, on the other hand, their application is often associated with safety issues due to cytotoxicity and immune response.

8.5.1.1 Retrovirus

Retrovirus has been primarily utilized through *ex vivo* strategies. When using a retrovirus as a vector, the plasmid DNA carried is fully integrated into the genome of the host cell, so that the transgene can be transmitted to each daughter cell after cell division, thus allowing for long-term gene expression. However, retroviruses have shown oncogenic potential by inserting random sequences in the host cell genome, thus inhibiting the expression of oncosuppressor genes and/or favoring the transcription of oncogenes (Hacein-Bey-Abina et al. 2003). In addition, retroviruses are able to transfect actively replicating cells only at the time of the inoculation (Robbins, Tahara, and Ghivizzani 1998).

Although retrovirus has been extensively investigated as a vector for gene therapy of bone and cartilage disorders, as well as IDD, major concerns regarding its biological activity have limited its application, allowing other viral vectors to be developed.

8.5.1.2 Adenovirus

Adenovirus has been widely adopted in gene therapy through *in vivo* strategies. Adenoviral vectors are double-stranded DNA viruses capable of infecting non-proliferating cells both *in vivo* and *in vitro*, thus leading to high levels of gene expression. Among the 47 known human serotypes, serotypes 2 and 5 are the most commonly utilized as viral vectors for gene therapy. The main advantages of adenoviruses are the high efficiency in transferring the target gene into host cells, the relatively simple manipulability (which needs the envelope E1 gene to be removed in order to modulate adenoviral gene expression), and the lack of genome integration into host cell DNA, as the viral genetic material remains episomial, which limits the risk of insertional mutagenesis. On the other hand, when compared to other vectors, adenoviruses are associated with a gradual decrease in transgene expression due to the immune response against viral proteins, which leads to adenoviral episome degradation, thus abating its expression (Yang et al. 1994). The same immune reaction triggers inflammation when injected in the joint space and in other sites (McCoy et al. 1995; Nita et al. 1996). In addition, as the viral episome is not integrated into the host cell genome, it does not undergo replication during cellular division. For these reasons, it has not been possible to maintain a transgene expression for more than 12 weeks in most musculoskeletal tissues in which adenoviral vectors have been inoculated. However, the inherent avascularity and the immune privilege that characterize the IVD might allow adenoviral vectors to prolong transgene expression. Indeed, Nishida et al. (2000) have demonstrated that intradiscal transgene expression markers were detectable up to 1 year after adenoviral transfection in a rabbit lumbar disc.

Despite the promising role of adenoviral vectors in IDD gene therapy, significant concerns about cytotoxicity and immune-mediated side effects on the surrounding central nervous system structures and the neurological sequelae associated with them still limit their application (Driesse et al. 2000, Tripathy et al. 1996; Wallach et al. 2006).

8.5.1.3 Adeno-Associated Virus

AAV is a single-stranded DNA parvovirus that is characterized by high transfection efficacy in both proliferating and nonproliferating cells and lack of causal connections with any known disease in humans or mammals (Afione et al. 1999). In addition, AAV integrates its genome in a specific site located on chromosome 19 without altering the preexisting sequence. AAV intrinsic genome contains only two genes, namely, Rep and Cap, which cannot be expressed after transfection unless a helper virus is present, thereby meaning that viral antigen expression remains blocked and immune response is not triggered.

However, AAV capacity to carry exogenous DNA is limited but still allows for including genetic sequences encoding for most growth factors.

Lattermann et al. (2005) investigated the role of AAV in IDD gene therapy by transfecting both rabbit and human NP cells *in vivo* and *in vitro*, respectively. The *in vivo* analysis showed low transgene expression markers at 2 weeks after intradiscal inoculation, while levels rose at 4 weeks and remained substantially constant at up to 6 weeks. However, the maximum transgene expression rate after transfection with the AAV-luciferase vector was ~50% of the same expression rate measured after utilizing an adenoviral vector (Lattermann et al. 2005).

Subsequent studies adopting different AAV vectors have demonstrated that the vector potency and tropism may vary significantly depending on which AAV serotype has been used, thus suggesting that AAV might be a considerable alternative to other viral vectors in regard to IDD gene therapy (Gao, Vandenberghe, and Wilson 2005).

In this regard, Mern et al. investigated NP cell-specific AAV serotypes by targeting human NP cells with different self-complementary AAV serotypes (scAAV). Cells were harvested *ex vivo* from patients undergoing spine surgery, cultured, and then transduced with either scAAV-1, -2, -3, -4, -5, -6, or -8 carrying a green fluorescent protein (GFP) gene. Among these vectors, scAAV-2, -6, and -3 showed a high and sustained GFP expression and a transduction efficiency of ≥89% in all three cases. However, scAAV-2 and -3 diminished NP cell viability by 25% and 10%, respectively, while scAAV-6 did not affect the expression of catabolic factors, proinflammatory cytokines, and matrix proteins. This ultimately suggests that AAV-6 could be the most suitable AAV serotype for IDD gene therapy due to the association with sustained transgene expression and lack of immune reaction (Mern and Thome 2015).

8.5.2 NONVIRAL SYSTEMS

Nonviral vectors are immunologically inert, easily constructible plasmid-based systems including liposomes, microbubble-enhanced ultrasound, synthetic polymers, gene guns, naked DNA, nucleofection (as discussed in Section 8.7.1), and

DNA–ligand complexes. Although biologically inert and thus safe, generally, nonviral systems are not able to guarantee a sustained transgene expression due to low transfection efficacy, which would be insufficient to treat chronic conditions such as IDD (Blanquer, Grijpma, and Poot 2015).

However, Nishida et al. (2006) demonstrated that intradiscal injection of plasmid DNA mixed with ultrasonography contrast agent (phospholipid-stabilized microbubbles), followed by local ultrasound irradiation, increased transgene uptake and expression by NP cells *in vivo*, which could be measured up to 24 weeks.

Chung et al. transfected ovine NP cells *in vitro* with a human telomerase reverse transcriptase (hTERT) mixed with lipofectamine, a nonviral carrier constituted by polycationic lipids. Telomerase, which is committed to extending telomeres and thus prolonging cellular lifespan, led to diminished cell senescence and increased type I and II collagen once transduced in NP cells for up to 9 months (Chung et al. 2007).

Morrey et al. compared different nonviral vectors loaded with marker genes and then transfected to human NP cells *in vitro*. Among them, the histone polypeptide/lipid-based vector LT1 showed both the lowest toxicity and the highest transfection efficacy compared to the other vectors (Morrey et al. 2008).

Recently, May et al. reported the successful use of nonviral transfection of bovine and human AF and NP cells with the commercially available plasmid pCMV6-AC-GFP by electroporation *in vitro*; in one group, the plasmid carried a marker gene, with another one containing the human growth and differentiation factor (GDF)-6 cDNA. Results showed a successful cell transfection after applying two pulses of 1400 V for 20 milliseconds, with a higher transfection rate in AF than in NP cells, both in human and bovine species. GDF-6 showed to be sustainedly upregulated for up to 14 days, after which its expression progressively decreased. Interestingly, no increase in aggrecan or type II collagen was reported (May et al. 2017).

8.6 RNA INTERFERENCE

RNA interference (RNAi) is a recently developed strategy to silence the expression of specific genes using small interfering RNA (siRNA) sequences. siRNAs are able to inhibit gene expression by binding in a site- and sequence-specific manner to a targeted mRNA, which undergoes degradation or is simply not translated into the corresponding protein.

Theoretically, catabolic genes encoding for products involved in disc matrix breakdown might be targeted and silenced by siRNA: this hypothesis has been successfully tested *in vitro*, even though siRNA sequences showed to have a significantly short half-life. In order to overcome this limitation, Kakutani et al. (2006) utilized DNA vector–siRNA complexes that were able to inhibit targeted gene expression in rat and human NP cells *in vitro* up to 3 weeks.

The first *in vivo* evidence of the application of RNAi for IDD came from Suzuki et al., who used two reporter luciferase plasmids (namely Firefly and Renilla) and unmodified siRNA duplexes targeting Firefly luciferase (in order to evaluate RNAi efficacy) and Fas ligand (Fas ligand [FasL], as a marker of endogenous gene RNAi). These complexes were cotransfected into rat coccygeal discs using the ultrasound gene transfer technique, which resulted in sustained gene expression

Gene Delivery for Intervertebral Disc

inhibition for up to 24 weeks, after which the inhibition rate was still 80% and 53% for Firefly and FasL expression, respectively, when compared to the control group (Suzuki et al. 2009).

The role of Fas was further investigated by Han et al. (2009), who treated rat NP cells with Fas siRNA *in vitro*. The Fas receptor is involved in maintaining the immune privilege of IVD (as discussed in Section 8.7) as well as in degenerative cell apoptosis: thus, its inhibition through RNAi resulted in a diminished apoptotic NP cell death in this study (Han et al. 2009).

Seki et al. suppressed ADAMTS-5 (a disintegrin and MMP with thrombospondin-like repeats-5) expression through a single intradiscal injection of the corresponding siRNA in a rabbit annular needle-puncture IDD model, so as to assess the efficacy of ADAMTS-5 RNAi *in vivo*. This MMP is mainly involved in disc matrix breakdown in IDD: results from this study showed an increase in signal intensity within the injected NP at magnetic resonance imaging (MRI) and diminished histological evidence of matrix degradation, without notable disc height restoration though (Seki et al. 2009).

Recently, Yamada et al. targeted caspase 3, a major apoptotic marker that is upregulated in mechanical overload-induced IDD, with a corresponding siRNA sequence both *in vitro* and *in vivo* using cultured human NP cells and a rabbit IDD model respectively. The *in vitro* assay showed an improvement in cell viability and a decrease in MMP-3 and -13 and ADAMTS-4 and -5 levels, thus resulting in a higher proteoglycan synthesis by NP cells supplemented with caspase 3 siRNA compared to controls after compressive loading. The *in vivo* study required animals to undergo IDD through the application of an external compression model: 1 week after the compression session, caspase 3 siRNA was injected intradiscally and rabbits were sacrificed at 8 or 16 weeks after the injection. Then, MRI, immunohistochemistry, histological assays, and TUNEL assay were performed, showing a significant inhibition of the degenerative process (Yamada et al. 2014).

This demonstrates that RNAi might be a beneficial approach to both inhibit specific catabolic gene expression and to analyze the role of peculiar gene expression profiles in IDD.

8.7 PROOF OF PRINCIPLE OF A GENE THERAPY FOR IDD

The first attempt to adopt gene therapy in order to counteract IDD was made by Wehling et al. (1997), who used a retroviral-mediated transfer of two exogenous genes, namely, bacterial enzyme β-galactosidase (lacZ) and human IL-1 receptor antagonist (IL-1Ra) to intervertebral endplate chondrocytes harvested from bovine coccygeal bone. While lacZ transgene expression rate was only 1%, IL-1Ra expression was sustainedly high for up to 48 hours. This study concluded that an *ex vivo* strategy requiring cell harvesting from the endplate, transfer of exogenous genes to cells *in vitro*, and then reinjection back into the disc could be effective in treating IDD (Wehling et al. 1997).

Shortly after, Nishida et al. performed the first *in vivo* intradiscal gene transfer by directly injecting into rabbit IVDs an adenoviral vector carrying the lacZ gene mixed with saline solution. The animals did not show any sign of immune reaction, which was confirmed histologically up to 3 months, and, in addition, transgene expression rate was found to be highly sustained at 3 months after injection and at 1 year, as reported by a subsequent study (Nishida et al. 1998).

In order to seek more deeply for the possible immunogenicity of the adenoviral vector, the authors injected subcutaneously the transgene-vector complex weeks before the intradiscal delivery and found no down-regulation of the luciferase gene marker expression later inoculated into rabbit IVDs. This "immune privilege" within the disc was supposed to be associated with the constitutive expression of FasL, which, in other immunoprivileged organs such as the testes and the retina, leads FasL-positive T cells to apoptosis, thus aborting the immune reaction.

The presence of FasL within the IVD and, more specifically, on NP cells was then demonstrated by Takada et al. (2002), thus providing a solid basis to the disc immune privilege theory.

8.7.1 Anabolic Factors

After having demonstrated the feasibility of an intradiscal gene therapy, Nishida et al. (1999) managed to transduce a cDNA encoding for human TGF-β1 *via* an adenoviral vector (Ad/TGF-β1) *in vivo* by directly injecting it into healthy rabbit lumbar IVDs. At 1 week after transduction, a 30-fold increase in active TGF-β1 synthesis and a 5-fold increase in total TGF-β1 expression in injected discs were recorded. In addition, proteoglycan synthesis increased up to twofold in the transduced discs when compared to controls, which were injected with an adenoviral vector carrying a marker gene, and did not show any elevation in TGF-β1 synthesis and proteoglycan production. As previously described, no local or systemic immune response was detected in this study, which confirmed not only the possibility to safely transfect IVD cells with potentially therapeutic genes but also that this approach could realistically result in an enhancement of NP matrix (Nishida et al. 1999).

Subsequent studies showed similar results after human IVD cell transfection with Ad/TGF-β1 *in vitro*. Particularly, Moon et al. (2000) reported that transduced human NP cells showed a threefold increase in proteoglycan synthesis when compared to controls transfected with marker genes. Furthermore, cells treated with exogenous TGF-β1 also showed a significant elevation of proteoglycan levels, even if lower than in the transfected group. Notably, these remarkable results were obtained with a viral load of between 25 and 70 multiplicity of infection (MOI), while the presumed viral load to transfect 100% of the cells was expected to be 150 MOI: this suggests that transduced cells may influence non-transduced ones *via* a paracrine effect (Moon et al. 2000).

As mentioned earlier, BMP-2 and IGF-1 have previously showed efficacy in increasing the anabolic activity within the IVD after exogenous administration, which led researchers to construct adenoviral vectors carrying corresponding genes and to test them on human IVD cells *in vitro*, resulting in an increase in proteoglycan synthesis (Moon et al. 2008).

An additional anabolic factor that has been investigated in the last decade for its possible role in IDD is GDF-5. It has been shown to promote human mesenchymal stem cell (hMSC) differentiation toward a chondrocyte-like phenotype and to exert a regenerative effect on NP cells both *in vitro* and *in vivo* using a rabbit IDD model. Wang et al. (2004) were the first to investigate the role of GDF-5 by transfecting rabbit and human IVD cells with an adenoviral vector loaded with the GDF-5 gene *in vitro*, which resulted in enhanced cell proliferation.

Similar findings were obtained by Liang et al. (2010), who constructed an adenoviral vector carrying the GDF-5 gene and transfected NP cells using a mouse IDD model *in vivo*. The transfected group showed an increased disc height and preserved histological aspect, with a sustained GAG and DNA content when compared to control groups (Liang et al. 2010).

Given these results, Cui et al. investigated the transduction of GDF-5 using a novel nonviral delivery system, namely, nucleofection. This technology combines specific solutions paired with electrical pulses, which drive the internalization of the cDNA into the cell nucleus by electroporation. In this study, NP cells were harvested from mice lumbar discs, cultured, and treated with either exogenous GDF-5 protein or GDF-5 cDNA by nucleofection. GDF-5 showed to enhance cell proliferation, GAG production, and aggrecan and type II collagen gene expression, while MMP-3 levels were notably diminished. These results demonstrate that GDF-5 has a potential regenerative role in the IVD both as a protein and as a plasmid cDNA and that nucleofection, whose application on IVD cells was reported by this study for the first time, could be a successful nonviral delivery system for IDD gene therapy (Cui et al. 2008).

An interesting combination of nonviral transgene delivery and stem cell therapy was shown by Bucher et al. (2013): the authors isolated hMSCs from human bone marrow aspirates and transfected them with human GDF-5 cDNA by nucleofection. Transfected cells were then cultured in alginate beads and showed increased expression of aggrecan, Sox9, and KRT19 (a marker of discogenesis), as well as an increased proteoglycan synthesis. Transfected cells were subsequently injected in a bovine IDD organ culture model, where they led to a partial recovery of the GAG/DNA ratio, which was significantly lower in the nontransfected group (Bucher et al. 2013).

8.7.2 ANTICATABOLIC FACTORS

Increasing proteoglycan and collagen content can be obtained not only by enhancing their synthesis utilizing anabolic factors but also by reducing their breakdown through the inhibition of catabolic activity. Among all the molecules involved, matrix MMPs seem to be primarily responsible for matrix degradation in degenerating discs and in IVDs put under significant hydrostatic pressure and other types of stress (Sobajima et al. 2005b).

Given that matrix degradation has appeared to be associated with an imbalance between MMP-3 and its physiological inhibitor, namely, tissue inhibitor of MMP-1 (TIMP-1) (Kanemoto et al. 1996), Wallach et al. (2003) transfected human disc cells cultured in 3D pellets with adenoviral vectors loaded with TIMP-1 gene and reported an increase of proteoglycan synthesis of up to fivefold compared to controls.

TIMP-1 use for IDD gene therapy *in vivo* was firstly reported by Leckie et al. (2012), who performed the intradiscal injection of an AAV vectors carrying either TIMP-1 or BMP-2 genes in a rabbit annulotomy IDD model. MRI scans were obtained at 0, 6, and 12 weeks and showed clear evidence of IDD in the control group, while the treated group was characterized by a lower grade of degeneration. Animals were sacrificed at 12 weeks and histological, biochemical, and biomechanical assays were performed. Histological analysis confirmed MRI findings, while the treated group showed better viscoelastic properties and lower C-telopeptide of collagen type II (CTX-II, a serum marker of this animal IDD model) levels than the control group (Leckie et al. 2012).

An additional and significant degenerative trigger is inflammation: it has been previously reported that proinflammatory cytokines, such as IL-1, TNF-α, and IL-6 effectively, activate catabolic factors and promote IDD. On the other hand, mechanical stress upregulates p38 mitogen-activated protein kinase (MAPK), which physiologically inhibits the catabolic effect mediated by the aforementioned cytokines (Studer et al. 2007).

Studer et al. (2008) reported that p38 MAPK diminished IL-1- and TNF-α-induced proteoglycan breakdown in rabbit NP cells *in vitro*, without exerting any effect on the anabolic activity mediated by TGF-β.

In addition, the inhibitory effect on MMP-3 expression without altering TIMP-1 levels may be a further mechanism by which the p38 MAPK pathway influences disc inflammation and metabolism.

An emerging factor that is mainly involved in the up-regulation of the aforementioned proinflammatory cytokines and that has been reported to be associated with aging-dependent IDD is nuclear factor kappa-light-chain-enhancer of activated B cells (NF-κB). Nasto et al. (2012) demonstrated that inhibiting NF-κB activation actually resulted in increased proteoglycan content and reduced cell loss in progeroid mice IVD discs, which intrinsically show a permanently enhanced NF-κB activity.

For these reasons, p38 MAPK and NF-κB may be effectively considered as additional targets for IDD gene therapy in the future.

8.7.3 Transcription Factors

Growth factor synthesis results from a complex and widely integrated molecular cascade that ultimately involves transcription factors, which operate by activating targeted gene expression under certain conditions.

Sox9 is a chondrogenic transcription factor that enhances type II collagen synthesis and production of other proteins that are essential to the IVD matrix.

This assumption led Paul et al. (2003) to transfect human IVD cells *in vitro* and rabbit IVD cells *in vivo* with an adenoviral vector expressing Sox9 gene (AdSox9). Human IVD cells showed higher levels of type II collagen production while rabbit IVD cells after intradiscal AdSox9 injection maintained a chondrocytic phenotype and the NP preserved it original architecture for up to 5 weeks compared to control discs (Paul et al. 2003).

LIM mineralization protein-1 (LMP-1) is an essential regulator of osteogenic differentiation and BMP synthesis in different musculoskeletal tissues, including the IVD (Boden et al. 1998a). The first attempt to utilize an LMP-1-based gene therapy was made by Boden et al. (1998b), who tried to replace iliac crest bone grafts for spinal fusion with grafts composed of nonosteoinductive devitalized bone matrix and bone marrow cells transfected with a pCMV2 vector expressing LMP-1. Results showed enhanced spinal fusion in all segments treated with the transfected graft while no fusion occurred in adjacent segments not expressing LMP-1 (Boden et al. 1998b). Yoon et al. (2004) delivered an adenoviral vector carrying LMP-1 gene (AdLMP-1) to rabbit IVD cells both *in vitro* and *in vivo*: cultured cells showed significantly higher levels of aggrecan, GAG, and BMP-2 and -7; comparable results were recorded *in vivo*. These results demonstrate that, curiously, LMP-1 shows a chondrogenic effect on IVD cells rather that an osteogenic effect, as it appears for other cell types.

8.7.4 Cell Survival and Apoptosis

It has been previously described that increased apoptosis of NP cells within the IVD is one of the major mechanisms involved in IDD. Programmed cell death is triggered by several stressful stimuli that activate intrinsic and extrinsic pathways and ultimately converge on the up-regulation of caspase-3, which then activates the apoptotic cascade. While caspase-3 has been successfully inhibited by RNAi, as described earlier, Ma et al. (2015) investigated the transfection of human NP cells *in vitro* with a lentivirus carrying the survivin gene. The latter is an inhibitor of apoptosis (IAP) that exerts its effects by binding to caspase-9 or preventing second mitochondria-derived activator of caspases (SMAC) from blocking other IAPs, thus down-regulating apoptosis. However, results showed a marked alteration of cell morphology, with a reduction in cytoplasm content and cell shrinkage, while no significant antiapoptotic effects were observed, even if survivin transgene expression remained stable and sustained (Ma et al. 2015).

Apoptosis can be also induced in the presence of the proinflammatory cytokine TNF-α, whose expression has been found to be elevated in IDD. Zhang et al. investigated the role of IGF-1 on rabbit NP cells *in vitro* through transfection with an adenoviral vector expressing recombinant human IGF-1 (Ad-hIGF-1). While it is known that this anabolic factor is able to enhance NP cell proliferation, the authors explored the possibility that it could inhibit TNF-α-mediated apoptosis as well. Cell cultures were treated with TNF-α and then divided in two groups, only one receiving Ad-hIGF-1. Cells transfected with Ad-hIGF-1 showed a sustained IGF-1 expression and low apoptotic rate, while the other group showed enhanced apoptosis with typical associated cell morphological alterations, thus suggesting that IGF-1 could realistically inhibit TNF-α-induced apoptosis (Zhang et al. 2014).

An additional apoptotic marker that has been recently designated as a target for IDD gene therapy is C/EBP homologous protein (CHOP), whose expression is triggered by endoplasmic reticulum stress provoked by biomechanical stress exerted on AF cells, thus leading to apoptosis. Zhang et al. (2014) combined the use of a lentiviral vector and RNAi by the introduction in the viral genome of a short hairpin RNA (shRNA) designed to bind and inhibit CHOP expression. Rat AF cells were cultured and transduced *in vitro* and then underwent cyclic tension loading; the experimental group showed decreased apoptosis and prolonged CHOP inhibition. Subsequently, the authors tested the effectiveness of this vector *in vivo* by an intradiscal injection in a rat IDD model. At 7 weeks after injection, MRI scans showed a higher intradiscal intensity in the experimental group than in the control group; corresponding findings were detected at the histological assay, where the injected group demonstrated no or mild signs of IDD. In addition, apoptosis was significantly decreased in the experimental group (Zhang et al. 2011).

Another attempt to extend cell lifespan has been performed by upregulating telomerase activity, as mentioned previously. Wu et al. (2014) reported the use of a lentiviral vector bearing hTERT gene on freshly isolated human NP cells *in vitro*. Transgene expression was detected up to 210 days after transfection and resulted in increased telomere length, delayed cell senescence, and enhanced cell proliferation (Wu et al. 2014).

On the other hand, Shi et al. (2015) have investigated the effect of an AAV loaded with hTERT gene in a canine IDD model by injection of the vector 4 weeks after needle-puncture-induced IDD; animals were followed for 12 weeks and then sacrificed. In the experimental group, no significant change in disc height or relative grayscale index at MRI was seen. Injected discs showed retained biomechanical features, disc morphology, and higher proteoglycan content (Shi et al. 2015).

8.7.5 Multiple Targets

As a single-agent gene therapy was successful, Moon et al. (2008) tested the transfection with multiple growth factor-encoding genes by the administration of a "gene therapy cocktail" to cultured human IVD cells. Adenoviral constructs carrying TGF-β1, IGF-1, and BMP-2 genes showed a synergistic effect when combined together, resulting in a higher proteoglycan synthesis rate compared to single-agent therapy (Moon et al. 2008).

In order to identify the most potent among the BMPs, Zhang et al. (2005) evaluated the effect of different BMP subtypes by transfecting bovine articular chondrocytes with corresponding genes *in vitro*. Among BMP-2, -4, -5, -7, -10, and -13, subtypes 2 and 7 resulted in a higher proteoglycan and collagen synthesis compared to other subtypes (Zhang et al. 2005).

The increasingly expanding knowledge of IDD pathophysiology and the numerous factors and pathways involved thus suggested the possibility to transduce multiple transgenes at the same time with a multigene therapy.

Liu et al. (2010) compared the transfection of monkey NP cells *in vitro* with an AAV vector loaded with either both connective tissue growth factor (CTGF) and TIMP-1 genes or with a single transgene (CTGF or TIMP-1). Single gene transduction led to enhanced proteoglycan and collagen synthesis, which was significantly higher with combined transduction of both transgenes (Liu et al. 2010).

A similar study design was adopted by Xi et al. (2013), who utilized AAV vectors carrying human vascular endothelial growth factor165 (hVEGF165) and TGF-β1 to transfect rabbit AF cells *in vitro*. Cotransfection with both vectors resulted in an increased type I collagen synthesis when compared to transfection with a single vector; the data also suggested that hVEGF165 could exert a synergistic effect on TGF-β1 by promoting collagen production (Xi et al. 2013).

The increased efficacy of a transgene cotransfection was further demonstrated by Ren et al. (2015), who used AAV vectors to transduce BM-7 and Sox9 genes, notable for their chondrogenetic effects, in human NP cells cultured under degenerative conditions. Type II collagen mRNA levels were significantly higher in the cotransfection group than in the single-vector transfection group (Ren et al. 2015).

Recently, Liu et al. (2016) constructed a lentiviral vector carrying TGF-β3, CTGF, and TIMP-1 genes at the same time and examined its role in possibly reversing IDD in a rabbit annular puncture IDD model *in vivo*. In comparison with untreated controls, MRI scans at 16 and 20 weeks postoperatively showed increased NP hydration, while reverse transcription polymerase chain reaction (RT-PCR) and western blot assay demonstrated higher levels of aggrecan and type II collagen in the experimental group (Liu et al. 2016).

Notably, Yue et al. (2016) were the first to report the effects of a lentivirus simultaneously carrying survivin, TGF-β3 and TIMP-1 genes, in order to modulate cell apoptosis, anabolic, and anticatabolic functions with a single vector. The study was performed *in vivo* using a rabbit IDD model: 3 weeks after annular puncture, the lentiviral vector was injected into the NP and animals were then sacrificed at 12 weeks after injection. The experimental group showed less evidence of IDD at MRI as well as diminished cell apoptosis and increased levels of aggrecan and type II collagen when compared to untreated controls (Yue et al. 2016).

8.8 *IN VIVO* EFFICACY AND FEASIBILITY

In order to effectively assess the feasibility and the efficacy of a gene therapy in preventing IDD and to progress toward the design of human clinical trials, it is crucial to test this approach on reliable animal IDD models that closely resemble the human aspects of the disease. Most researchers have used a rabbit model in which IDD was induced by a scalpel blade annular injury with subsequent disc herniation and prompt degeneration.

Sobajima et al. (2005) proposed a slowly progressive and reproducible animal model of IDD, which was accomplished by stabbing the anterolateral AF with a 16-gauge needle to a depth of 5mm. This approach did not cause NP herniation, while IDD occurred gradually over 24 weeks as defined by MRI and histological and radiographic assays, thus better resembling the progressivity of the human condition. On a molecular level, RT-PCR demonstrated that, at 3 weeks, aggrecan and type IIa collagen mRNA levels diminished significantly, while BMP-2 and -7 and IGF-1 mRNA decreased from week 3 to 6 but later increased at week 12 and 24 up to the preoperative level. On the other hand, TIMP-1 expression was 10% of the preoperative level at week 3 and stayed low throughout the entire degenerative process. The authors then adopted this model to test an adenoviral vector carrying the BMP-2 gene, which was directly injected into degenerating IVDs. T2-weighted MRI and X-ray assays concluded that BMP-2 gene transfer actually delayed IDD progression when compared to saline-injected control IVDs at 12 weeks after gene transfection. This beneficial effect has been likely due to sustained BMP-2 transgene expression (which was confirmed by enzyme-linked immunosorbent assay) and to an increase in disc proteoglycan and water content, as suggested by MRI (Sobajima et al. 2005).

8.9 SAFETY CONCERNS

Despite having repeatedly shown efficacy and sustained response under experimental conditions, considerable safety concerns regarding gene therapy still limit its application in a wide-scale clinical setting.

This issue is particularly relevant considering the nature of IDD, which is a chronic and progressive process, therefore needing a prolonged and reliable treatment; moreover, the anatomical proximity of IVDs and critical neurological and vascular structures make this concern even more significant due to possible and irreversible damage to them.

As described earlier, viral cytotoxicity and mutagenicity are among the main risks associated with gene therapy: this has led to the development of safer viral vectors, most notably AAV, as well as engineered nonviral vectors.

Safety concerns regarding intradiscally injected adenoviral vectors were firstly evidenced by Wallach et al. (2006), who demonstrated that a quantitatively inappropriate or directly injected viral load could cause paraplegia and significant histological alterations in the spinal cord subarachnoid space of rabbits treated with a high concentration of the adenoviral vector.

Later, Levicoff et al. (2008) analyzed the possible effects of a misplacement of the genetic material with leakage inside the spinal epidural space using either an adenoviral vector or an AAV vector in a rabbit model; animals were divided in groups receiving a normal dose or a high dose of viral load. After 6 weeks, some animals treated with a high-dose adenoviral-based therapy showed significant signs of toxicity, including paraplegia, sensory loss, and death. Histological analysis showed signs of transgene expression in the meninges, the dorsal roots, and in the spinal cord parenchyma, with evidence of inflammation and fibrosis in these structures. In contrast, the AAV-treated group did not show any clinical or histological side effect, independent of dosage (Levicoff et al. 2008).

These findings have led to additional efforts to better delineate the possible role of AAV as a vector for IDD gene therapy, due to its favorable safety profile and transfection efficacy, as described previously.

8.10 REGULATING TRANSGENE EXPRESSION

In order to prevent side effects caused by uncontrolled transgene expression, exogenous methods to modulate positively or negatively the synthesis of target proteins until reaching desired levels have been proposed. These act by binding to promoter regions on the vector genome, thus activating ("on" systems) or inhibiting ("off" systems) transgene expression.

In this regard, one of the major transgene regulators is tetracycline, which has been used to design tetracycline (tet)-inducible systems for modulating viral transgene expression. Chtarto et al. (2003) firstly reported the successful use of a tet-on system, which was integrated into AAV construct and characterized by an up-regulation of transgene expression after administration of tetracycline derivatives.

Vadala et al. (2007) described for the first time the application of a tet-inducible system for intradiscal gene therapy. In this study, human NP cells were transfected *in vitro* with an adenoviral vector carrying a marker gene and cultured in presence of tetracycline, which showed to reversibly inhibit transgene expression. The suitability of a tet-off system for down-regulating transgene expression would be of great value especially if applied to intradiscal gene therapy, since tetracycline can cross the blood–brain barrier and thus inhibit transgene expression in the epidural or intradural spaces in case of unexpected leakages (Vadala, Sowa, and Smith et al. 2007).

However, the ability to up-regulate intradiscal transgene expression through a tet-on system has demonstrated poor feasibility due to insufficient tetracycline concentration within the IVD after oral administration (Alp et al. 2006).

Thereby, novel transgene regulation systems have been investigated, including heat shock proteins, metallothionine, steroid regulatory promoters, tissue-specific inducible promoters, and ecdysone receptor-based systems (Graham 2002).

The latter was adopted in a study conducted by Sowa et al. (2011), who investigated the use of an AAV vector containing a novel control system, namely, RheoSwitch, which uses a specific promoter region that enables transgene expression only in the presence of an activator ligand. The vector carrying a GFP marker gene was transfected in rabbit NP cells *in vitro*; increasing concentrations of the activator ligand were added and fluorescence was then assessed during and after its removal. Results showed that fluorescence was time and dose dependent: it diminished 24 hours after removal of the ligand and was hardly detectable after 48 hours. Further, this regulation system has been tested *in vivo* using healthy rabbits that received an intradiscal injection of the virus. GFP expression was reported to be high in animals treated with both virus and activator ligand, while animals receiving either one of them did not show any appreciable fluorescence. In addition, no GFP expression was detected in surrounding tissues, thus possibly designating RheoSwitch as a reliable regulation system for IDD gene therapy (Sowa et al. 2011).

8.11 CONCLUSION

Recent advances in understanding IVD physiology as well as biochemical pathways and pathophysiology involved in the development of IDD have raised the possibility to target specific factors driving the degenerative process in order to delay or, theoretically, reverse it. Critical molecule expression regulating disc anabolism, catabolism, and cell survival has been found to be altered in IDD, and gene therapy seems to be a considerable tool to effectively modulate gene expression through sustained delivery of potentially therapeutic transgenes.

Multiple studies, both *in vitro* and *in vivo*, have demonstrated the feasibility and the beneficial potential of a gene therapy for IDD. Several and diverse molecules and growth factors have been targeted, using either different viral vectors, nonviral vectors, or RNAi, all yielding promising and significant results in terms of IVD structure preservation, increased matrix anabolism, and reduced cell loss.

However, major concerns regard the safety of this approach, especially when involving viral vectors, which can have detrimental effects on surrounding structures as well as systemic consequences; these limitations are being overcome through the development of transgene regulation systems and identification of vectors with safer profiles. Nonetheless, transgene construction and vector engineering are expensive and time-consuming procedures that might question the real need for a clinical translation.

In addition, further studies are needed to better comprehend the effect of a gene therapy for IDD over a long-term period, which has not been consented by past animal studies, whose duration was relatively short.

To date, the major shortcoming of any therapy for IDD is the relatively scarce knowledge of the exact mechanisms and pathways that lead to disc degeneration. Furthermore, as disc cell activity is critically influenced by biomechanical stimuli, it is crucial to deeply investigate the mechanobiology of these cell populations, in order to better define the correct conditions for the application of a treatment which, ideally, could prevent or delay IDD.

REFERENCES

Afione, S.A., J. Wang, S. Walsh et al. 1999. "Delayed expression of adeno-associated virus vector DNA." *Intervirology* 42 (4):213–20.

Alp, E., R.K. Koc, A.C. Durak et al. 2006. "Doxycycline plus streptomycin versus ciprofloxacin plus rifampicin in spinal brucellosis [ISRCTN31053647]." *BMC Infect Dis* 6:72.

An, H.S., K. Takegami, H. Kamada et al. 2005. "Intradiscal administration of osteogenic protein-1 increases intervertebral disc height and proteoglycan content in the nucleus pulposus in normal adolescent rabbits." *Spine (Phila Pa 1976)* 30 (1):25–31; discussion 31–22.

Annunen, S., P. Paassilta, J. Lohiniva et al. 1999. "An allele of COL9A2 associated with intervertebral disc disease." *Science* 285 (5426):409–12.

Battie, M.C., T. Videman, and E. Parent. 2004. "Lumbar disc degeneration: Epidemiology and genetic influences." *Spine (Phila Pa 1976)* 29 (23):2679–90.

Blanquer, S.B., D.W. Grijpma, and A.A. Poot. 2015. "Delivery systems for the treatment of degenerated intervertebral discs." *Adv Drug Deliv Rev* 84:172–87.

Boden, S.D., Y. Liu, G.A. Hair et al. 1998a. "LMP-1, a LIM-domain protein, mediates BMP-6 effects on bone formation." *Endocrinology* 139 (12):5125–34.

Boden, S.D., L. Titus, G. Hair et al. 1998b. "Lumbar spine fusion by local gene therapy with a cDNA encoding a novel osteoinductive protein (LMP-1)." *Spine (Phila Pa 1976)* 23 (23):2486–92.

Bruehlmann, S.B., J.B. Rattner, J.R. Matyas, and N.A. Duncan. 2002. "Regional variations in the cellular matrix of the annulus fibrosus of the intervertebral disc." *J Anat* 201 (2):159–71.

Bucher, C., A. Gazdhar, L.M. Benneker, T. Geiser, and B. Gantenbein-Ritter. 2013. "Nonviral gene delivery of growth and differentiation factor 5 to human mesenchymal stem cells injected into a 3D bovine intervertebral disc organ culture system." *Stem Cells Int* 2013:326828.

Cassidy, J.J., A. Hiltner, and E. Baer. 1989. "Hierarchical structure of the intervertebral disc." *Connect Tissue Res* 23 (1):75–88.

Chtarto, A., H.U. Bender, C.O. Hanemann et al. 2003. "Tetracycline-inducible transgene expression mediated by a single AAV vector." *Gene Ther* 10 (1):84–94.

Chung, S.A., A.Q. Wei, D.E. Connor et al. 2007. "Nucleus pulposus cellular longevity by telomerase gene therapy." *Spine (Phila Pa 1976)* 32 (11):1188–96.

Cui, M., Y. Wan, D.G. Anderson et al. 2008. "Mouse growth and differentiation factor-5 protein and DNA therapy potentiates intervertebral disc cell aggregation and chondrogenic gene expression." *Spine J* 8 (2):287–95.

Driesse, M.J., M.C. Esandi, J.M. Kros et al. 2000. "Intra-CSF administered recombinant adenovirus causes an immune response-mediated toxicity." *Gene Ther* 7 (16):1401–9.

Ellman, M.B., J. Kim, H.S. An et al. 2013. "Lactoferricin enhances BMP7-stimulated anabolic pathways in intervertebral disc cells." *Gene* 524 (2):282–91.

Evans, C.H., and P.D. Robbins. 1995. "Possible orthopaedic applications of gene therapy." *J Bone Joint Surg Am* 77 (7):1103–14.

Gao, G., L.H. Vandenberghe, and J.M. Wilson. 2005. "New recombinant serotypes of AAV vectors." *Curr Gene Ther* 5 (3):285–97.

Graham, L.D. 2002. "Ecdysone-controlled expression of transgenes." *Exp Opin Biol Ther* 2 (5):525–35.

Hacein-Bey-Abina, S., C. Von Kalle, M. Schmidt et al. 2003. "LMO2-associated clonal T cell proliferation in two patients after gene therapy for SCID-X1." *Science* 302 (5644):415–9.

Han, D., Y. Ding, S.L. Liu et al. 2009. "Double role of Fas ligand in the apoptosis of intervertebral disc cells *in vitro*." *Acta Biochim Biophys Sin (Shanghai)* 41 (11):938–47.

Hart, L.G., R.A. Deyo, and D.C. Cherkin. 1995. "Physician office visits for low back pain. Frequency, clinical evaluation, and treatment patterns from a U.S. national survey." *Spine (Phila Pa 1976)* 20 (1):11–9.

Humzah, M.D., and R.W. Soames. 1988. "Human intervertebral disc: Structure and function." *Anat Rec* 220 (4):337–56.

Kakutani, K., K. Nishida, K. Uno et al. 2006. "Prolonged down regulation of specific gene expression in nucleus pulposus cell mediated by RNA interference *in vitro*." *J Orthop Res* 24 (6):1271–8.

Kanemoto, M., S. Hukuda, Y. Komiya, A. Katsuura, and J. Nishioka. 1996. "Immunohistochemical study of matrix metalloproteinase-3 and tissue inhibitor of metalloproteinase-1 human intervertebral discs." *Spine (Phila Pa 1976)* 21 (1):1–8.

Kawaguchi, Y., R. Osada, M. Kanamori et al. 1999. "Association between an aggrecan gene polymorphism and lumbar disc degeneration." *Spine (Phila Pa 1976)* 24 (23):2456–60.

Kim, D.J., S.H. Moon, H. Kim et al. 2003. "Bone morphogenetic protein-2 facilitates expression of chondrogenic, not osteogenic, phenotype of human intervertebral disc cells." *Spine (Phila Pa 1976)* 28 (24):2679–84.

Lattermann, C., W.M. Oxner, X. Xiao et al. 2005. "The adeno associated viral vector as a strategy for intradiscal gene transfer in immune competent and pre-exposed rabbits." *Spine (Phila Pa 1976)* 30 (5):497–504.

Leckie, S.K., B.P. Bechara, R.A. Hartman et al. 2012. "Injection of AAV2-BMP2 and AAV2-TIMP1 into the nucleus pulposus slows the course of intervertebral disc degeneration in an *in vivo* rabbit model." *Spine J* 12 (1):7–20.

Levicoff, E.A., J.S. Kim, S. Sobajima et al. 2008. "Safety assessment of intradiscal gene therapy II: Effect of dosing and vector choice." *Spine (Phila Pa 1976)* 33 (14):1509–16; discussion 1517.

Li, J., S.T. Yoon, and W.C. Hutton. 2004. "Effect of bone morphogenetic protein-2 (BMP-2) on matrix production, other BMPs, and BMP receptors in rat intervertebral disc cells." *J Spinal Disord Tech* 17 (5):423–8.

Liang, H., S.Y. Ma, G. Feng, F.H. Shen, and X. Joshua Li. 2010. "Therapeutic effects of adenovirus-mediated growth and differentiation factor-5 in a mice disc degeneration model induced by annulus needle puncture." *Spine J* 10 (1):32–41.

Liu, Y., J. Kong, B.H. Chen, and Y.G. Hu. 2010. "Combined expression of CTGF and tissue inhibitor of metalloprotease-1 promotes synthesis of proteoglycan and collagen type II in rhesus monkey lumbar intervertebral disc cells *in vitro*." *Chin Med J (Engl)* 123 (15):2082–7.

Liu, Y., T. Yu, X.X. Ma et al. 2016. "Lentivirus-mediated TGF-beta3, CTGF and TIMP1 gene transduction as a gene therapy for intervertebral disc degeneration in an *in vivo* rabbit model." *Exp Ther Med* 11 (4):c1399–404.

Lively, M.W. 2002. "Sports medicine approach to low back pain." *South Med J* 95 (6):642–6.

Luo, X., R. Pietrobon, S.X. Sun, G.G. Liu, and L. Hey. 2004. "Estimates and patterns of direct health care expenditures among individuals with back pain in the United States." *Spine (Phila Pa 1976)* 29 (1):79–86.

Ma, X., Y. Lin, K. Yang, B. Yue, H. Xiang, and B. Chen. 2015. "Effect of lentivirus-mediated survivin transfection on the morphology and apoptosis of nucleus pulposus cells derived from degenerative human disc *in vitro*." *Int J Mol Med* 36 (1):186–94.

Masuda, K., Y. Imai, M. Okuma et al. 2006. "Osteogenic protein-1 injection into a degenerated disc induces the restoration of disc height and structural changes in the rabbit annular puncture model." *Spine (Phila Pa 1976)* 31 (7):742–54.

Masuda, K., T.R. Oegema, Jr., and H.S. An. 2004. "Growth factors and treatment of intervertebral disc degeneration." *Spine (Phila Pa 1976)* 29 (23):2757–69.

May, R.D., A. Tekari, D.A. Frauchiger et al. 2017. "Efficient nonviral transfection of primary intervertebral disc cells by electroporation for tissue engineering application." *Tissue Eng Part C Methods* 23 (1):30–7.

McCoy, R.D., B.L. Davidson, B.J. Roessler, G.B. Huffnagle, and R.H. Simon. 1995. "Expression of human interleukin-1 receptor antagonist in mouse lungs using a recombinant adenovirus: Effects on vector-induced inflammation." *Gene Ther* 2 (7):437–42.

Mern, D.S., and C. Thome. 2015. "Identification and characterization of human nucleus pulposus cell specific serotypes of adeno-associated virus for gene therapeutic approaches of intervertebral disc disorders." *BMC Musculoskelet Disord* 16:341.

Moon, S.H., K. Nishida, L.G. Gilbertson et al. 2008. "Biologic response of human intervertebral disc cells to gene therapy cocktail." *Spine (Phila Pa 1976)* 33 (17):1850–5.

Moon, S.H., K. Nishida, L.G. Gilbertson, P.D. Robbins, and J.D. Kang. 2000. *Proteoglycan Synthesis in Human Intervertebral Disc Cells Cultured in Alginate Beads; Exogenous TGF-beta1 Versus Adenovirus-Mediated Gene Transfer of TGF-beta1 cDNA*. Orlando, FL: Orthopaedic Research Society. 1061.

Morrey, M.E., P.A. Anderson, G. Chambers, and R. Paul. 2008. "Optimizing nonviral-mediated transfection of human intervertebral disc chondrocytes." *Spine J* 8 (5):796–803.

Nasto, L.A., K. Ngo, A.S. Leme et al. 2014. "Investigating the role of DNA damage in tobacco smoking-induced spine degeneration." *Spine J* 14 (3):416–23.

Nasto, L.A., H.Y. Seo, A.R. Robinson et al. 2012. "ISSLS prize winner: Inhibition of NF-kappaB activity ameliorates age-associated disc degeneration in a mouse model of accelerated aging." *Spine (Phila Pa 1976)* 37 (21):1819–25.

Nishida, K., M. Doita, T. Takada et al. 2006. "Sustained transgene expression in intervertebral disc cells *in vivo* mediated by microbubble-enhanced ultrasound gene therapy." *Spine (Phila Pa 1976)* 31 (13):1415–9.

Nishida, K., L. Gilbertson, S.H. Moon et al. 2000. *Immune-Privilege of the Intervertebral Disc: Long-Term Transgene Expression following Direct Adenovirus-Mediated Gene Transfer*. Adelaide, Australia: International Society for the Study of the Lumbar Spine.

Nishida, K., J.D. Kang, L.G. Gilbertson et al. 1999. "Modulation of the biologic activity of the rabbit intervertebral disc by gene therapy: An *in vivo* study of adenovirus-mediated transfer of the human transforming growth factor beta 1 encoding gene." *Spine (Phila Pa 1976)* 24 (23):2419–25.

Nishida, K., J.D. Kang, J.K. Suh et al. 1998. "Adenovirus-mediated gene transfer to nucleus pulposus cells. Implications for the treatment of intervertebral disc degeneration." *Spine (Phila Pa 1976)* 23 (22):2437–42; discussion 2443.

Nita, I., S.C. Ghivizzani, J. Galea-Lauri et al. 1996. "Direct gene delivery to synovium. An evaluation of potential vectors *in vitro* and *in vivo*." *Arthritis Rheum* 39 (5):820–8.

Osada, R., H. Ohshima, H. Ishihara et al. 1996. "Autocrine/paracrine mechanism of insulin-like growth factor-1 secretion, and the effect of insulin-like growth factor-1 on proteoglycan synthesis in bovine intervertebral discs." *J Orthop Res* 14 (5):690–9.

Paassilta, P., J. Lohiniva, H.H. Goring et al. 2001. "Identification of a novel common genetic risk factor for lumbar disk disease." *JAMA* 285 (14):1843–9.

Paul, R., R.C. Haydon et al. 2003. "Potential use of Sox9 gene therapy for intervertebral degenerative disc disease." *Spine (Phila Pa 1976)* 28 (8):755–63.

Ren, X.F., Z.Z. Diao, Y.M. Xi et al. 2015. "Adeno-associated virus-mediated BMP-7 and SOX9 *in vitro* co-transfection of human degenerative intervertebral disc cells." *Genet Mol Res* 14 (2):3736–44.

Robbins, P.D., H. Tahara, and S.C. Ghivizzani. 1998. "Viral vectors for gene therapy." *Trends Biotechnol* 16 (1):35–40.

Roughley, P.J., L.I. Melching, T.F. Heathfield, R.H. Pearce, and J.S. Mort. 2006. "The structure and degradation of aggrecan in human intervertebral disc." *Eur Spine J* 15 (Suppl 3):S326–32.

Samartzis, D., J. Karppinen, D. Chan, K.D. Luk, and K.M. Cheung. 2012. "The association of lumbar intervertebral disc degeneration on magnetic resonance imaging with body mass index in overweight and obese adults: A population-based study." *Arthritis Rheum* 64 (5):1488–96.

Seki, S., Y. Asanuma-Abe, K. Masuda et al. 2009. "Effect of small interference RNA (siRNA) for ADAMTS5 on intervertebral disc degeneration in the rabbit annular needle-puncture model." *Arthritis Res Ther* 11 (6):R166.

Shi, Z., T. Gu, H. Xin et al. 2015. "Intervention of rAAV-hTERT-transducted nucleus pulposus cells in early stage of intervertebral disc degeneration: A study in canine model." *Tissue Eng Part A* 21 (15–16):2186–94.

Sivan, S.S., E. Wachtel, and P. Roughley. 2014. "Structure, function, aging and turnover of aggrecan in the intervertebral disc." *Biochim Biophys Acta* 1840 (10):3181–9.

Skrzypiec, D., M. Tarala, P. Pollintine, P. Dolan, and M.A. Adams. 2007. "When are intervertebral discs stronger than their adjacent vertebrae?" *Spine (Phila Pa 1976)* 32 (22):2455–61.

Sobajima, S., J.F. Kompel, J.S. Kim et al. 2005a. "A slowly progressive and reproducible animal model of intervertebral disc degeneration characterized by MRI, X-ray, and histology." *Spine (Phila Pa 1976)* 30 (1):15–24.

Sobajima, S., A.L. Shimer, R.C. Chadderdon et al. 2005b. "Quantitative analysis of gene expression in a rabbit model of intervertebral disc degeneration by real-time polymerase chain reaction." *Spine J* 5 (1):14–23.

Sowa, G., E. Westrick, C. Pacek et al. 2011. "*In vitro* and *in vivo* testing of a novel regulatory system for gene therapy for intervertebral disc degeneration." *Spine (Phila Pa 1976)* 36 (10):E623–8.

Studer, R.K., A.M. Aboka, L.G. Gilbertson et al. 2007. "p38 MAPK inhibition in nucleus pulposus cells: A potential target for treating intervertebral disc degeneration." *Spine (Phila Pa 1976)* 32 (25):2827–33.

Studer, R.K., L.G. Gilbertson, H. Georgescu et al. 2008. "p38 MAPK inhibition modulates rabbit nucleus pulposus cell response to IL-1." *J Orthop Res* 26 (7):991–8.

Suzuki, T., K. Nishida, K. Kakutani et al. 2009. "Sustained long-term RNA interference in nucleus pulposus cells *in vivo* mediated by unmodified small interfering RNA." *Eur Spine J* 18 (2):263–70.

Takada, T., K. Nishida, M. Doita, and M. Kurosaka. 2002. "Fas ligand exists on intervertebral disc cells: A potential molecular mechanism for immune privilege of the disc." *Spine (Phila Pa 1976)* 27 (14):1526–30.

Takahashi, M., H. Haro, Y. Wakabayashi et al. 2001. "The association of degeneration of the intervertebral disc with 5a/6a polymorphism in the promoter of the human matrix metalloproteinase-3 gene." *J Bone Joint Surg Br* 83 (4):491–5.

Takegami, K., E.J. Thonar, H.S. An, H. Kamada, and K. Masuda. 2002. "Osteogenic protein-1 enhances matrix replenishment by intervertebral disc cells previously exposed to interleukin-1." *Spine (Phila Pa 1976)* 27 (12):1318–25.

Thompson, J.P., T.R. Oegema, Jr., and D.S. Bradford. 1991. "Stimulation of mature canine intervertebral disc by growth factors." *Spine (Phila Pa 1976)* 16 (3):253–60.

Tripathy, S.K., H.B. Black, E. Goldwasser, and J.M. Leiden. 1996. "Immune responses to transgene-encoded proteins limit the stability of gene expression after injection of replication-defective adenovirus vectors." *Nat Med* 2 (5):545–50.

Trout, J.J., J.A. Buckwalter, K.C. Moore, and S.K. Landas. 1982. "Ultrastructure of the human intervertebral disc. I. Changes in notochordal cells with age." *Tissue Cell* 14 (2):359–69.

Urban, J.P., and J.F. McMullin. 1985. "Swelling pressure of the inervertebral disc: Influence of proteoglycan and collagen contents." *Biorheology* 22 (2):145–57.

Urban, J.P., and J.F. McMullin. 1988. "Swelling pressure of the lumbar intervertebral discs: Influence of age, spinal level, composition, and degeneration." *Spine (Phila Pa 1976)* 13 (2):179–87.

Urban, J.P., S. Smith, and J.C. Fairbank. 2004. "Nutrition of the intervertebral disc." *Spine (Phila Pa 1976)* 29 (23):2700–9.

Vadala, G., G.A. Sowa, and J.D. Kang. 2007. "Gene therapy for disc degeneration." *Exp Opin Biol Ther* 7 (2):185–96.

Vadala, G., G.A. Sowa, L. Smith et al. 2007. "Regulation of transgene expression using an inducible system for improved safety of intervertebral disc gene therapy." *Spine (Phila Pa 1976)* 32 (13):1381–7.

Videman, T., J. Leppavuori, J. Kaprio et al. 1998. "Intragenic polymorphisms of the vitamin D receptor gene associated with intervertebral disc degeneration." *Spine (Phila Pa 1976)* 23 (23):2477–85.

Vos, T., A.D. Flaxman, M. Naghavi et al. 2012. "Years lived with disability (YLDs) for 1160 sequelae of 289 diseases and injuries 1990–2010: A systematic analysis for the Global Burden of Disease Study 2010." *Lancet* 380 (9859):2163–96.

Wallach, C.J., J.S. Kim, S. Sobajima et al. 2006. "Safety assessment of intradiscal gene transfer: A pilot study." *Spine J* 6 (2):107–12.

Wallach, C.J., S. Sobajima, Y. Watanabe et al. 2003. "Gene transfer of the catabolic inhibitor TIMP-1 increases measured proteoglycans in cells from degenerated human intervertebral discs." *Spine (Phila Pa 1976)* 28 (20):2331–7.

Wang, H., M. Kroeber, M. Hanke et al. 2004. "Release of active and depot GDF-5 after adenovirus-mediated overexpression stimulates rabbit and human intervertebral disc cells." *J Mol Med (Berl)* 82 (2):126–34.

Wehling, P., K.P. Schultz, P.D. Robbins, C.H. Evans, and J.A. Reinecke. 1997. "Transfer of genes to chondrocytic cells of the lumbar spine. Proposal for a treatment strategy of spinal disorders by local gene therapy." *Spine (Phila Pa 1976)* 22 (10):1092–7.

Winn, S.R., H. Uludag, and J.O. Hollinger. 1999. "Carrier systems for bone morphogenetic proteins." *Clin Orthop Relat Res* () Suppl): S95–106.

Woods, B.I., N. Vo, G. Sowa, and J.D. Kang. 2011. "Gene therapy for intervertebral disk degeneration." *Orthop Clin North Am* 42 (4):563–74, ix.

Wu, J., D. Wang, D. Ruan et al. 2014. "Prolonged expansion of human nucleus pulposus cells expressing human telomerase reverse transcriptase mediated by lentiviral vector." *J Orthop Res* 32 (1):159–66.

Xi, Y.M., Y.F. Dong, Z.J. Wang et al. 2013. "Co-transfection of adeno-associated virus-mediated human vascular endothelial growth factor165 and transforming growth factor-beta1 into annulus fibrosus cells of rabbit degenerative intervertebral discs." *Genet Mol Res* 12 (4):4895–908.

Yamada, K., H. Sudo, K. Iwasaki et al. 2014. "Caspase 3 silencing inhibits biomechanical overload-induced intervertebral disk degeneration." *Am J Pathol* 184 (3):753–64.

Yang, Y., F.A. Nunes, K. Berencsi et al. 1994. "Cellular immunity to viral antigens limits E1-deleted adenoviruses for gene therapy." *Proc Natl Acad Sci U S A* 91 (10):4407–11.

Yoon, S.T., J.S. Park, K.S. Kim et al. 2004. "ISSLS prize winner: LMP-1 upregulates intervertebral disc cell production of proteoglycans and BMPs *in vitro* and *in vivo*." *Spine (Phila Pa 1976)* 29 (23):2603–11.

Yue, B., Y. Lin, X. Ma et al. 2016. "Survivin-TGFB3-TIMP1 gene therapy via lentivirus vector slows the course of intervertebral disc degeneration in an *in vivo* rabbit model." *Spine (Phila Pa 1976)* 41 (11):926–34.

Zhang, C.C., G.P. Cui, J.G. Hu et al. 2014. "Effects of adenoviral vector expressing hIGF-1 on apoptosis in nucleus pulposus cells *in vitro*." *Int J Mol Med* 33 (2):401–5.

Zhang, Y., Z. Li, E.J. Thonar et al. 2005. "Transduced bovine articular chondrocytes affect the metabolism of cocultured nucleus pulposus cells *in vitro*: Implications for chondrocyte transplantation into the intervertebral disc." *Spine (Phila Pa 1976)* 30 (23):2601–7.

Zhang, Y.H., C.Q. Zhao, L.S. Jiang, and L.Y. Dai. 2011. "Lentiviral shRNA silencing of CHOP inhibits apoptosis induced by cyclic stretch in rat annular cells and attenuates disc degeneration in rats." *Apoptosis* 16 (6):594–605.

Index

A

ADCT, see Autologous disc chondrocyte transplant
Adeno-associated virus (AAV), 235, 237
Adenoviral vector carrying LMP-1 gene (AdLMP-1), 243
Adenoviral vector expressing recombinant human IGF-1 (Ad-hIGF-1), 243
Adipose-tissue-derived MSCs (AT-MSCs), 106
Adipose-tissue-derived stem cells (ADSCs), 107
Adjacent segment disease, 105
Adult stem cells for intervertebral disc repair, 103–135
   adipose-tissue-derived MSCs, 106
   adipose-tissue-derived stem cells, 107
   adjacent segment disease, 105
   adult stem cell-based therapies, 108–110
   adult stem cells, 106–108
   autologous disc chondrocyte transplant, 110
   best source of adult stem cells, 122–123
   bone-marrow-derived MSCs, 106
   challenges, 117–125
   clinical studies, 110–117
   committed cell-based therapies, 105–106
   delivery of adult stem cells, 124–125
   first signs of degeneration, 104
   induced pluripotent stem cells, 107
   intervertebral disc and its degeneration, 103–105
   IVD-derived MSCs, 107
   mechanism of action of transplanted adult stem cells, 120
   mesenchymal precursor cells, 117
   optimal dose of adult stem cells, 123–124
   predifferentiation of adult stem cells, 121–122
   survival of transplanted adult stem cells, 117–120
   target patient profile, 125
AF, see Annulus fibrosus
Aggrecan (ACAN), 113, 120, 184
Alginate, 141
Animal models and imaging of intervertebral disc (IVD) degeneration, 19–66
   anatomy of IVD (comparison of various animal models), 29–32
   animal models of spontaneously occurring DDD, 35–37
   biomechanics of IVD, 33–34
   cellular population of IVD, 32–33
   characteristics of IVD among species, 29–34
   chemonucleolysis, 43
   chondroid metamorphosis, 35
   DDD induction technique, 34–45
   experimentally induced disc disease, 38–45
   imaging of IVD degeneration and perspectives, 46–51
   in vivo models of DDD, 21–28
   MRI, new developments in, 51–53
   nucleotomy, 41
   potential applications of different models, 45–46
   surgical approach to IVD, 45
Annulus fibrosus (AF), 20, 68, 103
Annulus fibrosus (AF), materials for, 145–148
   natural materials, 146
   synthetic materials, 147–148
AT-MSC, see Adipose-tissue-derived MSCs
Autologous disc chondrocyte transplant (ADCT), 110

B

Basic fibroblast growth factor (bFGF), 202
BDNF, see Brain-derived neurotrophic factor
Biacuplasty, 8
Biomolecular strategies, 10
Bone marrow (BM), 160
Bone-marrow-derived MSCs (BM-MSCs), 106
Bone morphogenetic protein (BMP), 203, 234
Brain-derived neurotrophic factor (BDNF), 202

C

Cartilaginous end plate (CEP), 69, 184, 232
CBT, see Cognitive behavioral therapy
C/EBP homologous protein (CHOP), 243
Cell delivery materials, 137–153
   alginate, 141
   annulus fibrosus, materials for, 145–148
   chitosan, 141
   gellan gum-based hydrogels, 142
   hydrogels, 140
   methacrylated gellan gum hydrogels, 142
   natural materials, 140–142, 146
   nucleus pulposus, materials for, 138–145
   polyethylene glycol, 140, 145
   polylactides/glycolides, 142, 145
   synthetic materials, 142–145, 147–148
Cell recruitment for intervertebral disc, 155–182
   cell migration in IVD (new paradigm), 169–173
   cell recruitment for endogenous repair/regeneration, 159–165

cell recruitment in IVD, 171–173
cell viability, 158
endothelial progenitor cells, 160
hematopoietic stem cells, 160
immune cell migration, 164–165, 168
intracellular adhesion molecule, 164
IVD anatomy, 156–158
IVD degeneration, 158–159
IVD degeneration (involved cytokines and chemokines), 169–171
macrophage inflammatory protein, 169
motivation, 155–156
stem cell homing and migration, 160–164
stem cell recruitment (chemokine delivery systems), 165–168
strategies to enhance cell migration, 165–168
vascular cell adhesion molecule, 161, 164
vascular endothelial growth factor, 162
CFU-F, *see* Colony-forming unit-fibroblast
Chemokine receptors (CCR), 162
Chemonucleolysis, 12, 43
Chitosan, 141
Chondroid metamorphosis, 35
Chondroitinase ABC, 12, 43
Chymopapain, 43
Clinics, intervertebral disc degeneration in, 1–17
annular tears, 6
biacuplasty, 8
biomolecular strategies, 10
chemonucleolysis, 12
clinical manifestations of disc degeneration, 3–4
cognitive behavioral therapy, 7
degenerative disc disease, 1
disc bulge, 5
discogenic pain, 1
electrothermal ablation of ramus communicans, 9
endplate abnormalities, 6
epidural glucocorticoid injections, 11
fracture, 3
gene vector systems, 10–11
imaging studies, intervertebral disc degeneration in, 4–6
importance of disc degeneration and LBP, 2–3
intradiscal electrothermal therapy, 8
intradiscal steroid injections, 9
Lasègue's test, 4
lumbar arthroplasty, 10
lumbar discectomy, 11
lumbar disc herniation, 5
nerve root pain, 4
nucleoplasty, 13
operative treatment, 9–10
passive physiotherapy modalities, 8
percutaneous intradiscal radiofrequency thermocoagulation, 8
percutaneous laser disc decompression, 12
percutaneous techniques, 8–9

prognosis, 7
protrusion, 5
provocative discography, 6
radicular compression, 1
radicular pain, 4
regenerative techniques, 10–11
relation between disc degeneration and pain, 6–7
sciatica, prognosis of, 7
sciatica, therapeutic challenges for, 11–13
sequestration, 5
sequestrectomy, 12
stem cells, 11
straight leg raising test, 4
therapeutic challenges for discogenic LBP, 7–11
thermal annular procedures, 8
tissue-engineering technology, 11
total disc replacement, 10
Cognitive behavioral therapy (CBT), 7
Collagen, 70–71
Colony-forming unit-fibroblast (CFU-F), 113
Complementary DNA (cDNA), 235
Computed tomography (CT), 68
Cytokines, 71

D

Degenerative disc disease (DDD), 1, 20
induction technique, 34–45
in vivo models of, 21–28
Dekompressor, 13
Disc immune privilege theory, 240
Discogenic pain, 1
Dorsal root ganglion (DRG), 187, 202

E

Endothelial progenitor cells (EPCs), 160
Enzyme-linked immunosorbent assay (ELISA), 88
Enzymes, 72
Epidural glucocorticoid injections, 11
Extracellular matrix (ECM), 20, 69, 137

F

Fibroblasts, 196, 201
First signs of degeneration, 104
Fracture, 3

G

GAG, *see* Glycosaminoglycan
Gellan gum-based hydrogels, 142
Gene delivery for intervertebral disc, 231–253
adeno-associated virus, 235, 237
adenoviral vector expressing recombinant human IGF-1, 243

Index

adenovirus, 236–237
anabolic factors, 240–241
anticatabolic factors, 241–242
bone morphogenetic protein, 234
C/EBP homologous protein, 243
cell survival and apoptosis, 243–244
disc immune privilege theory, 240
gene delivery systems and strategies, 235–238
gene therapy for IDD, 234–235
growth and differentiation factor, 238, 241
human mesenchymal stem cell differentiation, 240
immune privilege, 240
intervertebral disc degeneration, 232–233
in vivo efficacy and feasibility, 245
luciferase plasmids, 238
modulating disc cell activity, 233–234
multiple targets, 244–245
nonviral systems, 237–238
proof of principle of gene therapy for IDD, 239–245
regulating transgene expression, 246–247
retrovirus, 236
RNA interference, 238–239
safety concerns, 246
second mitochondria-derived activator of caspases, 243
self-complementary AAV serotypes, 237
transcription factors, 242–243
viral systems, 236–237
Gene vector systems, 10–11
Glycosaminoglycan (GAG), 70, 85, 232
Granulocyte-macrophage colony-stimulating factor (GM-CSF), 202
Green fluorescent protein (GFP), 173, 237
Growth and differentiation factor (GDF), 203, 238, 241

H

Hematopoietic stem cells (HSCs), 160
Human mesenchymal stem cell (hMSC) differentiation, 240
Hydrogels
 natural-origin, 140
 synthetic, 142

I

ICAM, *see* Intracellular adhesion molecule
IDET, *see* Intradiscal electrothermal therapy
Imaging of IVD degeneration, 46–51; *see also* Animal models and imaging of intervertebral disc degeneration
 magnetic resonance imaging, 49–51
 nuclear medicine, 49
 physiopathology context, 46–47
 studies, 4–6
 X-ray imaging, 47–49
Immune cell migration, 164–165
Immune privilege, 236, 240
Immunohistochemistry (IHC), 86
Immunomodulation, 183–230
 basic fibroblast growth factor, 202
 bioactive molecules, 204–209
 bone morphogenetic protein, 203
 brain-derived neurotrophic factor, 202
 cell-based therapies, 212–214
 dorsal root ganglion, 187, 202
 endogenous therapies, 212–213
 exogenous stem cell delivery, 213–214
 gene therapy, 211
 granulocyte-macrophage colony-stimulating factor, 202
 growth differentiation factor, 203
 IL-1β, 194–195
 immune cell activation, 197–202
 immunogenic phenotype of IVD cell populations and induced immune cell response, 185–187
 inflammation mediators, 188–193
 key proinflammatory molecules in IVD degeneration and associated inflammation, 187–196
 macrophages, 201–202
 microRNAs, 196–197
 mitogen-activated protein kinase, 194
 molecular therapy (clinical trials), 203–210
 molecular therapy (*in vivo* and *ex vivo* studies), 210–211
 monocyte chemoattractant protein, 196
 nerve growth factor, 195, 202
 strategies for immunomodulation of degenerated IVD, 202–211
 T cells, 200
 toll-like receptors, 195–196
 tumor necrosis factor-α, 185, 187–194
 VEGF, 202
Induced pluripotent stem cells (iPSCs), 107
In situ zymography (ISZ), 89
Interferon (IFN)-γ, 185
Interleukin (IL)-1 α/β, 158
Intervertebral disc (IVD), 20
 anatomy (comparison of various animal models), 29–32
 biomechanics of, 33–34
 cellular population of, 32–33
 characteristics among species, 29–34
 -derived MSCs (IVD-MSCs), 107
 description of, 103
 surgical approach to, 45
Intervertebral disc degeneration (IDD), gene therapy for, 234–235, 239–245
Intervertebral disc (IVD) degeneration, imaging of, 46–51
 magnetic resonance imaging, 49–51
 nuclear medicine, 49

physiopathology context, 46–47
X-ray imaging, 47–49
Intracellular adhesion molecule (ICAM), 164
Intradiscal electrothermal therapy (IDET), 8

L

Lasègue's test, 4
LDH, *see* Lumbar disc herniation
Low back pain (LBP), 20
   importance of disc degeneration and, 2–3
   specific causes, 2
   statistics on, 184
Low back pain (LBP), discogenic (therapeutic challenges for), 7–11
   biomolecular strategies, 10
   conservative treatment, 7–8
   gene vector systems, 10–11
   lumbar arthroplasty, 10
   operative treatment, 9–10
   percutaneous techniques, 8–9
   regenerative techniques, 10–11
   stem cells, 11
   tissue-engineering technology, 11
   total disc replacement, 10
Luciferase plasmids, 238
Lumbar arthroplasty, 10
Lumbar discectomy, 11
Lumbar disc herniation (LDH), 5, 68

M

Macrophage inflammatory protein (MIP), 169
Macrophages, 201–202
Magnetic resonance imaging (MRI), 3, 51–53, 109
MAPK, *see* Mitogen-activated protein kinase
Matrix metalloproteinase 3 (MMP-3), 39, 242
MCP, *see* Monocyte chemoattractant protein
Mechanobiology, 72–73, 92
Mesenchymal precursor cells (MPCs), 117
Mesenchymal stem cells (MSCs), 138, 160
   adipose-tissue-derived, 106
   bone-marrow-derived, 106
   IVD-derived, 107
   varied locations of, 107
Methacrylated gellan gum (GG-MA) hydrogels, 142
Methylene blue, 9
MicroRNAs, 196–197
MIP, *see* Macrophage inflammatory protein
Mitogen-activated protein kinase (MAPK), 194, 242
MMP-3, *see* Matrix metalloproteinase, 3
Monocyte chemoattractant protein (MCP), 196
MPCs, *see* Mesenchymal precursor cells
MRI, *see* Magnetic resonance imaging
MSCs, *see* Mesenchymal stem cells
Multiplicity of infection (MOI), 240

N

Nerve growth factor (NGF), 195, 202
Nerve root pain, 4
Nuclear medicine, 49
Nucleoplasty, 13
Nucleotomy, 41
Nucleus pulposus (NP), 29, 68, 103, 232
Nucleus pulposus (NP), materials for, 138–145
   alginate, 141
   chitosan, 141
   gellan gum-based hydrogels, 142
   hydrogels, 140
   methacrylated gellan gum hydrogels, 142
   natural materials, 140–142
   polyethylene glycol, 140, 145
   polylactides/glycolides, 142, 145
   synthetic materials, 142–145

O

Osteogenic protein-1 (OP-1), 234
Ozone, 12

P

PCR, *see* Polymerase chain reaction
PEG, *see* Polyethylene glycol
Percutaneous intradiscal radiofrequency thermocoagulation (PIRFT), 8
Percutaneous laser disc decompression (PLDD), 12
Pfirrmann grades, 6
Platelet-rich plasma (PRP), 203
Polyethylene glycol (PEG), 140, 145
Polylactides/glycolides (PLA/PGA), 142, 145
Polymerase chain reaction (PCR), 87
Proteoglycans, 70, 85–86
Provocative discography, 6

R

Radicular compression, 1
Radicular pain, 4
Randomized controlled trials (RCTs), 8
RNA interference (RNAi), 238–239

S

scAAV, *see* Self-complementary AAV serotypes
Sciatica, prognosis of, 7
Sciatica, therapeutic challenges for, 11–13
   chemonucleolysis, 12
   conservative management, 11
   epidural glucocorticoid injections, 11
   lumbar discectomy, 11
   nucleoplasty, 13
   percutaneous laser disc decompression, 12

Second mitochondria-derived activator of caspases (SMAC), 243
Self-complementary AAV serotypes (scAAV), 237
Sequestrectomy, 12
Small interfering RNA (siRNA) sequences, 238
Sox-9, 113
Stem cell(s), 11; *see also* Adult stem cells for intervertebral disc repair
   delivery, exogenous, 213–214
   hematopoietic, 160
   homing and migration, 160–164
   recruitment (chemokine delivery systems), 165–168
Straight leg raising test, 4
Stromal cell derived factor-1α (SDF-1α), 162

T

T cells, 200
Terminal deoxynucleotidyl transferase dUTP nick end labeling (TUNEL), 89
TGF, *see* Transforming growth factor
Tissue-engineering technology, 11
Tissue inhibitor of metalloproteinases (TIMP)-1, 184, 211, 241
Toll-like receptors (TLRs), 195–196
Total disc replacement, 10
Transforming growth factor (TGF), 88, 234
Tumor necrosis factor (TNF)-α, 158, 187–194
TUNEL, *see* Terminal deoxynucleotidyl transferase dUTP nick end labeling
Type II collagen, 113, 120

V

Vascular cell adhesion molecule (VCAM), 161, 164
Vascular endothelial growth factor (VEGF), 162, 202
Very-late antigen-4 (VLA-4), 161

W

Whole organ cultures, IVD (model choice), 67–101
   cell viability, 89–90
   changes in human IVD degeneration, 69–73
   collagen, 70–71, 86–87
   cytokines, 71, 87–89
   enzyme-linked immunosorbent assay, 88
   enzymes, 72, 89
   established whole-organ culture models, 73–84
   ex vivo IVD organ culture, monitoring degenerative changes in, 85–92
   glycosaminoglycan, 70, 85
   immunohistochemistry, 86
   in situ zymography, 89
   macroscopic evaluation, 91–92
   mechanobiology, 72–73, 92
   polymerase chain reaction, 87
   proteoglycans, 70, 85–86
   terminal deoxynucleotidyl transferase dUTP nick end labeling, 89
   transforming growth factor, 88

X

X-ray imaging, 47–49

PGSTL 05/01/2018